CAUSAL MODELS
IN THE
SOCIAL SCIENCES

List of Contributors

Albert Ando

H. M. Blalock, Jr.

Herbert L. Costner

Otis Dudley Duncan

Marcus Felson

Franklin M. Fisher

Albert N. Halter

Michael T. Hannan

Jerald R. Herting

Karl G. Jöreskog

Tjalling C. Koopmans

Kenneth C. Land

Robert L. Linn

Robert Mason

Herbert A. Simon

Robert H. Strotz

John L. Sullivan

Charles E. Werts

H. O. A. Wold

Sewall Wright

CAUSAL MODELS IN THE SOCIAL SCIENCES

Second Edition

Edited by

H. M. Blalock, Jr.

1985

ALDINE
Publishing Company
New York

About the Editor

H. M. Blalock, Jr., is Professor, Department of Sociology, University of Washington, Seattle. Dr. Blalock has authored and/or edited eighteen books and has been a major contributor to numerous academic and professional journals. He was recipient of the 1973 ASA Samuel Stouffer Prize and is a fellow of the American Statistical Association, American Academy of Arts and Sciences, and a member of the National Academy of Sciences. He is Past-President of the American Sociological Association.

Copyright © 1985 H. M. Blalock, Jr.
All rights reserved. No part of this publication may be reproduced or transmitted in any form or by any means, electronic or mechanical, including photocopy, recording, or any information storage and retrieval system, without permission in writing from the publisher.

Aldine Publishing Company
200 Saw Mill River Road
Hawthorne, New York 10532

Library of Congress Cataloging in Publication Data

Main entry under title:
Causal models in the social sciences.

Bibliography: p.
  1. Social sciences—Mathematical models—Addresses, essays, lectures. 2. Social sciences—Methodology—Addresses, essays, lectures. I. Blalock, Hubert M.
H61.25.C38 1985   300'.724   84-24258
ISBN 0-202-30313-6 (lib. bdg.)
ISBN 0-202-30314-4 (pbk.)

Printed in the United States of America
10 9 8 7 6 5 4 3 2 1

Contents

Preface

There is a growing literature on causal models and structural systems of equations that crosscuts a number of different fields. However, much of this material is widely scattered throughout the journal literature and varies considerably in terms of both level of difficulty and substantive application. It therefore seemed wise to attempt to capture the essential flavor of the main developments of this methodological approach through an edited volume that may be used to supplement textual materials of a more integrated nature.

The most systematic discussions of this general approach have appeared in the econometrics literature, where several general texts are available. However, many of these discussions are too technical for most sociologists, political scientists, and others who lack strong backgrounds in mathematics. In editing this volume, I have attempted to integrate a few of the less technical papers written by econometricians such as Koopmans, Wold, Strotz, and Fisher with discussions of causal approaches in the biological sciences and with relatively more exploratory treatments by sociologists and other social scientists.

It is assumed that the reader has some rudimentary familiarity with the subject, as can be obtained by reading nontechnical discussions such as contained in my *Causal Inferences in Nonexperimental Research* (Chapel Hill: University of North Carolina Press, 1964). Those readers who wish to study the subject seriously should learn the basic elements of matrix algebra and

ix

then take up treatments of simultaneous-equation approaches in textbooks on econometrics. Many of the papers in this volume can be understood by those who have been exposed to simple least-squares procedures but who lack training in mathematical statistics. However, there can be no getting around the fact that many of the most important methodological problems we face in the social sciences involve highly technical issues. One of my major purposes in editing this volume is to convince the reader that there *are* ways of handling the kinds of complexities that have been loosely discussed in the verbal literature, but that we cannot expect to obtain definitive answers without first mastering at least some of this rapidly accumulating technical literature.

In thinking about how to revise the original volume, which was published in 1971, I was tempted to cut back severely on the earlier articles and to expand considerably in terms of much more recent discussions. To do so, however, would risk giving the student the impression that important ideas are developed full-bloom and without benefit of a diffusion of knowledge across disciplinary boundaries, even though such a diffusion process may be erratic and very uneven at times. I therefore decided to retain many of these important earlier articles, while making the volume more compact. But much has been written in the past dozen or so years, and as I attempted to collect those papers likely to be of greatest interest to sociologists and political scientists, it became obvious that a second, companion volume would be needed. Since a large proportion of these more recent contributions deal with panel or experimental designs, I decided that this would be the focus in a second volume on design issues. Several of the papers in the original volume were thus shifted over to this second volume. This particular revised version of the original volume, however, contains three new papers: one, by Land and Felson, that suggests strategies for conducting sensitivity analyses of identifying assumptions, and two others, by Herting and by Herting and Costner, that are concerned with LISREL procedures for dealing with multiple-indicator models.

H. M. Blalock, Jr.

CAUSAL MODELS
IN THE
SOCIAL SCIENCES

SIMPLE RECURSIVE MODELS AND PATH ANALYSIS

I

Part I focuses on the kinds of relatively simple causal models that have served for many years as first approximations for political scientists and sociologists. These models all involve one-way causation and can be handled by what are referred to as *recursive systems of equations*. Such models provide a heuristic device for broadening the scope of simple regression approaches that commonly focus on a single dependent variable and a set of "predictors." In fact, they justify the procedure of treating each such equation as separate from the rest so that its coefficients can be estimated by ordinary least squares. As we shall see in Part II, there are numerous instances where more complex kinds of estimating procedures must be used and indeed many situations where there will be too many unknowns in the system to permit any estimation at all. These latter kinds of complications do not arise in the case of one-way causation, provided we are willing to make certain kinds of simplifying assumptions about the omitted variables.

Sociologists and political scientists, in particular, are well aware of the hiatus or gap between our verbal theories, on the one hand, and our research techniques, on the other. The causal modeling approach offers a systematic way out of this impasse, although at the same time it makes one well aware of the limitations of the inference procedures involved. In effect, it provides a set of rules for making causal inferences on the basis of empirical interrelationships. These rules alone can never provide infallible guides nor can they assure one that any particular causal explanation is correct. But they can give us a systematic way of building block upon block, so that our theories can become cumulative and so that alternative explanations that are not consistent with the data can be rejected.

The notion of recursive equations is fundamental. The basic idea is that variables can be hierarchically arranged in terms of their causal priorities in such a way that it becomes possible to neglect variables that are clearly dependent on a given subset of variables. Suppose we are considering four variables X_1, X_2, X_3, and X_4. If we are willing to assume that X_4 does not affect X_1, X_2, and X_3, then regardless of the influence that any of these latter variables may have on X_4, we are justified in ignoring X_4 when considering the interrelationships among these first three variables. Similarly, if we assume that X_3 does not influence X_1 and X_2, we are justified in ignoring X_3 in studying their interrelationship. In fact, if we were to introduce X_3 into their relationship, say by relating X_2 and X_1 controlling for X_3, we might be badly misled. Finally, if we are willing to assume that X_2 cannot affect X_1, then we may write the following set of equations:

$$X_1 = e_1$$

$$X_2 = b_{21}X_1 + e_2$$

$$X_3 = b_{31}X_1 + b_{32}X_2 + e_3$$

3

and

$$X_4 = b_{41}X_1 + b_{42}X_2 + b_{43}X_3 + e_4$$

where the e_i are disturbance terms representing the effects of all omitted variables. Obviously, this basic idea can be extended to any number of variables as long as we are willing to assume such a causal ordering.

The recursive models appropriate to one-way causation, as well as the more general models discussed in Part II, assume perfect measurement in all variables that appear as "independent" variables somewhere in the causal system. Obviously, this assumption is unrealistic in the case of all research, but it is especially so in the case of the social sciences. Nevertheless, it is useful to ignore problems of measurement error when studying other kinds of complications in the models. Approaches to the study of measurement errors and their implications will be considered later in Part III. There are other kinds of elaborations, however, that cannot be extensively discussed in this single volume. These include nonlinear or nonadditive models, the handling of correlated disturbance terms that are especially likely in the case of time-series and ecological data, and the use of truly dynamic formulations that explicitly involve the time dimension. The reader who wishes to pursue these additional topics will find extensive bibliographies suggested in many of the chapters in this volume.

Much of the literature on causal models and simultaneous equations consists of elaborations on this basic kind of model. As we shall see in Part II the simple recursive model can be replaced by one in which there are sets or blocks of variables that are recursively related, but where one can allow for reciprocal causation within blocks. In these more complex models, however, there may be too many unknowns in the system, in which case we refer to the system as being *underidentified*. Furthermore, as soon as we allow for reciprocal causation, we are forced to think more carefully about our assumptions concerning the disturbance terms e_i. In order to justify the use of ordinary least squares, we commonly assume that these disturbances are uncorrelated with the *independent* variables in each equation as well as with each other. It can be shown that in the case of the more general nonrecursive system, this kind of assumption breaks down, and we must therefore find alternatives to ordinary least-squares estimating procedures.

A whole host of problems revolves around the very simple assumptions about these error terms commonly made in connection with recursive systems, and these involve the very core issues of causal inferences. We ordinarily conceive of these disturbances as having been brought about by variables that have been omitted from the theoretical system. In the case of recursive systems, the use of ordinary least squares therefore requires the assumption that the aggregate effect of omitted variables that affect the dependent variable in any given equation must be such that the error term is uncorrelated with each of the independent variables. For example, in the case of the equation for X_4, we assume that omitted variables that affect X_4 are uncorrelated in the aggregate with X_1, X_2, and X_3. This is a somewhat weaker form of the assumption that "other things" are equal, in that we do not need to assume that all relevant variables are literally constant. They are merely assumed not to disturb the basic relationships within the system.

In the first chapter of Part I, Simon shows how this assumption about the disturbance terms is utilized to provide the rationale behind our more intuitive ideas about spurious correlations. He notes that whenever one is not satisfied by this particular simplifying assumption, he must expand the theoretical system to include those variables thought to be violating the assumption about the disturbance terms. My own chapter (2) on four-variable models extends the Simon approach by showing its implications for the use of control variables.

The chapters by Wright (3) and Duncan (4) are concerned with an approach to causal model building referred to as "path analysis," following the terminology of Sewall Wright, who introduced the procedure into the biometric literature over 60 years ago. Most of the applications of path analysis that have appeared in the social science literature have involved recursive models and exactly the same basic assumptions as have been discussed in the first two chapters. The terminology, notation, and heuristics of path analysis, however, differ somewhat from what has been referred to as the *Simon – Blalock approach* by some authors, and therefore certain of these differences need to be noted.

The emphasis in path analysis is on tracing causal paths and in decomposing total correlations into component parts attributed to simple and compound paths between any two variables. This places a greater emphasis on estimating the numerical values of each path coefficient rather than testing models by examining the behaviors of partial correlations, but in this respect there is no really important difference between the two approaches. Indeed, one can rather easily decompose regression coefficients in much the same fashion so as to compare the relative magnitudes of direct effects, indirect effects, and effects of variables that may be creating spurious relationships among some of the variables in the causal system [1].

One of the principal features of path analysis is that the emphasis is placed on variables that have been standardized in terms of standard-deviation units that are appropriate for any given sample, thereby permitting direct comparisons across different kinds of variables. Wright's chapter points out that path coefficients, which are standardized measures, have certain advantages over the unstandardized coefficients, which Wright has termed *path regression coefficients.* Thus, one of the issues that divides those who utilize "path analysis" from those who have followed Simon or who have been trained in the econometric tradition is that of whether standardized coefficients are preferable to unstandardized ones. Duncan's discussion of sociological applications of path analysis utilizes the standardized measures, whereas in my chapter on causal inferences and closed populations (5), I argue in favor of the unstandardized measures. Thus, the differences between perspectives partly involve the question of the *kinds* of measures to utilize, but the models and assumptions involved are fundamentally the same.

There is also another difference that may be important in terms of the psychological process of theory building. The path-analytic perspective involves more explicit attention to the "error" or "residual" components of the equation. These terms are conceptualized as though they were due to a distinct though unmeasured variable, thus making each variable completely determined by the remaining "variables" in the system. The resulting algebraic manipulations and formulas therefore appear different from those utilized by Simon and in connection with ordinary least-squares procedures, but it can be shown that the two systems are mathematically equivalent, provided one begins with the same models and assumptions.

This difference in the way equations are written down, as well as the pathtracing algorithms that have been developed out of the path-analytic tradition, offers the social scientist a dual approach that may afford more insights than could be attained through the use of either approach alone. In particular, the use of curved double-headed arrows to represent unexplained correlations among exogenous variables is a practical heuristic device for adding reality to a causal system. Thus, although the two approaches are formally equivalent, there are many instances in which it is useful to retain the distinction. This point will be clarified in Part III where we turn our attention to multiple indicators and various kinds of measurement errors.

REFERENCE

[1] Namboodiri, N. K., Carter, L. F., and Blalock, H. M. *Applied Multivariate Analysis and Experimental Designs.* New York: McGraw-Hill, 1975.

Spurious Correlation: A Causal Interpretation*

Herbert A. Simon

1

Even in the first course in statistics, the slogan "Correlation is no proof of causation!" is imprinted firmly in the mind of the aspiring statistician or social scientist. It is possible that he leaves the course (and many subsequent courses) with no very clear ideas as to what *is* proved by correlation, but he never ceases to be on guard against "spurious" correlation, that master of imposture who is always representing himself as "true" correlation.

The very distinction between "true" and "spurious" correlation appears to imply that while correlation in general may be no proof of causation, "true" correlation does constitute such proof. If this is what is intended by the adjective "true," are there any operational means for distinguishing between true correlations, which do imply causation, and spurious correlations, which do not?

A generation or more ago, the concept of *spurious correlation* was examined by a number of statisticians, and in particular by G. U. Yule [8]. More recently, important contributions to our understanding of the phenomenon have been made by Hans Zeisel [9] and by Patricia L. Kendall and Paul F. Lazarsfeld [1]. Essentially, all these treatments deal with the three-variable

* Reprinted by permission of the author and publisher from the *Journal of the American Statistical Association* 49 (1954): 467–479.

case—the clarification of the relation between two variables by the introduction of a third. Generalizations to n variables are indicated but not examined in detail.

Meanwhile, the main stream of statistical research has been diverted into somewhat different (but closely related) directions by Frisch's work on confluence analysis and the subsequent exploration of the "identification problem" and of "structural relations" at the hands of Haavelmo, Hurwicz, Koopmans, Marschak, and many others.[1] This work has been carried on at a level of great generality. It has now reached a point where it can be used to illuminate the concept of spurious correlation in the three-variable case. The bridge from the identification problem to the problem of spurious correlation is built by constructing a precise and operationally meaningful definition of causality—or, more specifically, of causal ordering among variables in a model.[2]

STATEMENT OF THE PROBLEM

How do we ordinarily make causal inferences from data on correlations? We begin with a set of observations of a pair of variables, x and y. We compute the coefficient of correlation, r_{xy}, between the variables, and whenever this coefficient is significantly different from zero, we wish to know what we can conclude as to the causal relation between the two variables. If we are suspicious that the observed correlation may derive from "spurious" causes, we introduce a third variable, z, that, we conjecture, may account for this observed correlation. We next compute the partial correlation, $r_{xy \cdot z}$, between x and y with z "held constant," and compare this with the zero-order correlation, r_{xy}. If $r_{xy \cdot z}$ is close to zero, while r_{xy} is not, we conclude that either: (a) z is an intervening variable—the causal effect of x on y (or vice versa) operates through z; or (b) the correlation between x and y results from the joint causal effect of z on both those variables, and hence this correlation is spurious. It will be noted that in case (a), we do not know

---

[1] See Koopmans [2] for a survey and references to the literature.

[2] Simon [6] and [7]. See also Orcutt [4] and [5]. I should like, without elaborating it here, to insert the caveat that the concept of *causal ordering* employed in this chapter does not in any way solve the "problem of Hume" nor contradict his assertion that all we can ever observe are covariations. If we employ an ontological definition of cause—one based on the notion of the "necessary" connecton of events—then correlation cannot, of course, prove causation. But neither can anything else prove causation, and hence we can have no basis for distinguishing "true" from "spurious" correlation. If we wish to retain the latter distinction (and working scientists have not shown that they are able to get along without it), and if at the same time we wish to remain empiricists, then the term *cause* must be defined in a way that does not entail objectionable ontological consequences. That is the course we shall pursue here.

whether the causal arrow should run from x to y or from y to x (via z in both cases); and in any event, the correlations do not tell us whether we have case (a) or case (b).

The problem may be clarified by a pair of specific examples adapted from Zeisel [9, pp. 192–195].[3]

1. The data consist of measurements of three variables in a number of groups of people: x is the percentage of members of the group that is married, y is the average number of pounds of candy consumed per month per member, z is the average age of members of the group. A high (negative) correlation, r_{xy}, was observed between marital status and amount of candy consumed. But there was also a high (negative) correlation, r_{yz}, between candy consumption and age; and a high (positive) correlation, r_{xz}, between marital status and age. However, when age was held constant, the correlation $r_{xy \cdot z}$, between marital status and candy consumption was nearly zero. By our previous analysis, either age is an intervening variable between marital status and candy consumption; or the correlation between marital status and candy consumption is spurious, being a joint effect caused by the variation in age. "Common sense"—the nature of which we will examine below in detail—tells us that the latter explanation is the correct one.

2. The data consist again of measurements of three variables in a number of groups of people: x is the percentage of female employees who are married, y is the average number of absences per week per employee, z is the average number of hours of housework performed per week per employee.[4] A high (positive) correlation, r_{xy}, was observed between marriage and absenteeism. However, when the amount of housework, z, was held constant, the correlation $r_{xy \cdot z}$ was virtually zero. In this case, by applying again some common sense notions about the direction of causation, we reach the conclusion that z is an intervening variable between x and y: that is, that marriage results in a higher average amount of housework performed, and this, in turn, in more absenteeism.

Now what is bothersome about these two examples is that the same statistical evidence, so far as the coefficients of correlation are concerned, has been used to reach entirely different conclusions in the two cases. In the first case, we concluded that the correlation between x and y was spurious; in the second case, that there was a true causal relationship, mediated by the intervening variable z. Clearly, it was not the statistical evidence, but the "common sense" assumptions added afterward, that permitted us to draw these distinct conclusions.

---

[3] Reference to the original source will show that in this and the following example, we have changed the variables from attributes to continuous variables for purposes of exposition.

[4] Zeisel [9], pp. 191–192.

CAUSAL RELATIONS

In investigating spurious correlation, we are interested in learning whether the relation between two variables persists or disappears when we introduce a third variable. Throughout this chapter (as in all ordinary correlation analyses), we assume that the relations in question are linear and without loss of generality, that the variables are measured from their respective means.

Now suppose we have a system of three variables whose behavior is determined by some set of linear mechanisms. In general, we will need three mechanisms, each represented by an equation — three equations to determine the three variables. One such set of mechanisms would be that in which each of the variables *directly influenced* the other two. That is, in one equation, x would appear as the dependent variable, y and z as independent variables; in the second equation, y would appear as the dependent variable, x and z as the independent variables; in the third equation, z as dependent variable, x and y as independent variables.[5]

$$x + a_{12}y + a_{13}z = u_1 \tag{1.1}$$

(I) $$a_{21}x + \quad y + a_{23}z = u_2 \tag{1.2}$$

$$a_{31}x + a_{32}y + \quad z = u_3 \tag{1.3}$$

where the u's are "error" terms that measure the net effects of all other variables (those not introduced explicitly) upon the system. We refer to $A = \|a_{ij}\|$ as the *coefficient matrix* of the system.

Next, let us suppose that not all the variables directly influence all the others — that some independent variables are absent from some of the equations. This is equivalent to saying that some of the elements of the coefficient matrix are zero. By way of specific example, let us assume that $a_{31} = a_{32} = a_{21} = 0$. Then the equation system (I) reduces to:

$$x + a_{12}y + a_{13}z = u_1 \tag{1.4}$$

(II) $$y + a_{23}z = u_2 \tag{1.5}$$

$$z = u_3 \tag{1.6}$$

---

[5] The question of how we distinguish between "dependent" and "independent" variables is discussed in Simon [7] and will receive further attention in this chapter.

By examining the equations (II), we see that a change in u_3 will change the value of z directly, and the values of x and y indirectly; a change in u_2 will change y directly and x indirectly, but will leave z unchanged; a change in u_1 will change only x. Then we may say that y *is causally dependent on z* in (II), and that x is causally dependent on y and z.

If x and y were correlated, we would say that the correlation was genuine in the case of the system (II), for $a_{12} \neq 0$. Suppose, instead, that the system were (III):

$$x + a_{13}z = u_1 \tag{1.7}$$

(III) $$y + a_{23}z = u_2 \tag{1.8}$$

$$z = u_3 \tag{1.9}$$

In this case, we would regard the correlation between x and y as spurious, because it is due solely to the influence of z on the variables x and y. Systems (II) and (III) are, of course, not the only possible cases, and we shall need to consider others later.

THE *A PRIORI* ASSUMPTIONS

We shall show that the decision that a partial correlation is or is not spurious (does not or does indicate a causal ordering) can in general only be reached if *a priori* assumptions are made that certain *other* causal relations do *not* hold among the variables. This is the meaning of the "common sense" assumptions mentioned earlier. Let us make this more precise.

Apart from any statistical evidence, we are prepared to assert in the first example of Section 1 that the age of a person does *not* depend upon either his candy consumption or his marital status. Hence, z cannot be causally dependent upon either x or y. This is a genuine empirical assumption, since the variable "chronological age" really stands, in these equations, as a surrogate for physiological and sociological age. Nevertheless, it is an assumption that we are quite prepared to make on evidence apart from the statistics presented. Similarly, in the second example of Statement of the Problem, we are prepared to assert (on grounds of other empirical knowledge) that marital status is not causally dependent upon either amount of housework or absenteeism.[6]

---

[6] Since these are empirical assumptions, it is conceivable that they are wrong, and indeed, we can imagine mechanisms that would reverse the causal ordering in the second example. What is argued here is that these assumptions, right or wrong, are implicit in the determination of whether the correlation is true or spurious.

The need for such *a priori* assumption follows from considerations of elementary algebra. We have seen that whether a correlation is genuine or spurious depends on which of the coefficients, a_{ij}, of A are zero, and which are non-zero. But these coefficients are not observable nor are the "error" terms, u_1, u_2 and u_3. What we observe is a sample of values of x, y, and z.

Hence, from the standpoint of the problem of statistical estimation, we must regard the $3n$ sample values of x, y, and z as numbers given by observation, and the $3n$ error terms, u_i, together with the six coefficients, a_{ij}, as variables to be estimated. But then we have $(3n + 6)$ variables ($3n$ u's and six a's) and only $3n$ equations (three for each sample point). Speaking roughly in "equation-counting" terms, we need six more equations, and we depend on the *a priori* assumptions to provide these additional relations.

The *a priori* assumptions we commonly employ are of two kinds:

1. *A priori* assumptions that certain variables are not directly dependent on certain others. Sometimes such assumptions come from knowledge of the time sequence of events. That is, we make the general assumption about the world that if y precedes x in time, then $a_{21} = 0$ — x does not directly influence y.

2. *A priori* assumptions that the errors are uncorrelated — that is, that "all other" variables influencing x are uncorrelated with "all other" variables influencing y, and so on. Writing $E(u_i u_j)$ for the expected value of $u_i u_j$, this gives us the three additional equations:

$$E(u_1 u_2) = 0 \qquad E(u_1 u_3) = 0 \qquad E(u_2 u_3) = 0$$

Again, it must be emphasized that these assumptions are *"a priori"* only in the sense that they are not derived from the statistical data from which the correlations among x, y, and z are computed. The assumptions are clearly empirical.

As a matter of fact, it is precisely because we are unwilling to make the analogous empirical assumptions in the two-variable case (the correlation between x and y alone) that the problem of spurious correlation arises at all. For consider the two-variable system:

$$x + b_{12}y = v_1 \tag{1.10}$$

(IV)

$$y = v_2 \tag{1.11}$$

We suppose that y precedes x in time, so that we are willing to set $b_{21} = 0$ by an assumption of type (1). Then, if we make the type (2) assumption that

$E(v_1 v_2) = 0$, we can immediately obtain a unique estimate of b_{12}. For multiplying the two equations and taking expected values we get:

$$E(xy) + b_{12}E(y^2) = E(v_1 v_2) = 0 \qquad (1.12)$$

when

$$b_{12} = -\frac{E(xy)}{E(y^2)} = -\frac{\sigma_y}{\sigma_x} r_{xy} \qquad (1.13)$$

It follows immediately that (sampling questions aside) b_{12} will be zero or non-zero as r_{12} is zero or non-zero. *Hence, correlation is proof of causation in the two-variable case if we are willing to make the assumptions of time precedence and noncorrelation of the error terms.*

If we suspect the correlation to be spurious, we look for a common component, z, of v_1 and v_2 that might account for their correlation:

$$v_1 \equiv u_1 - a_{13}z \qquad (1.14a)$$

$$v_2 \equiv u_2 - a_{23}z \qquad (1.14b)$$

Substitution of these relations in (IV) brings us back immediately to systems like (II). This substitution replaces the unobservable v's by unobservable u's. Hence, we are not relieved of the necessity of postulating independence of the errors. We are more willing to make these assumptions in the three-variable case because we have explicitly removed from the error term the component z, which we suspect is the source, if any, of the correlation of the v's.

Stated otherwise, introduction of the third variable, z, to test the genuineness or spuriousness of the correlation between x and y, is a method for determining whether in fact the v's of the original two-variable system were uncorrelated. But the test can be carried out only on the assumption that the unobservable error terms of the three variable system are uncorrelated. If we suspect this to be false, we must further enlarge the system by introduction of a fourth variable, and so on, until we obtain a system we are willing to regard as "complete" in this sense.

Summarizing our analysis we conclude that:

1. Our task is to determine which of the six off-diagonal matrix coefficients in a system like (I) are zero.
2. But we are confronted with a system containing a total of nine vari-

ables (six coefficients and three unobservable errors) and only three equations.

3. Hence, we must obtain six more relations by making certain *a priori assumptions*. (*a*) Three of these relations may be obtained, from considerations of time precedence of variables or analogous evidence, in the form of direct assumptions that three of the a_{ij} are zero. (*b*) Three more relations may be obtained by assuming the errors to be uncorrelated.

SPURIOUS CORRELATION

Before proceeding with the algebra, it may be helpful to look a little more closely at the matrix of coefficients in systems like (I), (II), and (III), disregarding the numerical values of the coefficients but considering only whether they are nonvanishing (X), or vanishing (0). An example of such a matrix would be

$$
\begin{Vmatrix}
X & 0 & 0 \\
X & X & X \\
0 & 0 & X
\end{Vmatrix}
$$

In this case, x and z both influence y but not each other, and y influences neither x nor z. Moreover, a change in $u_2 - u_1$ and u_3 being constant — will change y but not x or z; a change in u_1 will change x and y but not z; a change in u_3 will change z and y but not x. Hence, the causal ordering may be depicted thus:

In this case, the correlation between x and y is "true" and not spurious.

Since there are six off-diagonal elements in the matrix, there are $2^6 = 64$ possible configurations of X's and 0's. The *a priori* assumptions (1), however, require 0's in three specified cells, and hence for each such set of assumptions there are only $2^3 = 8$ possible distinct configurations. If (to make a definite assumption) x does not depend on y, then there are three possible orderings of the variables (z, x, y; x, z, y; x, y, z), and consequently $3 \cdot 8 = 24$ possible configurations, but these 24 configurations are not all distinct. For example, the one depicted above is consistent with either the ordering (z, x, y) or the ordering (x, z, y).

Still assuming that x does not depend on y, we will be interested, in particular, in the following configurations:

$$
\begin{Vmatrix} X & 0 & 0 \\ X & X & X \\ 0 & 0 & X \end{Vmatrix} \quad
\begin{Vmatrix} X & 0 & X \\ X & X & 0 \\ 0 & 0 & X \end{Vmatrix} \quad
\begin{Vmatrix} X & 0 & 0 \\ X & X & 0 \\ X & 0 & X \end{Vmatrix}
$$
$$
\quad (\alpha) \qquad\qquad (\beta) \qquad\qquad (\gamma)
$$

$$
\begin{Vmatrix} X & 0 & X \\ 0 & X & X \\ 0 & 0 & X \end{Vmatrix} \quad
\begin{Vmatrix} X & 0 & 0 \\ 0 & X & X \\ X & 0 & X \end{Vmatrix}
$$
$$
\quad (\delta) \qquad\qquad (\epsilon)
$$

In Case α, either x may precede z or z, x. In Cases β and δ, z precedes x; in Cases γ and ϵ, x precedes z. The causal orderings that may be inferred are:

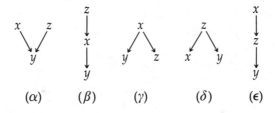

$$
(\alpha) \qquad (\beta) \qquad (\gamma) \qquad (\delta) \qquad (\epsilon)
$$

The two cases we were confronted with in our earlier examples were δ and ϵ, respectively. Hence, δ is the case of spurious correlation due to z; ϵ the case of true correlation with z as an intervening variable.

We come now to the question of which of the matrices that are consistent with the assumed time precedence is the correct one. Suppose, for definiteness, that z precedes x and x precedes y. Then $a_{12} = a_{31} = a_{32} = 0$; and the system (I) reduces to:

$$
x \quad + a_{13}z = u_1 \tag{1.15}
$$

$$
a_{21}x + y + a_{23}z = u_2 \tag{1.16}
$$

$$
z = u_3 \tag{1.17}
$$

Next, we assume the errors to be uncorrelated:

$$
E(u_1 u_2) = E(u_1 u_3) = E(u_2 u_3) = 0 \tag{1.18}
$$

Multiplying equations $(1.15) - (1.17)$ by pairs and taking expected values we get:

$$a_{21}E(x^2) + E(xy) + a_{23}E(xz) + a_{13}[a_{21}E(xz) + E(yz) + a_{23}E(z^2)]$$
$$= E(u_1u_2) = 0 \quad (1.19)$$

$$E(xz) + a_{13}E(z^2) = E(u_1u_3) = 0 \quad (1.20)$$

$$a_{21}E(xz) + E(yz) + a_{23}E(z^2) = E(u_2u_3) = 0 \quad (1.21)$$

Because of (1.21), the terms in the bracket of (1.19) vanish, giving:

$$a_{21}E(x^2) + E(xy) + a_{23}E(xz) \equiv 0 \quad (1.22)$$

Solving for $E(xz)$, $E(yz)$ and $E(xy)$ we find:

$$E(xz) = -a_{13}E(z^2) \quad (1.23)$$

$$E(yz) = (a_{13}a_{21} - a_{23})E(z^2) \quad (1.24)$$

$$E(xy) = a_{13}a_{23}E(z^2) - a_{21}E(x^2) \quad (1.25)$$

Case α: Now in the matrix of case α above, we have $a_{13} = 0$. Hence:

$$E(xz) = 0 \quad (1.26a)$$

$$E(yz) = -a_{23}E(z^2) \quad (1.26b)$$

$$E(xy) = -a_{21}E(x^2) \quad (1.26c)$$

Case β: In this case, $a_{23} = 0$. Hence,

$$E(xz) = -a_{13}E(z^2) \quad (1.27a)$$

$$E(yz) = a_{13}a_{21}E(z^2) \quad (1.27b)$$

$$E(xy) = -a_{21}E(x^2) \quad (1.27c)$$

from which it also follows that:

$$E(xy) = E(x^2)\frac{E(yz)}{E(xz)} \qquad (1.28)$$

Case δ: In this case, $a_{21} = 0$. Hence,

$$E(xz) = - a_{13}E(z^2) \qquad (1.29a)$$

$$E(yz) = - a_{23}E(z^2) \qquad (1.29b)$$

$$E(xy) = a_{13}a_{23}E(z^2) \qquad (1.29c)$$

and we deduce also that:

$$E(xy) = \frac{E(xz)E(yz)}{E(z^2)} \qquad (1.30)$$

We have now proved that $a_{13} = 0$ implies (1.26a); that $a_{23} = 0$ implies (1.28); and that $a_{21} = 0$ implies (1.30). We shall show that the converse also holds.

To prove that (1.26a) implies $a_{13} = 0$, we need only set the left-hand side of (1.23) equal to zero.

To prove that (1.28) implies that $a_{23} = 0$, we substitute in (1.28) the values of the cross-products from (1.23) $-$ (1.25). After some simplification, we obtain:

$$a_{23}[E(x^2) - a_{13}{}^2E(z^2)] = 0 \qquad (1.31)$$

Now since, from (1.15)

$$E(x^2) - E(u_1{}^2) + 2a_{13}E(zu_1) = a_{13}{}^2E(z^2) \qquad (1.32)$$

and since, by multiplying (1.17) by u_1, we can show that $E(zu_1) = 0$, the second factor of (1.31) can vanish only in case $E(u_1{}^2) = 0$. Excluding this degenerate case, we conclude that $a_{23} = 0$.

To prove that (1.30) implies that $a_{21} = 0$, we proceed in a similar manner, obtaining:

$$a_{21}[E(x^2) - a_{13}{}^2 E(z^2)] = 0 \qquad\qquad (1.33)$$

from which we can conclude that $a_{21} = 0$.

We can summarize the results as follows:

1. If $E(xz) = 0$, $E(yz) \neq 0$, $E(xy) \neq 0$, we have Case α
2. If none of the cross-products is zero, and

$$E(xy) = E(x^2)\,\frac{E(yz)}{E(xz)}$$

we have Case β.

3. If none of the cross-products is zero, and

$$E(xy) = \frac{E(xz)E(yz)}{E(z^2)}$$

we have Case δ.

We can combine these conditions to find the conditions that two or more of the coefficients a_{13}, a_{23}, a_{21} vanish:

4. If $a_{13} = a_{23} = 0$, we find that:

$$E(xz) = 0, \; E(yz) = 0 \qquad \text{Call this Case } (\alpha\beta)$$

5. If $a_{13} = a_{21} = 0$, we find that:

$$E(xz) = 0, \; E(xy) = 0 \qquad \text{Call this Case } (\alpha\delta)$$

6. If $a_{23} = a_{21} = 0$, we find that:

$$E(yz) = 0, \; E(xy) = 0 \qquad \text{Call this Case } (\beta\delta)$$

7. If $a_{13} = a_{23} = a_{21} = 0$, then

$$E(xz) = E(yz) = E(xy) = 0 \qquad \text{Call this Case } (\alpha\beta\delta)$$

8. If none of the conditions (1)–(7) are satisfied, then all three coefficients a_{13}, a_{23}, a_{21} are non-zero. Thus, by observing which of the conditions (1)–(8) are satisfied by the expected values of the cross products, we can determine what the causal ordering is of the variables.[7]

We can see also, from this analysis, why the vanishing of the partial correlation of x and y is evidence for the spuriousness of the zero-order correlation between x and y. For the numerator of the partial correlation coefficient $r_{xy \cdot z}$, we have:

$$N(r_{xy \cdot z}) = \frac{E(xy)}{\sqrt{E(x^2)E(y^2)}} - \frac{E(xz)E(yz)}{E(z^2)\sqrt{E(x^2)E(y^2)}} \tag{1.34}$$

We see that the condition for Case δ is precisely that $r_{xy \cdot z}$ vanish while none of the coefficients, r_{xy}, r_{xz}, r_{yz} vanish. From this, we conclude that the first illustrative example of Section 1 falls in Case δ, as previously asserted. A similar analysis shows that the second illustrative example of Section 1 falls in Case ϵ.

In summary, our procedure for interpreting, by introduction of an additional variable z, the correlation between x and y consists in making the six *a priori* assumptions described earlier; estimating the expected values, $E(xy)$, $E(xz)$, and $E(yz)$; and determining from their values which of the eight enumerated cases holds. Each case corresponds to a specified arrangement of zero and non-zero elements in the coefficient matrix and hence to a definite causal ordering of the variables.

THE CASE OF EXPERIMENTATION

In the latter two sections, we have treated u_1, u_2, and u_3 as random variables. The causal ordering among x, y, and z can also be determined without *a priori* assumptions in the case where u_1, u_2, and u_3 are controlled by an experimenter. For simplicity of illustration, we assume there is time precedence among the variables. Then the matrix is triangular, so that $a_{ij} \neq 0$ implies $a_{ji} = 0$; and $a_{ij} \neq 0$, $a_{jk} \neq 0$ implies $a_{ki} = 0$.

Under the given assumptions, at least three of the off-diagonal a's in (I) must vanish, and the equations and variables can be reordered so that all the nonvanishing coefficients lie on or below the diagonal. If (with this ordering) u_2 or u_3 are varied, at least the variable determined by the first equation

---

[7] Of course, the expected values are not, strictly speaking, observables except in a probability sense. However, we do not wish to go into sampling questions here and simply assume that we have good estimates of the expected values.

will remain constant (since it depends only on u_1). Similarly, if u_3 is varied, the variables determined by the first and second equations will remain constant.

In this way, we discover which variables are determined by which equations. Further, if varying u_i causes a particular variable other than the ith to change in value, this variable must be causally dependent on the ith.

Suppose, for example, that variation in u_1 brings about a change in x and y, variation in u_2 a change in y, and variation in u_3 a change in x, y, and z. Then we know that y is causally dependent upon x and z and x upon z. But this is precisely the Case β treated previously under the assumption that the u's were stochastic variables.

CONCLUSION

In this chapter I have clarified the logical processes and assumptions that are involved in the usual procedures for testing whether a correlation between two variables is true or spurious. These procedures begin by imbedding the relation between the two variables in a larger three-variable system that is assumed to be self-contained, except for stochastic disturbances or parameters controlled by an experimenter.

Since the coefficients in the three-variable system will not in general be identifiable, and since the determination of the causal ordering implies identifiability, the test for spuriousness of the correlation requires additional assumptions to be made. These assumptions are usually of two kinds. The first, ordinarily made explicit, are assumptions that certain variables do *not* have a causal influence on certain others. These assumptions reduce the number of degrees of freedom of the system of coefficients by implying that three specified coefficients are zero.

The second type of assumption, more often implicit than explicit, is that the random disturbances associated with the three-variable system are uncorrelated. This assumption gives us a sufficient number of additional restrictions to secure the identifiability of the remaining coefficients, and hence to determine the causal ordering of the variables.

ACKNOWLEDGMENTS

I am indebted to Richard M. Cyert, Paul F. Lazarsfeld, Roy Radner, and T. C. Koopmans for valuable comments on earlier drafts of this chapter.

REFERENCES

[1] Kendall, Patricia L., and Lazarsfeld, Paul F. "Problems of Survey Analysis." In *Continuities in Social Research*, edited by R. K. Merton and P. F. Lazarsfeld. New York: The Free Press, 1950.

[2] Koopmans, Tjalling C. "Identification Problems in Economic Model Construction." *Econometrica* 17 (1949): 125–144. Reprinted as Chapter II in *Studies in Econometric Methods*. Cowles Commission Monograph 14.

[3] Koopmans, Tjalling C. "When Is an Equation System Complete for Statistical Purposes?" *Statistical Inference in Dynamic Economic Models*. Cowles Commission Monograph 10.

[4] Orcutt, Guy H. "Actions, Consequences, and Causal Relations." *The Review of Economics and Statistics* 34 (1952): 305–314.

[5] Orcutt, Guy H. "Toward Partial Redirection of Econometrics." *The Review of Economics and Statistics* 34 (1952): 195–213.

[6] Simon, Herbert A. "On the Definition of the Causal Relation." *The Journal of Philosophy* 49 (1952): 517–528.

[7] Simon, Herbert A. "Causal Ordering and Identifiability." *Studies in Econometric Methods*. Cowles Commission Monograph 14.

[8] Yule, G. Udny. *An Introduction to the Theory of Statistics*. 10th ed. London: Charles Griffin and Co., 1932. (Equivalent chapters will be found in all subsequent editions of Yule and Yule and Kendall, through the 14th.)

[9] Zeisel, Hans. *Say It With Figures*. New York: Harper and Brothers, 1947.

Four-Variable Causal Models and Partial Correlations*

H. M. Blalock, Jr.

2

In several recent papers, the writer has discussed the use of Simon's method for making causal inferences from correlational data.[1] Simon's method yields results that are consistent with those obtained through partial correlations, but it also forces one to make explicit assumptions about both outside disturbing influences and the nature of the various causal links among an entire set of variables. It thereby serves to keep our attention focused on the complete network of causal relationships rather than on a single dependent variable.

But Simon's method can also be used to provide a rationale for predicting what should happen to empirical intercorrelations under various control situations. Our major purpose in systematically treating the four-variable case is to gain certain insights into what happens to a zero-order correlation when controls are introduced for variables that are causally related to the original variables in different ways. A secondary purpose of the chapter is to

* Reprinted by permission of the publisher from the *American Journal of Sociology* 68: 182–194. Copyright 1962, The University of Chicago Press.

[1] See Blalock [1,2,3]. Discussions of the general method and the rationale for expressing causal relations in terms of simultaneous equations can be found in Simon [8,9].

provide a set of prediction equations for four-variable causal models for the convenience of the nonmathematically inclined social scientist, who may not wish to carry out the computations required by Simon's method.

Why focus on the four-variable case to study what happens in the controlling process? Since even the five-variable case will involve over a thousand distinct models, a systematic treatment of most of these possibilities would be out of the question. Why not merely make an exhaustive study of the three-variable case? First, the three-variable case has been more or less systematically treated in the literature, and there is no need to repeat these discussions.[2] But second, the three-variable case is too simple for our purposes. For example, we may wish to consider models in which a relationship between two variables could be partly spurious but also partly indirect through an intervening variable. Although in the present chapter we cannot discuss the important problem of *when* and *why* we control in such instances, we can at least deal with the question of *what* will happen if controls are made.[3]

RATIONALE AND LIMITATIONS
OF THE METHOD

We cannot attempt a formal definition of causality without becoming involved with issues that are outside our present focus. Metaphysically, it is difficult to think without the aid of notions such as *causes* or *forces*. According to Bunge, the essential idea behind causal thinking is that of some agent *producing* some change of state in a system [4]. It is not necessary that there be a definite temporal sequence or even a constant conjunction of events, though we do require that an effect can never precede its cause in time [4, p. 438]. Simon argues that the really essential aspect of a causal relationship is that it is asymmetrical in nature and not that it always involves temporal sequences [8, p. 51]. In view of the difficulties in actually demonstrating causal relationships empirically, Simon prefers to confine the notions of cause and effect to *models* of the real world that consist of only a finite number of explicitly defined variables [8, p. 51]. We shall follow Simon in this respect.

We begin by postulating a causal model involving a given number of

---

[2] See esp. Simon [9]; Nowak [7]; Hyman [6].

[3] Here we shall deal only with the sort of control situation in which one is interested in seeing whether or not a relationship between two variables disappears in controlling for antecedent or intervening variables. In such instances, we may suspect that there is no direct link between the two variables being related and that the relationship is either spurious or can be interpreted through an intervening variable.

variables X_i. We make certain simplifying assumptions about how these variables are interrelated and about the behavior of variables that have not been included in the causal network. In particular, we shall assume additive and linear models, as is commonly done in regression analyses. It should be specifically noted that the presence of nonlinear relationships may invalidate our conclusions. We also make the explicit assumption that variables that have been left out of the causal model create "error" terms that are essentially random. Whatever factors produce variations in one of the X_i should be uncorrelated with factors that give rise to disturbances in any of the remaining X's.

In nonexperimental situations, such as cross-sectional studies, this particular kind of assumption may not be so plausible as in the case of experimental designs in which randomization has been possible. Our only alternative in such nonexperimental studies is to bring as many outside disturbing influences as possible into the causal picture as explicit variables. This will ordinarily necessitate the use of a larger number of variables, and more complex causal models, than would be required to handle experimental data. And, unfortunately, the more variables we use, the simpler our assumptions usually have to be about *how* these variables are interrelated. The reader is therefore cautioned concerning the possibility of reaching erroneous conclusions with the method when one or more of these assumptions are violated.

Having committed ourselves on a particular set of variables, we can then define a causal relationship in terms of manipulations that might be carried out in an ideal experiment. We shall say that X is a direct cause of Y (written $X \rightarrow Y$) if and only if we can produce a change in the mean value of Y by changing X, holding constant *all* other variables that have been explicitly introduced into the system and that are not causally dependent upon Y. Since there is a finite number of such variables, and since they have been explicitly defined, there is no problem of determining whether or not one has controlled for all "relevant" variables.

The causal notion is confined to this particular model, as is the directness of the causal relationship. The introduction of a single additional variable might therefore change a causal relationship from direct to indirect, or even to a noncausal one. By an indirect causal relationship we mean one in which a change in X produces a change in certain other variables, which, in turn, affect Y. More precisely, we can say that X is an indirect cause of Y, within a particular causal model, if and only if it is possible to find certain variables $U, V, \ldots W$, explicitly included in the system, which are such that $X \rightarrow U \rightarrow V \rightarrow \ldots \rightarrow W \rightarrow Y$.

We now write a set of equations, which must hold simultaneously, and which express the various causal relationships mathematically. In a one-way linear causal system, these will be of the form

$$X_1 = e_1$$

$$X_2 = b_{21}X_1 + e_2$$

$$X_3 = b_{31 \cdot 2}X_1 + b_{32 \cdot 1}X_2 + e_3$$

$$X_4 = b_{41 \cdot 23}X_1 + b_{42 \cdot 13}X_2 + b_{43 \cdot 12}X_3 + e_4$$

These particular equations indicate that X_1 depends causally only on outside variables, the effects of which are represented by e_1. But X_2 depends on X_1 as well as outside factors, X_3 depends upon both X_1 and X_2, and finally X_4 depends upon all of the remaining X's. Such a system is referred to as "recursive," as contrasted with a possible set of equations in which there may be reciprocal causation in which X_1 depends upon X_2 and vice versa.[4] Recursive equations have the property that the various b's can all be estimated without bias by means of simple least-squares methods [10, p. 51].

A particular recursive set of equations involves the *assumption* that a given causal arrangement is appropriate. If some of the b's can be assumed to be equal to zero, certain restrictions will be imposed on the data if the equations are to be mutually consistent. For each of the b's set equal to zero, we impose the condition that the comparable partial correlation should also be zero, thus obtaining a prediction that can actually be tested from the data. But setting one of the b's equal to zero is equivalent to our postulating that there is no *direct* causal link between the two variables concerned. Thus, if we set $b_{31 \cdot 2} = 0$, we are saying that there is no direct causal link between X_1 and X_3, and that a partial correlation between X_1 and X_3 (in this case $r_{13 \cdot 2}$) should vanish. Similarly, if $b_{42 \cdot 13} = 0$, this means that we are assuming that X_2 is no longer a direct cause of X_4 and therefore $r_{24 \cdot 13}$ should be zero. But X_2 may of course be an indirect cause of X_4 through X_3.

Before considering four-variable causal models, one further caution should be introduced. There will always be a number of alternative models that will give the same predictions (e.g., the same vanishing partials) as any particular model under study. We can only proceed by *eliminating* inadequate models that give incorrect predictions, since it will ordinarily be impossible to rule out all the logical alternatives on the basis of the data at hand. In this sense, one can never "establish" any given causal model.

---

[4] See Wold and Juréen [10]. As Wold and Juréen point out, recursive systems can also be used to handle instances of reciprocal causation, provided that "feedback" is not more or less instantaneous. By lagging certain variables and collecting data at several points in time, we may take a given variable as independent at time t-1 and also as dependent at time t.

SPECIFIC FOUR-VARIABLE MODELS

Let us designate the four variables as X_1, X_2, X_3, and X_4. Since, for any given set of data, the variables can always be labeled arbitrarily, we shall suppose that X_4 cannot be a cause of the other three variables, that X_3 cannot cause either X_1 or X_2, and that X_2 cannot cause X_1. In effect, then, we are ruling out two-way causation, and we are supposing that, unless the appropriate arrows have been omitted, we are moving in the direction $X_1 \rightarrow X_2 \rightarrow X_3 \rightarrow X_4$ rather than the other way around.

With four variables, there will be six pairs of relationships. Since any given arrow may be either present or absent, there will be 2^6, or 64, possible causal situations. Certain of these models can be eliminated as either trivial or uninteresting, however. If all six arrows are present, Simon's method cannot be used. Cases involving no arrows or one arrow are completely trivial. In instances where there are only two arrows, these arrows will either by-pass one variable completely, in which case we are dealing with only three variables, or they will connect two completely unrelated pairs of variables. Therefore, we need be concerned only with situations in which there are three, four, or five arrows.

The various possible causal models involving three, four, and five arrows are given in Fig. 2.1. Before discussing some of these cases individually, we should first clarify the notation and organization used. The six possible models involving five arrows have been presented at the top of the figure. These models have been labeled in terms of the particular arrow omitted from the diagram. In model A, the arrow between X_1 and X_2 has been left out; in B, the arrow between X_1 and X_3 is missing; and, finally, in F we are dealing with the situation in which there is no direct link between X_3 and X_4.

Single numerical subscripts have been used to denote subtypes for the four-arrow cases. For example, in model A_1 not only is there no link between X_1 and X_2 (as is true for all A's), but there is also no arrow between X_1 and X_3. Similarly, in A_2 there are no arrows between X_1 and X_2 and also between X_2 and X_3. In A_3, there is no arrow between X_1 and X_4, and so on. Notice that when we come to the B series in four arrows, there is no B_1 because of the fact that if we use the same subscripts as in the A series, a subscript of 1 would indicate no link between X_1 and X_3, and this is already covered by B. But we have used the symbol B_2 for the B model involving no link between X_2 and X_3. When we come to the C series, there will be no need for either C_1 or C_2, since both of these cases have been handled previously. Finally, there will be only one four-arrow case in the E series (i.e., E_5) and none in the F series. There are $6 \times 5/2$, or 15, four-arrow cases in all.

Double numerical subscripts have been used in the three-arrow cases in analogous fashion. For example, the symbol A_{12} is associated with the three-arrow model, which represents a combination of A_1 and A_2, with

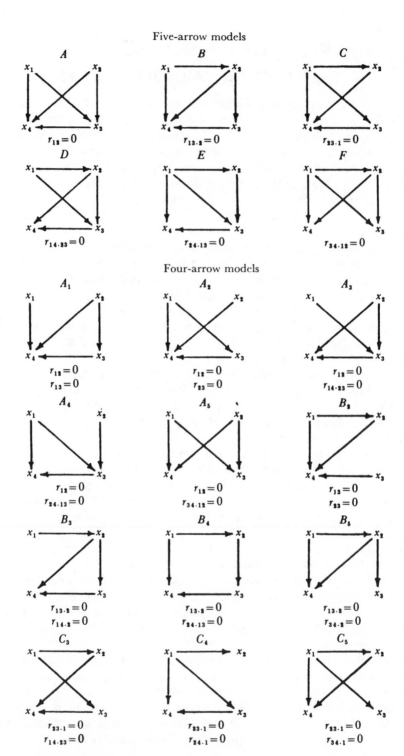

Figure 2.1. Prediction equations for four-variable causal models.

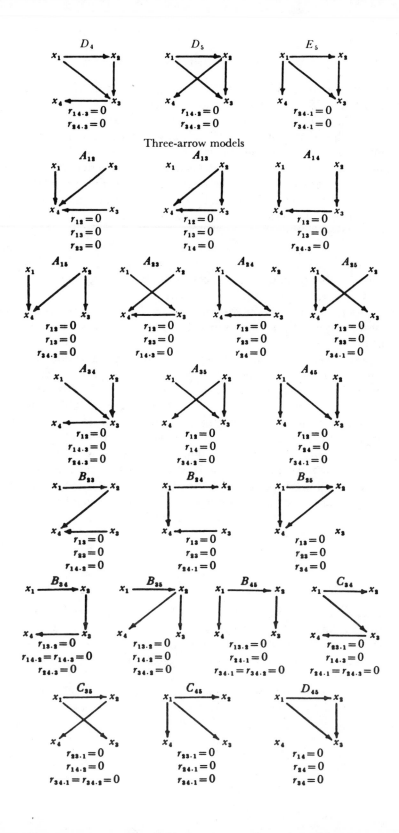

arrows missing between X_1 and X_2, X_1 and X_3, and X_2 and X_3. Likewise, B_{23} is a composite of B_2 and B_3, with no arrows between X_1 and X_3, X_2 and X_3, and X_1 and X_4. Once the reader has become familiar with this particular notational scheme, it should be relatively easy to pass back and forth from one model to another whenever comparisons are to be made.

Notice that we have one prediction equation for each five-arrow model, two equations for each four-arrow model, and three for every model involving only three arrows. This is in line with the fact that Simon's method yields a prediction equation for each pair of variables that have *not* been connected by an arrow. Each prediction equation involves the disappearance of some partial or zero-order correlation between a pair of unconnected variables. For example, if X_2 and X_4 have not been linked with an arrow, then either r_{24} or one of the partials, with X_1 and X_3 as controls, will vanish. Conceivably, all correlations between X_2 and X_4 (zero, first, and second order) may be zero. But they may not, and our problem becomes that of determining exactly which ones will vanish under a particular set of causal conditions.

As we shall see in the four-variable case, and as appears to be true more generally, certain higher-order partials can always be expected to disappear. These vanishing partials involve controls for *all* variables that are either antecedent to, or intervening between, the particular variables being related, but they do *not* involve controls for variables taken to be dependent upon *both* of these variables. Thus, if there were no arrow between X_1 and X_3, the value of $r_{13 \cdot 2}$ should be approximately zero, but if we were to control for X_4, we would not expect the partial to disappear. Conceivably, however, the *total* association between X_1 and X_3 might be zero. In relating X_3 and X_4, supposing no direct link between these variables, the value of $r_{34 \cdot 12}$ should be approximately zero. But we do not, as yet, know whether or not $r_{34 \cdot 1}$, $r_{34 \cdot 2}$, or r_{34} should also vanish.

The A Series

In all of the A-series models, there is no direct link between X_1 and X_2, the two variables that are taken as causally prior to the remaining variables X_3 and X_4. This will mean that under these models, the *total* association between X_1 and X_2 should be approximately zero, subject to the possibility of sampling error. Suppose we were to control for one or both of the two variables that are taken to be dependent upon X_1 and X_2. It can be seen that the resulting partials will *not* be equal to zero, except in those situations in which certain other arrows linking the first two variables with X_3 and X_4 have also been erased. An examination of the formula for a first-order partial will clarify the point:

$$r_{12\cdot3} = \frac{r_{12} - r_{13}r_{23}}{\sqrt{1 - r_{13}^2}\sqrt{1 - r_{23}^2}}$$

Since the denominator on the right cannot be greater than unity, it is clear that the partial will vanish only when the numerator is zero. Simon's method indicates that r_{12} should be zero, but we are asking whether or not $r_{12\cdot3}$ will also vanish. Evidently, this can occur only if either r_{13} or r_{23} is zero, that is, if X_3 is unrelated to one or the other of the first two variables. This, in fact, occurs in both A_1 and A_2, but since it does not always happen in the A series, we cannot generally conclude that the disappearance of r_{12} implies the vanishing of $r_{12\cdot3}$ or $r_{12\cdot4}$. This particular point should take on more meaning as we consider some of the remaining models.

Let us now turn our attention to four-arrow models in the A series. Notice that models A_1 and A_2 are basically similar, with the roles of variables X_1 and X_2 interchanged. Likewise, A_3 and A_4 are similar. It is because of the fact that there is no link between X_1 and X_2, making the question of asymmetry or temporal sequences irrelevant for these two variables, that they can be interchanged in such a manner. In A_1 and A_2, we have one of the independent variables being a direct cause of X_4 but completely unrelated to the remaining two variables. In A_1, for example, X_1 is in no way related to X_2 or X_3, and these zero-order correlations with X_1 vanish. If one were to relate X_1 to X_2, controlling for X_3, the resulting partial would also be zero, as we have just seen. But a control for X_4 would ordinarily produce a non-zero partial.

Similarly, both A_3 and A_4 represent situations in which an independent variable (i.e., X_1 or X_2) is a direct cause of only the "middle" variable in a set of three other variables. In relating this independent variable to X_4, one must control for *both* the remaining variables in order to have the association disappear. In A_3, for example, not only is the relationship between X_1 and X_4 indirect through X_3, but since X_2 is operating to produce a partly spurious relationship between X_3 and X_4, we must also control for X_2. In essence, once we have taken out the effects of X_2 on the relationship between X_3 and X_4, we can then handle the problem as though it involved only three variables, with X_3 intervening between X_1 and X_4.

In A_5, we have the case of a spurious relationship between X_3 and X_4 caused by the independent operation of X_1 and X_2. As may be expected, we would have to control for both of these variables in order to have the partial disappear. In this special case, where the two independent variables are completely unrelated, we find a very simple relationship among the correlations, namely,

$$r_{34} = r_{13}r_{14} + r_{23}r_{24}$$

If we examine the three-arrow cases in the A series, we notice an important fact that also applies to the remaining three-arrow models. The prediction equations will involve the disappearance of either first-order partials or zero-order coefficients. It can easily be shown that the appropriate second-order partials also disappear, but we can make use of "simplified" equations that require only that the lower-order partials disappear.

We can see numerically why this will occur. If we compare, for example, A_{14} with the four-arrow model A_4, we note that for the latter model, we have $r_{24 \cdot 13} = 0$, whereas we need only control for X_3 in A_{14} in order to have the relationship between X_2 and X_4 disappear. This is true because in A_{14} we also have $r_{13} = 0$ (see A_1). With both r_{12} and r_{13} being zero, we then know that $r_{12 \cdot 3}$ must be zero as well. But

$$r_{24 \cdot 31} = \frac{r_{24 \cdot 3} - r_{12 \cdot 3} r_{14 \cdot 3}}{\sqrt{1 - r_{12 \cdot 3}^2} \sqrt{1 - r_{14 \cdot 3}^2}}$$

Since from A_4 we know that $r_{24 \cdot 31}$ is zero, and since $r_{12 \cdot 3} = 0$, we must also have $r_{24 \cdot 3} = 0$. Thus, the combined facts that the second-order partial is zero and $r_{12} = r_{13} = 0$ imply that the first-order partial (with X_3 as control) will likewise be zero.

Notice what has happened when A_1 and A_4 have been combined into A_{14}. We have dropped the diagonal arrows and have reduced the situation to one in which there is a single independent variable X_1 operating on only one variable (in this case X_4) of a set of variables (X_2, X_3, and X_4), which are interrelated by only two arrows. When we now obtain the prediction equation linking X_2 and X_4, we find that we can actually ignore the effects of the first variable and still have the partial reduced to zero.

Similarly, in A_{15} we can ignore X_1 and still have $r_{34 \cdot 2}$ vanish. Since X_2 affects only X_4 in A_{23}, we likewise obtain the result that $r_{14 \cdot 3} = 0$. In A_{35}, we see that X_1 affects only X_3, and therefore $r_{34 \cdot 2}$ vanishes as well as $r_{34 \cdot 12}$. These facts can easily be verified by an examination of the appropriate formulas.

The particular phenomenon we are noting did not arise in the four-arrow models because of the fact that if only a single arrow connected any one variable to the others, there must have been *three* additional arrows linking the triad of remaining variables, making it impossible for Simon's method to yield a prediction equation among these latter variables. After we have examined the B series, we shall be in a position to make a more general assertion about this sort of situation found in the three-arrow models.

Several of the three-arrow situations are worthy of comment. Notice that in both A_{13} and A_{24} (as well as B_{25} and D_{45}) we have trivial cases in which one variable has been completely cut out and where we are essentially dealing with only three variables. In each of these instances, the prediction equations involve vanishing zero-order correlations with the isolated variable. But there are no predictions for the remaining three variables. Model A_{12} is another simple case in which we have three unrelated variables operating on a single dependent variable. In A_{34}, we have two independent variables causing X_3, which in turn causes X_4. In relating either of these former variables to X_4, we may merely control for X_3 in order to have the partial vanish. It will not be necessary to control for the remaining independent variable.

The B Series

In the B series, we have no direct link between X_1 and X_3, though we are now assuming that X_1 is a cause of X_2. A control for the intervening variable X_2 gives a zero partial between X_1 and X_3, but since X_4 is assumed not to cause either X_1 or X_3, directly or indirectly, an additional control for X_4 ordinarily produces a nonvanishing partial.

We see that B_2 involves basically the same type of situation as A_1 and A_2, with a variable causing X_4 completely independently of the other two variables. We again find that two total correlations (in this case, r_{13} and r_{23}) should disappear. Both $r_{13 \cdot 2}$ and $r_{23 \cdot 1}$ will also be zero because of the fact that X_3 is unrelated to *both* X_1 and X_2.

The situations represented in B_3 and B_5, as well as certain of the models appearing in later series, pose a more difficult problem. In B_3, the prediction equation tells us that it should be unnecessary to control for X_3 in order for the partial between X_1 and X_4 to vanish. Similarly, in B_5 the partial between X_3 and X_4 (controlling for X_2) is zero even without a control for X_1. An examination of the formula for $r_{14 \cdot 23}$ shows us that for B_3 the pair of prediction equations $r_{13 \cdot 2} = 0$ and $r_{14 \cdot 2} = 0$ automatically implies that $r_{14 \cdot 23}$ must also be zero since

$$r_{14 \cdot 23} = \frac{r_{14 \cdot 2} - r_{13 \cdot 2} r_{34 \cdot 2}}{\sqrt{1 - r_{13 \cdot 2}^2} \sqrt{1 - r_{34 \cdot 2}^2}}$$

Likewise in B_5, we can show that the equations $r_{13 \cdot 2} = 0$ and $r_{34 \cdot 2} = 0$ give us the result that $r_{34 \cdot 21}$ is also zero. Notice that in both B_3 and B_5 we have two prediction equations involving first-order partials using the *same control variable* (in each case, X_2). We shall find this again happening in certain other models (i.e., C_4, C_5, D_4, D_5, and E_5).

In B_3, we see that the link between X_2 and X_4 is both direct and indirect through X_3. The numerical value of r_{24} will reflect the influence of X_3 as well as the direct effect of X_2. It will therefore be unnecessary to control for X_3 in relating X_1 and X_4. Similarly, in B_5 the relationship between X_2 and X_4 is both direct and partly spurious, and the numerical value of r_{24} will reflect this fact. We thus need only control for X_2 in order to reduce the partial between X_3 and X_4 to zero.

There is an apparent similarity between both B_3 and B_5 and two of the models in the A series, namely, A_3 and A_4. In the latter two models, we have an independent variable connected to a triad of variables by an arrow to the middle variable of the triad. In relating this independent variable to X_4, the dependent variable of the triad, we found it necessary to control for *both* remaining variables in order for the partial to vanish. One might raise the question as to why it was necessary in A_3, for example, to control for X_2 as well as X_3, since the numerical value of r_{34} reflects the fact that the relationship between X_3 and X_4 is partly spurious. The essential difference between these two A models and B_3 and B_5, as well as certain other models in later series, seems to be that in both A_3 and A_4 we have two variables operating completely independently of each other on the remaining variables. Even though the diagrams may look similar to some of those for models in other series, the algebra shows that when one of these independent variables is related to the dependent variable, the effects of the other cannot be ignored if we expect the partial to vanish.

The remaining four-arrow case in the B series can be handled briefly. In B_4, the relationship between X_2 and X_4 is partly spurious and partly indirect through X_3. Controls for both X_1 and X_3 will therefore be necessary in order to reduce the relationship between X_2 and X_4 to zero. We have, here, an instance where the two control variables operate in different ways, a fact that can easily be overlooked if one does not develop the habit of drawing causal diagrams before attempting to interpret his findings.

Turning finally to the three-arrow models in the B series, we note only one new type of result. In both B_{34} and B_{45} we see instances in which *either* one of two first-order partials will disappear. In B_{34}, we have a simple causal chain in which X_1 causes X_2, X_2 causes X_3, and X_3 causes X_4. When we come to the correlation between X_1 and X_4, we obtain the very simple result that

$$r_{14} = r_{12} r_{23} r_{34}$$

A control for either X_2 or X_3, the two variables that are intermediate in the chain, will produce a zero partial between X_1 and X_4.

In B_{45}, we do not have a causal chain, but it again turns out that we can control for either of the two "middle" variables and still produce a zero partial between X_3 and X_4.[5] In this particular model we also get the simple

[5] By "middle variables" we do not necessarily mean X_2 and X_3 but whatever variables are connected to two of the remaining variables. By "end variables" we shall mean the two variables, in a simple sequence, that are connected directly to only one other variable. The type of results found in models B_{34} and B_{45}, as well as C_{34} and C_{35}, seem to hold in the general case. For example, in the k-variable causal chain, a control for any of the middle variables will produce a zero partial.

result that

$$r_{34} = r_{14}r_{12}r_{23}$$

In the case of the three-arrow models in the A series, we noted that none of the prediction equations involved second-order partials. We saw that, in instances where there was an independent variable affecting only one of the three remaining variables, the effects of this independent variable could be ignored and still have the appropriate partials reduced to zero. But actually in the A series, the single independent variable always operated on a dependent or intervening variable in the triad of remaining variables. In B_{34} and B_{35}, however, we have instances in which X_1 affects only X_2, the variable that would be taken as *independent* in the triad (X_2, X_3, X_4). Nevertheless, we still find it unnecessary to control for X_1 in order for the appropriate partials to disappear.

We thus seem to have a general rule that *whenever a given variable affects directly only one of the remaining variables, its effects on the partials for the latter variables can safely be ignored.* It appears as though this general rule will also hold for more than four variables, though to the writer's knowledge this has not been proven mathematically.

The Remaining Series

We can discuss the C, D, E, and F series much more briefly since there are no basically new problems posed by these models. In the C series, there is no direct link between X_2 and X_3, these variables being spuriously related through X_1. The value of $r_{23 \cdot 1}$ should be zero, although if we were to control for the dependent variable X_4 the value of $r_{23 \cdot 14}$ would ordinarily not be zero.

In C_3, we have the case in which X_1 and X_4 are related indirectly by means of two intervening variables, themselves not directly related, both of which must be controlled if the partial between X_1 and X_4 is to vanish. Models C_4 and C_5 are basically the same, with the roles of X_2 and X_3 being interchanged. In each case, X_1 causes both X_2 and X_3. But in C_4, there is no link between X_2 and X_4, whereas in C_5 there is no arrow from X_3 to X_4. Examining the relationship between X_4 and whichever of these variables is *not* linked to it with an arrow, we find that when we control for X_1, the partial disappears. No control for the remaining variable is needed.

Models C_{34} and C_{35} have prediction equations that are similar to those of B_{34} and B_{45}. When relating the two "end variables" we can obtain a zero partial by controlling for either of the variables falling intermediate in the sequence. Model C_{45} represents the special case where X_2, X_3, and X_4 are all spuriously related through X_1, so that a control for this latter variable produces vanishing partials among the former variables.

In the D series, X_1 and X_4 are only indirectly related through the intervening variables X_2 and X_3. In D_4, there is no need to control for X_2 in relating X_1 and X_4 since the indirect effects of X_1 on X_3 through X_2 will be taken into consideration in the value of r_{13}. Likewise, in this same model, we do not need to control for X_1 in order for the partial between X_2 and X_4 to disappear. Essentially the same kinds of control situations occur in D_5. No special comments are necessary concerning D_{45}, which represents the trivial case in which X_4 is completely unrelated to any of the remaining variables.

Model E is the situation in which the relationship between X_2 and X_4 is partly spurious and partly indirect but with the added complication that X_1 also directly affects X_3. In F, the relationship between X_3 and X_4 is spurious, owing to the effects of the two related variables X_1 and X_2. In E_5, we find another example of a situation in which it becomes unnecessary to control for more than one variable in order to have a partial disappear.

EXTENSIONS TO MORE THAN FOUR VARIABLES

Considering the present stage in the development of sociological theory, as well as the fact that a number of variables in any particular study will usually be only weakly related to most of the remainder, it will probably seldom be necessary to make use of models involving more than five or six variables. By using an alternative computing routine, which will be described elsewhere, it should not be too difficult to work out the prediction equations for any particular five-variable model, although even the six-variable case begins to involve rather tedious computations.

Certain relatively simple procedures can be suggested, however, for situations in which the investigator wishes to test the adequacy of any particular causal model but is not interested in determining exactly which of the lower-order partials will disappear. First, one can often eliminate certain variables, either because of their low correlations with other variables or because they appear in positions analogous, say, to that of X_3 in model B_3 or of X_1 in model B_5. Second, one may then single out the first four of the remaining variables, leaving out of the picture those variables that stand in dependent relationships to these four variables. The prediction equations among these four variables can then be found from Fig. 2.1. Here, we are making use of the fact that prediction equations never involve controls for variables that are dependent, directly or indirectly, upon *both* variables being interrelated.

Finally, it can be shown for the general case, assuming one-way causation, that we may reintroduce the remaining variables, writing a prediction equation for each arrow that has been omitted when these remaining variables are related to each other and to the four variables taken as causally prior. Each of these latter prediction equations will involve the disappearance of the highest-order partials, using controls for *all* variables except those taken to be causally dependent upon the two variables being related. In any particular case, certain lower-order partials may also vanish, but it may not be feasible to carry out the algebra to determine exactly which of these partials can be expected to do so under the assumed model.

CONCLUDING REMARKS

We have seen that Simon's method for making causal inferences from correlational data can be used to provide a systematic basis for predicting what should happen when we control under a given causal model. The four-variable situation has been taken up in detail, and certain suggestions have been made for handling larger numbers of variables. In discussing the various specific four-variable models, we have made sense intuitively out of the prediction equations derived by Simon's method. It is possible that many, if not all, of these equations might have been developed by using "common sense." The value of Simon's method, however, is in providing a rationale for whatever intuitive ideas we may have and in giving us a check whenever common sense might lead us astray.

Strictly speaking, Simon's method provides us with such a rationale only when we have made use of interval scales and linear models. Since a dichotomy can be considered a special case of an interval scale, the rationale would also seem appropriate for attribute data, though results with such data should be interpreted cautiously in view of possible peculiarities produced by extreme marginals or the use of arbitrary cut-points.[6] At present, there seems to be no comparable rationale for making causal inferences from rank-order correlations, but even in the absence of such a rationale, one might still wish to make causal inferences on the basis of a measure such as Kendall's tau. Provided he considered the study exploratory and provided he interpreted the results with extreme caution, he might thereby gain valuable theoretical insights that otherwise would have been lost.

The problems of evaluating sampling error and developing a satisfactory set of criteria for deciding on the adequacy of the goodness of fit of a particular model to a given set of data are too complex to be discussed in the present chapter. Provided one is willing to assume multivariate normality, he might make a series of significance tests for the disappearance of the various partials. However, it should be noted that (1) the researcher may often be on the wrong end of the test, actually wishing to accept the null hypothesis that a particular model is correct; and (2) the results of a signifi-

---

[6] For excellent discussions of causality in terms of attribute data, see Nowak [7] and Francis [5, Chapter iii]. It should be noted, however, that if the concept of *causality* is defined or conceived in terms of attributes (e.g., in terms of necessary and/or sufficient conditions for the presence of an attribute), one runs into difficulties in conceptualizing causal relationships among continuous variates.

cance test are dependent not only upon the degree to which a particular partial departs from zero, but upon the size of the sample as well.

Finally, it must be recalled that, since a number of different causal models all yield the same set of prediction equations, the mere fact that we cannot reject a particular model does not mean that we have established its validity.[7] We can, however, determine which models successfully resist elimination.

REFERENCES

[1] Blalock, H. M. "Correlation and Causality: The Multivariate Case." *Social Forces* 39 (1961): 246–251.

[2] Blalock, H. M. "Evaluating the Relative Importance of Variables." American Sociological Review 26 (1961): 866–874.

[3] Blalock, H. M. "Spuriousness versus Intervening Variables: The Problem of Temporal Sequences." *Social Forces* 40 (1962): 330–336.

[4] Bunge, M. "Causality, Chance, and Law." *American Scientist* 49 (1961): 432–448.

[5] Francis, R. G. *The Rhetoric of Science.* Minneapolis: University of Minnesota Press, 1961.

[6] Hyman, H. *Survey Design and Analysis.* Glencoe, Ill.: The Free Press, 1955.

[7] Nowak, S. "Some Problems of Causal Interpretation of Statistical Relationships." *Philosophy of Science* 27 (1960): 23–38.

[8] Simon, H. A. "Causal Ordering and Identifiability." In *Studies in Econometric Method,* edited by W. C. Hood and T. C. Koopmans, pp. 49–74. New York: John Wiley & Sons, 1953.

[9] Simon, H. A. *Models of Man.* New York: John Wiley & Sons, 1957.

[10] Wold, H., and Juréen, L. *Demand Analysis.* New York: John Wiley & Sons, 1953.

[7] Figure 2.1 has been set up in such a manner that it appears as though all causal models yield different predictions. But it must be remembered that the variables could have been reordered and that the particular variable labeled X_1 might have been taken as X_3 instead.

Path Coefficients and Path Regressions: Alternative or Complementary Concepts?*

Sewall Wright

3

INTRODUCTION

In a recent paper, Turner and Stevens [5] develop a modification of the method of path coefficients (Wright [9,13,15]). Following Tukey [5], they advocate systematic replacement in path analysis of the dimensionless path coefficients by the corresponding concrete path regressions. The purpose of the present chapter is to discuss this and other points that they raise.

1. The authors concur with Tukey in treating the standardized and concrete forms of correlational statistics as if they were alternative conceptions between which it is necessary to make a choice. It has always seemed to me that these should be looked upon as two aspects of a single theory corresponding to different modes of interpretation that, taken together, often give a deeper understanding of a situation than either can give by itself.

2. Even when the sole objectives of analysis are the concrete coefficients, actual path analysis takes a simpler and more homogeneous form in terms of the standardized ones. The application of the method to data usually requires algebraic manipulation of coefficients pertaining to unmeasured

* Reproduced from S. Wright, "Path coefficients and path regressions: alternative or complementary concepts?" *Biometrics* 16 (1960): 189–202. With permission from the author and The Biometric Society.

variables on the same basis as measured ones. As the former can only be dealt with in standardized form, homogeneity requires that all be so dealt with in the course of the algebra. It is such a simple matter to pass from either form to the other (in the cases in which standard deviations are available to all) that the economy of effort in using the concrete coefficients as far as possible, where these are the objectives, is usually outweighed by the loss of economy in other respects.

3. It is of first importance in path analysis to make use of all of the available data. This is not done by Turner and Stevens in most of their examples. The use of standardized coefficients leads naturally to the systematic expression of all of the available information in the form of equations to be solved simultaneously.

4. Turner and Stevens, again following Tukey, go into the direct treatment of reciprocal interaction between variables by path analysis. This interesting topic requires more extended discussion than is appropriate here.

REVIEW OF THE METHOD OF PATH ANALYSIS

Before taking up these points in detail, it seems necessary to review the method briefly to try to clear up certain misunderstandings.

The method is one for dealing with a system of interrelated variables. It is based on the construction of a qualitative diagram in which every included variable, measured or hypothetical, is represented (by arrows) either as *completely* determined by certain others (which may be represented as similarly determined) or as an *ultimate* factor. Each ultimate factor in the diagram must be connected (by lines with arrowheads at both ends) with each of the other ultimate factors to indicate possible correlation through still more remote unrepresented factors, except in cases in which it can safely be assumed that there is no correlation.

The necessary formal completeness of the diagram requires the introduction of a symbol for the array of unknown residual factors among those back of each variable that is not represented as one of the ultimate factors, unless it can safely be assumed that there is complete determination by the known factors. Such a residual factor can be assumed by definition to be uncorrelated with any of the other factors immediately back of the same variable but cannot be assumed to be independent of other variables in the system without careful consideration.

It is assumed here that all relations are linear: Nonlinear relations may sometimes be transformed systematically throughout a diagram into linear ones. Approximate results may be obtained without transformation where deviations from linearity are small within the range of actual variation. Thus, a product of uncorrelated variables XY may be treated as approximately additive $\delta(XY) = \overline{Y}\,\delta X + \overline{X}\,\delta Y$ if the coefficients of variability $\sigma_x/\overline{X}$ and $\sigma_y/\overline{Y}$ are not too large. If the latter are equal, the fraction of the variance that is excluded is less than one-half the squared coefficient of variability. It is also possible to deal rigorously with joint variability in restricted cases, but this extension will not be dealt with here.

The validity of the system requires that variables that enter into two or more relations in the system (as a common factor of two or more, or as an intermediary in a chain) act as if point variables. If one part of a composite variable (such as a total or average) is more significant in one relation and another part in another, the treatment of the variable as if it were a unit may lead to grossly erroneous results. Fortunately, the parts of a composite variable are often known to be so strongly correlated in their values, or in their action, or both, that they may be used to obtain approximate results. It cannot be emphasized too much, however, that apart from special extensions, the strict validity of the method depends on the properties of formally complete linear systems of point variables.

The primary purpose was stated in the first general account (Wright [9]) as follows:

> The present paper is an attempt to present a method of measuring the direct influence along each separate path in such a system and thus of finding the degree to which variation of a given effect is determined by each particular cause. The method depends on the combination of knowledge of the degrees of correlation among the variables in a system with such knowledge as may be possessed of the causal relations. In cases in which the causal relations are uncertain, the method can be used to find the logical consequences of any particular hypothesis in regard to them.

It was brought out here and later that the method "is by no means restricted to relations that can be described as ones of cause and effect. It can be applied to purely mathematical systems of linear relations and merges into the methods of multiple regression and multivariable vectorial analysis when applied to the symmetrical systems of relations that characterize these methods" [15].

The basic diagram in developing the theory is one in which a variable V_0 (Fig. 3.1) is represented as completely determined by a number of immediate factors $V_1, V_2, \ldots, V_m, V_u$, all of which, except the unknown residual V_u, are represented as intercorrelated. We are to consider the correlation of V_0 with any variable V_q. The latter must be represented as correlated with each of the factors of V_0, including the residual V_u if there is no reason to the contrary.

As all relations are assumed to be linear, we have:

$$V_0 = c_0 + c_{01}V_1 + c_{02}V_2 + \cdots + c_{0m}V_m + c_{0u}V_u \qquad (3.1)$$

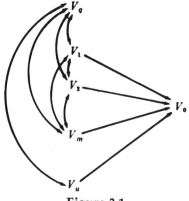

Figure 3.1

The coefficients, c_{01}, etc., are of the type of partial regression coefficients but are in a system that involves the residual V_u (unless there is known to be complete determination by the other immediate factors). It may involve other unmeasured hypothetical variables. The coefficients are thus ordinarily not deducible directly from the statistics in the way that is possible for a conventional partial regression coefficient such as $b_{01 \cdot 23}$ where V_2 and V_3 as well as V_0 and V_1 are measured variables. They have meaning only in connection with a specified diagram. The symbol, c_{01}, is used to distinguish such a quantity, defined as a *path regression coefficient* (Wright [9]), from the total regression coefficient b_{01}.

The path regression coefficient, c_{01}, measures the concrete contribution that V_1 is supposed to make *directly* to V_0 from the point of view represented in the diagram. If this correctly represents the causal relations, the path regression measures this contribution in an absolute sense, and its value can be used in the analysis of other populations (Wright [9, 1931]). Tukey [4] and Turner and Stevens [5] properly emphasize this virtue. The standardized *path coefficient* $p_{01} = c_{01}\sigma_1/\sigma_0$ obviously does not have this property, but it has other virtues including greater convenience in analysis. Let

$$X_0 = (V_0 - \overline{V}_0)/\sigma_0, \text{ etc.}$$

Then

$$X_0 = p_{01}X_1 + p_{02}X_2 + \cdots + p_{0m}X_m + p_{0u}X_u \tag{3.2}$$

In this standardized form, all correlation coefficients are reduced to product moments.

$$r_{0q} = (1/n) \sum x_0 x_q \tag{3.3}$$

$$= p_{01}r_{1q} + p_{02}r_{2q} + \cdots + p_{0m}r_{mq} + p_{0u}r_{uq} = \sum_{i=1}^{u} p_{0i}r_{iq}$$

If V_q is one of the immediate factors, for example, V_1, $r_{iu} = 0$,

$$r_{01} = p_{01} + p_{02}r_{12} + \cdots + p_{0m}r_{1m} \tag{3.4}$$

If V_q is V_0 itself,

$$r_{00} = p_{01}r_{01} + p_{02}r_{02} + \cdots + p_{0m}r_{0m} + p_{0u}^2 = 1 \tag{3.5}$$

$$r_{00} = \sum_{j=1}^{m} p_{0j}r_{0j} + p_{0u}^2 \text{ (where } V_j \text{ does not include } V_u \text{)}$$

The term, $\Sigma\, p_{0j}r_{0j}$, is the squared coefficient of correlation r_{0E}^2 with the best estimate of V_0 that can be made from immediate factors other than V_u (squared coefficient of multiple correlation) and $r_{0u}^2 = p_{0u}^2 (= 1 - \Sigma\, p_{0j}r_{0j})$ is the squared error of estimate. Thus, $r_{00} = r_{0E}^2 + r_{0u}^2$.

Returning to (3.3), which it may be noted does not depend on the assumption of normality of any of the variables, we note that it contains correlation coefficients that are capable of analysis by application of this formula to itself, if any of the immediate factors or V_q are represented as determined by more remote factors in a more extended diagram. The principle that is arrived at (for systems in which there are no paths that return on themselves) may be stated as follows: The correlation between any two variables in a properly constructed diagram of relations is equal to the sum of contributions pertaining to the paths by which one may trace from one to the other in the diagram without going back after going forward along an arrow and without passing through any variable twice in the same path. A coefficient pertaining to the whole path connecting two variables, and thus measuring the contribution of that path to the correlation, is known as a *compound path coefficient*. Its value is the product of the values of the coefficients pertaining to the elementary paths along its course. One, but not more than one of these, may pertain to a two-headed arrow without violating the rule against going back after going forward.

A uni-directional compound path coefficient may be indicated by listing the variables in order, from dependent to most remote independent, as subscripts. Thus, in Fig. 3.2, p_{013} pertains to the path $V_0 \leftarrow V_1 \leftarrow V_3$ and has the value $p_{01}p_{13}$. In a bi-directional compound path coefficient, it is convenient to list the variables in order from either end, but set off the ultimate

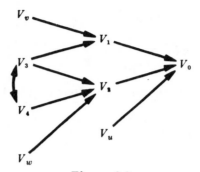

Figure 3.2

common factor or pair of factors by parentheses. Thus $p_{01(3)2}$ in Fig. 3.2 pertains to the path $V_0 \leftarrow V_1 \leftarrow V_3 \rightarrow V_2$, with value $p_{01}p_{13}p_{23}$ and $p_{01(34)2}$ pertains to the path $V_0 \leftarrow V_1 \leftarrow V_3 \leftrightarrow V_4 \rightarrow V_2$ with value $p_{01}p_{13}r_{34}p_{24}$. According to the rule above, $r_{02} = p_{02} + p_{01(3)2} + p_{01(34)2}$. In previous papers, the turning point in a compound path has been indicated by a typographically somewhat awkward dot or dash over the pertinent subscript or pair of subscripts.

CONCRETE VERSUS STANDARDIZED COEFFICIENTS

The comparison of the uses of standardized path coefficients and concrete path regressions can best be made in conjunction with a consideration of those of ordinary correlation and regression coefficients.

1. The coefficient of correlation is a useful statistic in providing a scale from -1 through 0 to $+1$ for comparing degrees of correlation.

It is not the only statistic that can provide such a scale, and these scales need not agree. There are statistics that agree at the three points referred to above without even rough agreement elsewhere (cf. r_{01}^3 with r_{01}), but, if one has become familiar with the situation implied by such values of r_{01} as .10, .50, .90, etc., the specification of the correlation coefficient conveys valuable information about the population that is under consideration. It is not a good reason to discard this coefficient because some have made the mistake of treating it as if it were an absolute property of the two variables. The term *path coefficient* was first used in an analysis of variability of amount of white in the spotted pattern of guinea pigs (Wright [8]). There was a correlation of $+.211 \pm .015$ between parent and offspring and one of $+.214 \pm .018$ between litter mates in a randomly bred strain. The fact that the corresponding correlations in another population (one tracing to a single mating after seven generations of brother–sister mating) were significantly different $(+.014 \pm .022$ and $+.069 \pm .028$, respectively) presented in a clear way a question for analysis for which the method of path coefficients seemed well adapted, not a reason for abandoning correlation coefficients.

Path coefficients resemble correlation coefficients in describing relations on an abstract scale. A diagram of functional relations, in which it is possible to assign path coefficients to each arrow, gives at a glance the relative direct contributions of variability of the immediate causal factors to variability of the effect in each case. They differ from correlation coefficients in that they may exceed $+1$ or -1 in absolute value. Such a value shows at a glance that direct action of the factor in question is tending to bring about greater variability than is actually observed. The direct effect must be offset by opposing correlated effects of other factors. As in the case of correlation coefficients, the fact that the same type of interpretative diagram yields different path coefficients when applied to different populations is valuable for comparative purposes.

2. The correlation coefficient seems to be the most useful parameter for supplementing the means and standard deviations of normally distributed variables in describing bivariate and multivariate distributions in mathematical form. Its value in this connection does not, however, mean that its usefulness otherwise is restricted to normally distributed variables as seems to be implied by Turner and Stevens.

The last point may be illustrated by citing one of the basic correlation arrays of population genetics, that for parent and offspring in a random breeding population, with respect to a single pair of alleles and no environmental complications.

Parent	Offspring			Total	Grade
	aa	Aa	AA		
AA	0	$q^2(1-q)$	q^3	q^2	$\alpha_0 + \alpha_1 + \alpha_2$
Aa	$q(1-q)^2$	$q(1-q)$	$q^2(1-q)$	$2q(1-q)$	$\alpha_0 + \alpha_1$
aa	$(1-q)^3$	$q(1-q)^2$	0	$(1-q)^2$	α_0
Total	$(1-q)^2$	$2q(1-q)$	q^2	1	

$$r_{OP} = \frac{(1-q)^2\alpha_1^2 + 2q(1-q)\alpha_1\alpha_2 + q^2\alpha_2^2}{(1-q)(2-q)\alpha_1^2 + 2q(1-q)\alpha_1\alpha_2 + q(1+q)\alpha_2^2}$$
$$= 1/2 \text{ if no dominance, } \alpha_2 = \alpha_1$$

As each variable takes only three discrete values, as the intervals are unequal unless dominance is wholly lacking ($\alpha_2 = \alpha_1$), and as gene frequency q may take any value between 0 and 1, making extreme asymmetry possible, this distribution is very far from being bivariate normal. Obviously, this correlation coefficient has no application as a parameter of such a distribution. It may, however, be used rigorously in all of the other respects discussed here. It may be noted that its value may be deduced rigorously from the population array $[(1 - q)a + qA]^2$, the assigned grades, and certain path coefficients that are obvious on inspection from a diagram representing the relation of parental and offspring genotypes under the Mendelian mechanism.

The correlation coefficients in a set of variables are statistical properties of the population in question that are independent of any point of view toward the relations among the variables. The dependence of calculated path coefficients on the point of view represented in a particular diagram, of course, restricts their use as parameters to this point of view.

3. Assuming linearity, the squared correlation coefficient measures the portion of the variance of either of the two variables, that is controlled directly or indirectly, by the other, in the sense that it gives the ratio of the variance of means of one for given values of the other to the total variance of the former $[r_{12}{}^2 = \sigma_{1(2)}{}^2/\sigma_1{}^2]$. Correspondingly, it gives the average portion of the variance of one that is lost at given values of the other $[\sigma_{1.2}{}^2 = \sigma_1{}^2(1 - r_{12}{}^2)]$. We are here using $\sigma_{1(2)}{}^2$ and $\sigma_{1.2}{}^2$ for the components of $\sigma_1{}^2$ that are dependent and independent respectively of variable V_2.

Equation (3.5), expressing complete determination of V_0 by its factors, can be expanded into the form.

$$r_{00} = 1 = \sum_{j=1}^{m} p_{0j}{}^2 + 2 \sum_{j,\,k=1}^{m} p_{0j}p_{0k}r_{jk} + p_{0u}{}^2, \qquad k > j \tag{3.6}$$

On multiplying both sides by $\sigma_0{}^2$, it may be seen that the squared path coefficients measure the portions of $\sigma_0{}^2$ that are determined directly by the indicated factors while other terms (which may be negative) measure correlational determination.

4. The correlation coefficient, r_{01}, measures the slope of the line of means of V_0 relative to V_i (or the converse) on standardized scales and merely needs to be multiplied by the proper ratio of standard deviations to express regression in concrete terms ($b_{01} = r_{01}\sigma_0/\sigma_1$, $b_{10} = r_{01}\sigma_1/\sigma_0$).

As already noted, the abstract path coefficients have the same relation to the concrete path regressions.

5. Another statistic of this family, the product moment,

$$M_{11}(V_1V_2) = \mathrm{cov}_{12} = (1/n) \sum_{1}^{n} (V_1 - \bar{V}_1)(V_2 - \bar{V}_2) = r_{12}\sigma_1\sigma_2$$

is useful on its own account in various ways. The product moment in a heterogeneous population may for example be analyzed into the sum of the product moment of the weighted means of the subpopulations and the average product moment within these (Wright [6]). It was because of this additive property, analogous to that of the squared standard deviation, that Fisher later renamed this quantity the covariance in analogy with his term variance (Fisher [1]) for the squared standard deviation.

Since the compound path coefficients analyze each correlation into additive contributions from each chain that connects the two variables, they merely need to be multiplied by the terminal standard deviations to give a similar analysis of the covariance.

6. In certain situations (including important ones in population genetics) the correlation coefficient can be interpreted as a probability. This may be analyzable into components in terms of compound path coefficients.

7. Finally, the formula for the correlation between linear functions is often useful and is the one that leads directly to path analysis. If

$$V_S = c_S + \Sigma\, c_{Si} V_i$$

$$\sigma_S^2 = \Sigma\, c_{Si}^2 \sigma_i^2 + 2\, \Sigma\, c_{Si} c_{Sj} \sigma_i \sigma_j r_{ij} \qquad j > i$$

$$p_{Si} = c_{Si} \sigma_i / \sigma_S \tag{3.7}$$

$$r_{ST} = \Sigma\, p_{Si} p_{Ti} + \Sigma\, p_{Si} p_{Tj} r_{ij} \qquad \text{all } i \text{ and } j$$

The most extensive applications of the method have been essentially of this sort, the deduction of correlations from known functional relations in population genetics (Wright [9,10,14]). The inverse problem, that of deducing path coefficients from known correlations and a given pattern of relations, depends on the solution of a system of simultaneous equations usually of higher degree than first and thus often requires rather tedious iteration.

We conclude that both the standardized and concrete coefficients for describing relationships between variables are useful, and that the rejection of either would impoverish the theory.

THE USE OF PATH COEFFICIENTS IN ANALYSIS

We come now to the point that even where concrete path regressions are the sole objectives, the analysis had best be carried out in terms of the standardized coefficients from which the desired concrete ones may be derived as the final step. We give below a number of simple systems. Numerical subscripts are used here for the variables that are supposed to be measured and literal ones for the hypothetical variables, including the residuals necessary for completion. All of the equations that can be written from the known correlations and from cases of complete determination are expressed in terms of path coefficients and residual correlations. They can all be written from inspection by tracing connecting paths. All of them can also be written in concrete terms by use of formulae given above. This is done in some of the cases. Parentheses enclose quantities that are inseparable in analysis restricted to concrete coefficients.

In this case, the first three equations are equally simple with concrete and standard coefficients and overdetermine two paths. One may (a) obtain a compromise solution as by the method of least squares; or (b) may attribute any inconsistency to correlations between V_u and the variables back of V_1 (which must compensate in such a way that $r_{1u} = 0$ as required by the definition of V_u as a residual); or (c) may assume that the measurements of V_1 are in error. The hypothesis of errors in either V_0 or V_2 does not resolve any inconsistency of equations (1), (2), and (3).

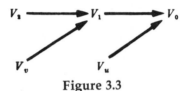

Figure 3.3

Standardized coefficients	Concrete coefficients
(1) $r_{12} = p_{12}$	$b_{12} = c_{12}$
(2) $r_{01} = p_{01}$	$b_{01} = c_{01}$
(3) $r_{02} = p_{01}p_{12}$	$b_{02} = c_{01}c_{12}$
(4) $r_{00} = 1 = p_{01}^2 + p_{0u}^2$	$\sigma_0^2 = c_{01}^2\sigma_1^2 + (c_{0u}^2\sigma_u^2)$
(5) $r_{11} = 1 = p_{12}^2 + p_{1v}^2$	$\sigma_1^2 = c_{12}^2\sigma_2^2 + (c_{1v}^2\sigma_v^2)$

The appropriate diagram and equations under (b) above are as follows:

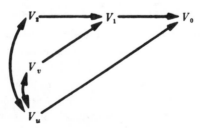

Figure 3.4

(1) $r_{12} = p_{12}$

(2) $r_{11} = 1 = p_{12}^2 + p_{1v}^2$

(3) $r_{01} = p_{01}$ (since $r_{1u} = 0$)

(4) $r_{00} = 1 = p_{01}^2 + p_{0u}^2$

(5) $r_{02} = p_{01}p_{12} + p_{0u}r_{2u}$

(6) $r_{1u} = 0 = p_{12}r_{2u} + p_{1v}r_{uv}$

A necessary condition for solution, that there be at least as many independent equations as paths, is met. This is not, in general, a sufficient condition since a system may be underdetermined in one part and overdetermined in another. In this case, however, the unknown path coefficients and correlations can be obtained in succession from the above equations.

The hypothesis that inconsistency of (1), (2), and (3) under Fig. 3.3 is due to errors of measurement of V_1 is represented in Fig. 3.5 and in the equations below.

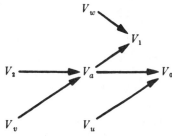

Figure 3.5

(1) $r_{01} = p_{0a}p_{1a}$

(2) $r_{12} = p_{1a}p_{a2}$

(3) $r_{02} = p_{0a}p_{a2}$

(4) $r_{00} = 1 = p_{0a}^2 + p_{0u}^2$

(5) $r_{11} = 1 = p_{1a}^2 + p_{1w}^2$

(6) $r_{aa} = 1 = p_{a2}^2 + p_{av}^2$

These again are easily solved. Turner and Stevens discuss the effects of errors of measurement but not by means of path analysis, which if attempted with concrete coefficients is encumbered with symbols of variances. Thus, equation (1) becomes

$$b_{01} = c_{0a}c_{1a}\sigma_a^2/\sigma_1^2 \quad \text{or} \quad \text{cov}_{01} = c_{0a}c_{1a}\sigma_a^2$$

There is, of course, indeterminancy if both (b) and (c) are assumed.

Encumbrance with unnecessary variances occurs wherever two variables trace to a third. The simplest case is that shown in Fig. 3.6.

There is overdetermination of two of the paths. Again, a compromise solution may be obtained if there is confidence in the diagram. If not, the latter may be revised to indicate a correlation, r_{uv}, between the residuals or

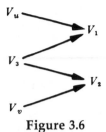

Figure 3.6

Standardized coefficients	Concrete coefficients
(1) $r_{13} = p_{13}$	$b_{13} = c_{13}$
(2) $r_{23} = p_{23}$	$b_{23} = c_{23}$
(3) $r_{12} = p_{13}p_{23}$	$b_{12} = c_{13}c_{23}\sigma_3^2/\sigma_2^2$
(4) $r_{11} = 1 = p_{13}^2 + p_{1u}^2$	$\sigma_1^2 = c_{13}^2\sigma_3^2 + (c_{1u}^2\sigma_u^2)$
(5) $r_{22} = 1 = p_{23}^2 + p_{2v}^2$	$\sigma_2^2 = c_{23}^2\sigma_3^2 + (c_{2v}^2\sigma_v^2)$

to indicate that there are errors of measurement in the intermediary V_3. A solution can be obtained under either of the latter two hypotheses (but not both at once).

A common situation is represented in Fig. 3.7.

The coefficients pertaining to all of the paths are readily calculated. This is a simple example of the patterns characteristic of multiple regression which are always easily solvable since only linear equations are involved. The equations (excluding [3.4]) can be written as simply in terms of the

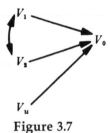

Figure 3.7

Standardized coefficients	Concrete coefficients
(1) $r_{01} = p_{01} + p_{02}r_{12}$	$b_{01} = c_{01} + c_{02}b_{21}$
(2) $r_{02} = p_{01}r_{12} + p_{02}$	$b_{02} = c_{01}b_{12} + c_{02}$
(3) r_{12} given	b_{21} and b_{12} given
(4) $r_{00} = 1 = p_{01}r_{01} + p_{02}r_{02} + p_{0u}^2$	

regressions as in terms of the standardized coefficients, but the former introduce an arbitrary asymmetry which is undesirable. Where only symmetrical patterns of this sort are being dealt with, the simplest procedure is, no doubt, to calculate the concrete coefficients directly from the ordinary normal equations of the method of least squares. This, however, takes us away from the sort of interpretive analysis that we are here considering, in which such symmetrical patterns are merely special cases.

Figure 3.8 is an example of another sort of symmetrical case.

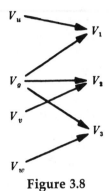

Figure 3.8

(1) $r_{12} = p_{1g}p_{2g}$ (4) $r_{11} = 1 = p_{1g}^2 + p_{1u}^2$
(2) $r_{13} = p_{1g}p_{3g}$ (5) $r_{22} = 1 = p_{2g}^2 + p_{2v}^2$
(3) $r_{23} = p_{2g}p_{3g}$ (6) $r_{33} = 1 = p_{3g}^2 + p_{3w}^2$

There are six paths and six equations that permit solution for the standardized coefficients. Concrete coefficients cannot be used in this case because of ignorance of the variance of the hypothetical general factor V_g.

This is a simple example of the conventional pattern of factor analysis with one general factor as proposed by Spearman [3]. There is overdeterminancy if there are more known variables. A solution that maximizes the sum of the squared path coefficients that relate the known variables to the general factor has been given by Hotelling [2]. Additional general factors may be postulated. The number of equations that can be written, given m known variables is $(1/2)m(m + 1)$. The number of coefficients to be determined with n common factors and m residuals is $m(n + 1)$. There may be exact determinancy, underdeterminancy, or overdeterminancy. Hotelling has shown that with any number of known variables there is complete determinancy by the same number of factors if the sums of the squared path coefficients relating to the factors are successively maximized. Other conventions for arriving at a unique solution have been given.

The first paper on path coefficients (as square roots of coefficients of direct determination) (Wright [7]) dealt with material of the sort to which factor analysis is applied (all possible correlations in a set of bone measurements in a rabbit population) but from a less symmetrical viewpoint. This has been developed in later papers (Wright [12,15]).

SUMMARY

The method of path coefficients and its more important pitfalls are reviewed briefly with reference to recent misunderstandings.

Reasons are discussed for looking upon standardized coefficients (correlations, path coefficients) and concrete ones (total and path regressions) as aspects of a single theory rather than as alternatives between which a choice should be made. They correspond to different modes of interpretation, which taken together give a deeper understanding of a situation than either can give itself.

It is brought out that even where the sole objectives of analysis are the concrete coefficients, actual path analysis takes a simpler and more homogeneous form in terms of the standardized ones, which can easily be converted into the concrete forms as the final step.

REFERENCES

[1] Fisher, R. A. "The Correlation between Relatives on the Supposition of Mendelian Inheritance." *Transactions of the Royal Society of Edinburgh* 52 (1918): 399–433.

[2] Hotelling, H. "Simplified Calculation of Principal Components." *Psychometrica* 1 (1936): 27–35.

[3] Spearman, C. "General intelligence, objectively determined and measured." *American Journal of Psychology* 15 (1904): 201–292.

[4] Tukey, J. W. "Causation, Regression and Path Analysis." In *Statistics and Mathematics in Biology*, edited by O. Kempthorne, T. A. Bancroft, J. W. Gowen, and J. L. Lush, Chapter 3, pp. 35–66. Ames, Iowa: Iowa State College Press, 1954.

[5] Turner, M. E., and Stevens, C. D. "The Regression Analysis of Causal Paths." *Biometrics* 15 (1959): 236–258.

[6] Wright, S. "The average correlation within Subgroups of a Population." *Journal of the Washington Academy of Science* 7 (1917): 532–535.

[7] Wright, S. "On the Nature of Size Factors." *Genetics* 3 (1918): 367–374.

[8] Wright, S. "The Relative Importance of Heredity and Environment in Determining the Piebald Pattern of Guinea Pigs." *Proceedings of the National Academy of Sciences* 6 (1920): 320–332.

[9] Wright, S. "Correlation and Causation." *Journal of Agricultural Research* 20 (1921): 557–585.

[10] Wright, S. "Evolution in Mendelian Populations." *Genetics* 16 (1931): 97–159.

[11] Wright, S. Statistical Methods in Biology. *Journal of the American Statistical Association Supplement; Papers and Proceedings of the 92nd Annual Meeting* 26 (1931): 155–163.

[12] Wright, S. "General, Group and Special Size Factors. *Genetics* 17 (1932): 603–619.

[13] Wright, S. "The Method of Path Coefficients." *Annals of Mathematical Statistics* 5 (1934): 161–215.

[14] Wright, S. "The Genetical Structure of Populations." *Annals of Eugenics* 15 (1951): 323–354.

[15] Wright, S. "The Interpretation of Multivariate Systems." *Statistics and Mathematics in Biology,* edited by O. Kempthorne, T. A. Bancroft, J. W. Gowen, and J. L. Lush, Chapter 2, pp. 11–33. Ames, Iowa: Iowa State College Press, 1954.

Path Analysis: Sociological Examples*

Otis Dudley Duncan

4

The long-standing interest of sociologists in causal interpretation of statistical relationships has been quickened by discussions focusing on linear
causal models. The basic work of bringing such models to the attention of
the discipline was done by Blalock [2], drawing upon the writings of Simon
[23, Chapter ii] and Wold and Juréen [31] in particular. The rationale of this
approach was strengthened when Costner and Leik [8] showed that "asymmetric causal models" of the kind proposed by Blalock afford a natural and
operational explication of the notion of "axiomatic deductive theory,"
which had been developed primarily by sociologists working with verbal
formulations. Most recently, Boudon [6] pointed out that the Simon–
Blalock type of model is a "special case" or "weak form" of path analysis (or
"dependence analysis," as Boudon prefers to call it). At the same time, he
noted that "convincing empirical illustrations are missing," since "moderately complicated causal structures with corresponding data are rather
scarce in the sociological literature." This chapter presents some examples
(in the form of reanalyses of published work) that may be interesting, if not
"convincing." It includes an exposition of some aspects of path technique,

* Reprinted by permission of the author and publisher from the *American Journal
of Sociology* 72: 1–16. Copyright 1966, The University of Chicago Press.

developing it in a way that may make it a little more accessible than some of the previous writings.

Path coefficients were used by the geneticist Sewall Wright as early as 1918, and the technique was expounded formally by him in a series of articles dating from the early 1920s. References to this literature, along with useful restatements and illustrations, will be found in Wright's papers of 1934, 1954, and 1960 [35;37, Chapter ii; 39]. The main application of path analysis has been in population genetics, where the method has proved to be a powerful aid to "axiomatic deductions." The assumptions are those of Mendelian inheritance, combined with path schemes representing specified systems of mating. The method allows the geneticist to ascertain the "coefficient of inbreeding," a quantity on which various statistical properties of a Mendelian population depend. It also yields a theoretical calculation of the genetic correlations among relatives of stated degrees of relationship. Most of Wright's expositions of this *direct* use of path coefficients are heavily mathematical [36]; an elementary treatment is given in the text by Li [19, Chapters xii–xiv; 20].

Apart from a few examples in Wright's own work, little use has been made of path coefficients in connection with the *inverse* problem of estimating the paths that may account for a set of observed correlations on the assumption of a particular formal or causal ordering of the variables involved. Of greatest substantive interest to sociologists may be an example relating to heredity and environment in the determination of intelligence [34]. Another highly suggestive study was a pioneer but neglected exercise in econometrics concerning prices and production of corn and hogs [33;35, pp. 192–204]. Although the subject matter is remote from sociological concerns, examples from studies in animal biology are instructive on methodological grounds [38,9]. If research workers have been slow to follow Wright's lead, the statisticians have done little better. There are only a few expositions in the statistical literature [24, Chapter iii; 16, Chapter xiv; 26;7;18, Chapter i; 21], some of which raise questions to which Wright has replied [39].

PATH DIAGRAMS AND THE BASIC THEOREM

We are concerned with linear, additive, asymmetric relationships among a set of variables that are conceived as being measurable on an interval scale, although some of them may not actually be measured or may even be purely hypothetical—for example, the "true" variables in measurement theory or the "factors" in factor analysis. In such a system, certain of the variables are represented to be dependent on others as linear functions. The remaining variables are assumed, for the analysis at hand, to be given. They may be correlated among themselves, but the explanation of their intercorrelation is not taken as problematical. Each "dependent" variable must be regarded explicitly as *completely* determined by some combination of variables in the system. In problems where complete determination by measured variables does not hold, a residual variable uncorrelated with other determining variables must be introduced.

Although it is not intrinsic to the method, the diagrammatic representation of such a system is of great value in thinking about its properties. A word of caution is necessary, however. Causal diagrams are appearing with increasing frequency in sociological publications. Most often, these have some kind of pictorial or mnemonic function without being isomorphic with the algebraic and statistical properties of the postulated system of variables — or, indeed, without having a counterpart in any clearly specified system of variables at all. Sometimes an investigator will post values of zero order or partial correlations, association coefficients, or other indications of the "strength" of relationship on such a diagram, without following any clearly defined and logically justified rules for entering such quantities into the analysis and its diagrammatic representation. In Blalock's work, by contrast, diagrams are employed in accordance with explicit rules for the representation of a system of equations. In general, however, he limits himself to the indication of the sign (positive or negative) of postulated or inferred direct relationships. In at least one instance [2] he inserts zero-order correlations into a diagram that looks very much like a causal diagram, although it is not intended to be such. This misleading practice should not be encouraged.

In path diagrams, we use one-way arrows leading from each determining variable to each variable dependent on it. Unanalyzed correlations between variables not dependent upon others in the system are shown by two-headed arrows, and the connecting line is drawn curved, rather than straight, to call attention to its distinction from the paths relating dependent to determining variables. The quantities entered on the diagram are symbolic or numerical values of *path coefficients*, or, in the case of the bidirectional correlations, the simple correlation coefficients.

Several of the properties of a path diagram are illustrated in Fig. 4.1. The original data, in the form of ten zero-order correlations, are from Turner's study of determinants of aspirations [25, pp. 49 and 52, Tables 11, 17, and 20]. The author does not provide a completely unequivocal formulation of the entire causal model shown here, but Fig. 4.1 appears to correspond to the model that he quite tentatively proposes. At one point, he states, "background affects ambition and ambition affects both IQ and class values; in addition . . . there is a lesser influence directly from background to class values, directly from background to IQ, and directly between IQ and class values [25, p. 107]." Elsewhere [25, pp. 54–61], he indicates that school rating operates in much the same fashion as (family) background. As for the relationship between the two, Turner notes, on the one hand, that "families may choose their place of residence," but also that "by introducing neighborhood, we may only be measuring family background more precisely [25, p. 61]." Hence, it seems that there is no firm assumption about the causal ordering within this pair of variables; but since these two precede the re-

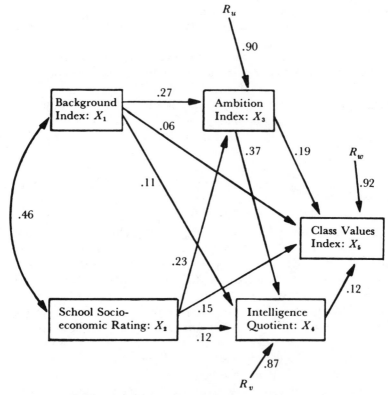

Figure 4.1. Causal model from Turner [25] with path coefficients estimated for male sample.

maining ones, it suffices to represent the link between X_1 and X_2 as merely a bidirectional correlation.

Allowing Turner to take responsibility for the causal ordering of the variables (assuming his statements are understood correctly) and deferring the question of how the path coefficients are estimated, let us see what the system represented by Fig. 4.1 is like. Each variable is taken to be in standard form; that is, if V_i is the ith variable as measured, then $X_i = (V_i - \overline{V}_i)/\sigma_{V_i}$. The same convention holds for the residuals, R_u, R_v, and R_w, to which a literal subscript is attached to indicate that these variables are not directly measured. The system represented in Figure 4.1 can now be written:

$$X_3 = p_{32}X_2 + p_{31}X_1 + p_{3u}R_u$$

$$X_4 = p_{43}X_3 + p_{42}X_2 + p_{41}X_1 + p_{4v}R_v \tag{4.1}$$

$$X_5 = p_{54}X_4 + p_{53}X_3 + p_{52}X_2 + p_{51}X_1 + p_{5w}R_w$$

The use of the symbol p for the path coefficient is perhaps obvious. Note that the order of the subscripts is significant, the convention being the same as that used for regression coefficients: The first subscript identifies the dependent variable; the second, the variable whose direct effect on the dependent variable is measured by the path coefficient. The order of subscripts is immaterial for correlations. But note that while $r_{42} = r_{24}$ and $r_{42 \cdot 13'} = r_{24 \cdot 13'}$, $p_{42} \neq p_{24}$; indeed p_{42} and p_{24} would never appear in the same system, given the restriction to recursive systems mentioned subsequently. Contrary to the practice in the case of partial regression and correlation coefficients, symbols for paths carry no secondary subscripts to identify the other variables assumed to affect the dependent variable. These will ordinarily be evident from the diagram or the equation system.

In one respect, the equation system (4.1) is less explicit than the diagram because the latter indicates what assumptions are made about residual factors. Each such factor is assumed by definition to be uncorrelated with any of the immediate determinants of the dependent variable to which it pertains. In Fig. 4.1, the residuals are also uncorrelated with each other, as in the Simon — Blalock development [2, p. 64; 6, p. 369]. We shall see later, however, that there are uses for models in which some residuals are intercorrelated, or in which a residual is correlated with variables antecedent to, but not immediate determinants of, the particular dependent variable to which it is attached. Where the assumption of uncorrelated residuals is made, deductions reached by the Simon – Blalock technique of expanding the product of two error variables agree with the results obtained by the formulas mentioned below, although path analysis involves relatively little use of the partial correlations that are a feature of their technique.

Equation system (4.1), as Blalock points out, is a recursive system. This discussion explicitly excludes nonrecursive systems, involving instantaneous reciprocal action of variables, although Wright has indicated ways of handling them in a path framework [40]. Thus, we shall not consider diagrams showing a direct or indirect feedback loop.

The principle that follows from equations in the form of (4.1) is that the correlation between any pair of variables can be written in terms of the paths leading from common antecedent variables. Consider r_{35}.

Since $X_3 = (V_3 - \bar{V}_3)/\sigma_3$ and $X_5 = (V_5 - \bar{V}_5)/\sigma_5$ we have $r_{35} = \Sigma (V_3 - \bar{V}_3) \times (V_5 - \bar{V}_5)/N\sigma_3\sigma_5 = \Sigma X_3 X_5 /N$.

We may expand this expression in either of two ways by substituting from (4.1) the expression for X_3 or the one for X_5. It is more convenient to expand the variable that appears later in the causal sequence:

$$r_{35} = \Sigma\, X_3 X_5 / N$$

$$= \frac{1}{N} \Sigma\, X_3(p_{54}X_4 + p_{53}X_3 + p_{52}X_2 + p_{51}X_1 + p_{5w}R_w) \qquad (4.2)$$

$$= p_{54}r_{34} + p_{53} + p_{52}r_{23} + p_{51}r_{13},$$

making use of the fact that $\Sigma\, X_3 X_3 / N = 1$ and the assumption that $r_{3w} = 0$, since X_3 is a factor of X_5. But the correlations on the right-hand side of (4.2) can be further analyzed by the same procedure; for example,

$$r_{34} = \frac{1}{N} \Sigma\, X_3 X_4 = \frac{1}{N} \Sigma\, X_3(p_{43}X_3 + p_{42}X_2 + p_{41}X_1 + p_{4v}R_v)$$
$$= p_{43} + p_{42}r_{23} + p_{41}r_{13} \qquad (4.3)$$

and

$$r_{32} = \frac{1}{N} \Sigma\, X_2 X_3 = \frac{1}{N} \Sigma\, X_2(p_{32}X_2 + p_{31}X_1 + p_{3u}R_u)$$
$$= p_{32} + p_{31}r_{12} \qquad (4.4)$$

Note that r_{12}, assumed as a datum, cannot be further analyzed so long as we retain the particular diagram of Fig. 4.1.

These manipulations illustrate the basic theorem of path analysis, which may be written in the general form:

$$r_{ij} = \sum_q p_{iq} r_{jq} \qquad (4.5)$$

where i and j denote two variables in the system, and the index q runs over all variables from which paths lead directly to X_i. Alternatively, we may expand (4.5) by successive applications of the formula itself to the r_{jq}. Thus, from (4.2), (4.3), (4.4), and a similar expansion of r_{13}, we obtain

$$r_{53} = p_{53} + p_{51}p_{31} + p_{51}r_{12}p_{32} + p_{52}p_{32} + p_{52}r_{12}p_{31} + p_{54}p_{42}p_{32}$$
$$+ p_{54}p_{42}r_{12}p_{31} + p_{54}p_{43} + p_{54}p_{41}p_{32}r_{12} + p_{54}p_{41}p_{31} \qquad (4.6)$$

Such expressions can be read directly from the diagram according to the following rule. Read *back* from variable i, then *forward* to variable j, forming the product of all paths along the traverse; then sum these products for all possible traverses. The same variable cannot be intersected more than once in a single traverse. In no case can one trace back having once started forward. The bidirectional correlation is used in tracing either forward or back, but if more than one bidirectional correlation appears in the diagram, only one can be used in a single traverse. The resulting expression, such as (4.6), may consist of a single direct path plus the sum of several compound paths representing all the indirect connections allowed by the diagram. The general formula (4.5) is likely to be the more useful in algebraic manipulation and calculation, the expansion on the pattern of (4.6) in appreciating the properties of the causal scheme. It is safer to depend on the algebra than on the verbal algorithm, at least until one has mastered the art of reading path diagrams.

An important special case of (4.5) is the formula for complete determination of X_i, obtained by setting $i = j$:

$$r_{ii} = 1 = \sum_q p_{iq} r_{iq} \tag{4.7}$$

or, upon expansion,

$$r_{ii} = \sum_q p_{iq}^2 + 2 \sum_{q,q'} p_{iq} r_{qq'} p_{iq'} \tag{4.8}$$

where the range of q and q' ($q' > q$) includes all variables, measured and unmeasured. A major use for (4.8) is the calculation of the residual path. Thus, we obtain p_{3u} in the system (4.1) from

$$p_{3u}^2 = 1 - p_{32}^2 - p_{31}^2 - 2p_{32} r_{12} p_{31} \tag{4.9}$$

The causal model shown in Figure 4.1 represents a special case of path analysis: one in which there are no unmeasured variables (other than residual factors), the residuals are uncorrelated, and each of the dependent variables is directly related to all the variables preceding it in the assumed causal sequence. In this case, path analysis amounts to a sequence of conventional regression analyses, and the basic theorem (4.5) becomes merely a compact statement of the normal equations of regression theory for variables in standard form. The path coefficients are then nothing other than the "beta coefficients" in a regression setup, and the usual apparatus for regression calculations may be employed [27, Chapter iii]. Thus, the paths in Figure 4.1 are obtained from the regression of X_3 on X_2 and X_1, setting $p_{32} = \beta_{32 \cdot 1}$ and $p_{31} = \beta_{31 \cdot 2}$; the regression of X_4 on X_3, X_2, and X_1, setting

$p_{43} = \beta_{43\cdot12}$, $p_{42} = \beta_{42\cdot13}$, and $p_{41} = \beta_{41\cdot23}$; and the regression of X_5 on the other four variables, setting $p_{54} = \beta_{54\cdot123}$, $p_{53} = \beta_{53\cdot124}$, and so on. Following the computing routine that inverts the matrix of intercorrelations of the independent variables, one obtains automatically the standard errors of the β coefficients (or b^*-coefficients, in the notation of Walker and Lev). In the present problem, with sample size exceeding 1000, the standard errors are small, varying between .027 and .032. All the β's are at least twice their standard errors and thus statistically significant.

In problems of this kind, Blalock [2, Chapter iii] has been preoccupied with the question of whether one or more path coefficients may be deleted without loss of information. As compared with his rather tedious search procedure, the procedure followed here seems more straightforward. Had some of the β's turned out both nonsignificant and negligible in magnitude, one could have erased the corresponding paths from the diagram and run the regressions over, retaining only those independent variables found to be statistically and substantively significant.

As statistical techniques, therefore, neither path analysis nor the Blalock–Simon procedure adds anything to conventional regression analysis as applied recursively to generate a system of equations, rather than a single equation. As a *pattern of interpretation*, however, path analysis is invaluable in making explicit the rationale for a set of regression calculations. One may not be wholly satisfied, for example, with the theoretical assumptions under lying the causal interpretation of Turner's data provided by Figure 4.1, and perhaps Turner himself would not be prepared to defend it in detail. The point is, however, that *any* causal interpretation of these data must rest on assumptions — at a minimum, the assumption as to ordering of the variables, but also assumptions about the unmeasured variables here represented as uncorrelated residual factors [2, pp. 46–47]. The great merit of the path scheme, then, is that it makes the assumptions explicit and tends to force the discussion to be at least internally consistent, so that mutually incompatible assumptions are not introduced surreptitiously into different parts of an argument extending over scores of pages. With the causal scheme made explicit, moreover, it is in a form that enables criticism to be sharply focused and hence potentially relevant not only to the interpretation at hand but also, perchance, to the conduct of future inquiry.

Another useful contribution of path analysis, even in the conventional regression framework, is that it provides a calculus for indirect effects, when the basic equations are expanded along the lines of (4.6). It is evident from the regression coefficients, for example, that the direct effect of school on class values is greater than that of background, but the opposite is true of the indirect effects. The pattern of indirect effects is hardly obvious without the aid of an explicit representation of the causal scheme. If one wishes a single summary measure of indirect effect, however, it is obtained as follows: indirect effect of X_2 on $X_5 = r_{52} - p_{52} = .28 - .15 = .13$; similarly, indirect effect of X_1 is $r_{51} - p_{51} = .24 - .06 = .18$. These summations of indirect effects include, in each case, the effects of one variable via its correlation with the other; hence, the two are not additive. Without commenting further on the substantive implications of the direct and indirect effects suggested by Turner's material, it may simply be noted that the investigator will usually want to scrutinize them carefully in terms of his theory.

DECOMPOSITION OF A DEPENDENT VARIABLE

Many of the variables studied in social research are (or may be regarded as) composite. Thus, population growth is the sum of natural increase and net migration; each of the latter may be further decomposed, natural increase being births minus deaths and net migration the difference between in- and out-migration. Where such a decomposition is available, it is of interest (1) to compute the relative contributions of the components to variation in the composite variable and (2) to ascertain how causes affecting the composite variable are transmitted via the respective components.

An example taken from work of Winsborough [30] illustrates the case of a variable with multiplicative components, rendered additive by taking logarithms. Studying variation in population density over the seventy-four community areas (omitting the central business district) of Chicago in 1940, Winsborough noted that density, defined as the ratio of population to area, can be written:

$$\frac{\text{Population}}{\text{Area}} = \frac{\text{Population}}{\text{Dwelling Units}} \times \frac{\text{Dwelling Units}}{\text{Structures}} \times \frac{\text{Structures}}{\text{Area}}$$

Let $V_0 = \log (\text{Population/Area})$, $V_1 = \log (\text{Population/Dwelling Units})$, $V_2 = \log (\text{Dwelling Units/Structures})$, and $V_3 = \log (\text{Structures/Area})$; then

$$V_0 = V_1 + V_2 + V_3$$

If each variable is expressed in standard form, we obtain,

$$\frac{V_0 - \bar{V}_0}{\sigma_0} = \frac{V_1 - \bar{V}_1}{\sigma_1} \cdot \frac{\sigma_1}{\sigma_0} + \frac{V_2 - \bar{V}_2}{\sigma_2} \cdot \frac{\sigma_2}{\sigma_0} + \frac{V_3 - \bar{V}_3}{\sigma_3} \cdot \frac{\sigma_3}{\sigma_0}$$

or

$$X_0 = p_{01}X_1 + p_{02}X_2 + p_{03}X_3$$

Table 4.1. Correlation matrix for logarithms of density and its components and two independent variables: Chicago community areas, 1940

Variable	X_1	X_2	X_3	W	Z
X_0 density (log)	−.419	.636	.923	−.663	−.390
X_1 persons per dwelling unit (log)	—	−.625	−.315	.296	.099
X_2 dwelling units per structure (log)	—	—	.305	−.594	−.466
X_3 structures per acre (log)	—	—	—	−.517	−.226
W distance from center	—	—	—	—	.549
Z recency of growth	—	—	—	—	—

Source: Winsborough, *op. cit.*, and unpublished data kindly supplied by the author.

where $X_0, \ldots, X_3$ are the variables in standard form and p_{01}, p_{02}, p_{03} are the path coefficients involved in the determination of X_0 by X_1, X_2, and X_3. Observe that the path coefficients can be computed in this kind of problem, where complete determination by measured variables holds as a consequence of definitions, without prior calculation of correlations:[1]

$$p_{01} = \sigma_1/\sigma_0 = .132 \quad \sigma_0 = .491 \quad \sigma_1 = .065$$

$$p_{02} = \sigma_2/\sigma_0 = .468 \qquad\qquad \sigma_2 = .230$$

$$p_{03} = \sigma_3/\sigma_0 = .821 \qquad\qquad \sigma_3 = .403$$

The intercorrelations of the components, shown in Table 4.1, are used to complete Figure 4.2a. The correlations of the dependent variable with its components may now be computed from the basic theorem, equation (4.5).

$$r_{01} = p_{01} + p_{02}r_{12} + p_{03}r_{13} = -.419$$

$$r_{02} = p_{01}r_{12} + p_{02} + p_{03}r_{23} = .636$$

and

$$r_{03} = p_{01}r_{13} + p_{02}r_{23} + p_{03} = .923$$

The analysis has not only turned up a clear ordering of the three components in terms of relative importance, as given by the path coefficients, it has

[1] Based on data kindly supplied by Winsborough.

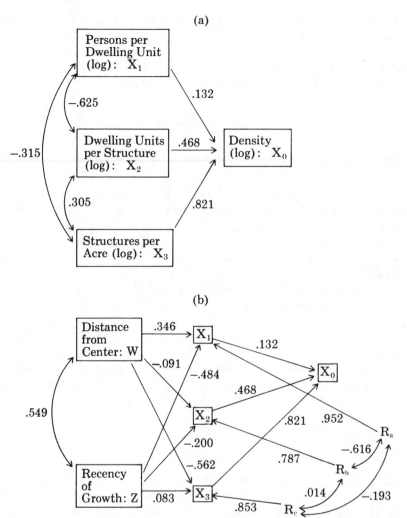

Figure 4.2. (a) Decomposition of log density (X_0) into components; (b) effects of distance and recency of growth on log density via components. (Source: Winsborough [30] and unpublished calculations kindly supplied by the author.)

also shown that one of the components is actually correlated negatively with the composite variable, owing to its negative correlations with the other two components.

Winsborough considered two independent variables as factors producing variation in density: distance from the city center and recency of growth (percentage of dwelling units built in 1920 or later). The diagram can be elaborated to indicate how these factors operate via the components of log density (see Figure 4.2b).

The first step is to compute the path coefficients for the relationships of each component to the two independent variables. (The requisite information is given in Table 4.1.) For example, the equations,

$$r_{1w} = p_{1w} + p_{1z}r_{zw}$$

$$r_{1z} = p_{1w}r_{zw} + p_{1z}$$

may be used to solve for p_{1z} and p_{1w}. (This is, of course, equivalent to computing the multiple regression of X_1 on W and Z, with all variables in standard form). Substantively, it is interesting that distance, W, has somewhat larger effects on each component of density than does recency of growth, Z, while the pattern of signs of the path coefficients is different for W and Z.

The two independent variables by no means account for all the variation in any of the components, as may be seen from the size of the residuals, p_{1a}, p_{2b}, and p_{3c}, these being computed from the formula (4.7) for complete determination. It is possible, nevertheless, for the independent variables to account for the intercorrelations of the components and, ideally, one would like to discover independent variables that would do just that. The relevant calculations concern the correlations between residuals. These are obtained from the basic theorem, equation (4.5), by writing, for example,

$$r_{23} = p_{2w}r_{3w} + p_{2z}r_{3z} + p_{2b}p_{c3}r_{bc}$$

which may be solved for $r_{bc} = .014$. In this setup, the correlations between residuals are merely the conventional second-order partial correlations; thus $r_{ab} = r_{12 \cdot wz}$, $r_{ac} = r_{13 \cdot wz}$, and $r_{bc} = r_{23 \cdot wz}$. Partial correlations, which otherwise have little utility in path analysis, turn out to be appropriate when the question at issue is whether a set of independent variables "explains" the correlation between two dependent variables. In the present example, while $r_{23} = .305$, we find $r_{bc} = r_{23 \cdot wz} = .014$. Thus, the correlation between the logarithms of dwelling units per structure (X_2) and structures per acre (X_3) is satisfactorily explained by the respective relationships of these two components to distance and recency of growth. The same is not true of the correlations involving persons per dwelling unit (X_1), but fortunately this is by far the least important component of density.

Although the correlations between residuals are required to complete the diagram and, in a sense, to evaluate the adequacy of the explanatory variables, they do not enter as such into the calculations bearing upon the final question: How are the effects of the independent variables transmitted to the dependent variable via its components? The most compact answer to this question is given by the equations

$$r_{0w} = p_{01}r_{1w} + p_{02}r_{2w} + p_{03}r_{3w}$$
$$= .039 - .278 - .424 = -.663$$

and

$$r_{0z} = p_{01}r_{1z} + p_{02}r_{2z} + p_{03}r_{3z}$$
$$= .013 - .218 - .185 = -.391$$

Density is negatively related to both distance and recency of growth, but the effects transmitted via the first component of density are positive (albeit quite small). Distance diminishes density primarily via its intermediate effect on structures per acre (X_3), secondarily via dwelling units per structure (X_2). The comparison is reversed for recency of growth, the less important of the two factors. More detailed interpretations can be obtained, as explained earlier, by expanding the correlations r_{1w}, r_{2w}, etc., using the basic theorem (4.5). For further substantive interpretation, the reader is referred to the source publication, which also offers an alternative derivation of the compound paths.

The density problem may well exemplify a general strategy too seldom employed in research: breaking a complex variable down into its components before initiating a search for its causes. One egregious error must, however, be avoided: that of treating components and causes on the same footing. By this route, one can arrive at the meaningless result that net migration is a more important "cause" of population growth than is change in manufacturing output. One must take strong exception to a causal scheme constructed on the premise, "If both demographic and economic variables help explain metropolitan growth, then we may gain understanding of growth processes by lumping the two together [29]." On the contrary, "understanding" would seem to require a clear distinction between demographic *components* of growth and economic *causes* that may affect growth via one or another of its components.

A CHAIN MODEL

Data reported by Hodge, Siegel, and Rossi [14] seem to fit well the model of a *simple causal chain* (see Fig. 4.3a). These authors give correlations between the occupational prestige ratings of four studies completed at widely separated dates: Counts (1925), Smith (1940), National Opinion

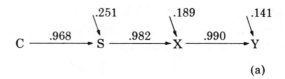

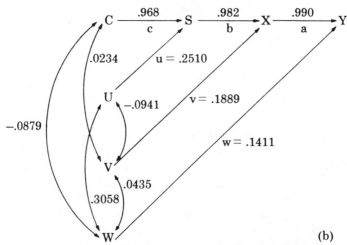

Figure 4.3. Causal chain. (a) Correlations taken from Hodge *et al.* [14] (C = Counts, 1925; S = Smith, 1940; X = NORC, 1947; Y = NORC, 1963); (b) intercorrelations of residuals implied by acceptance of chain hypothesis for the data in (a).

Research Center (NORC) (1947), and NORC replication (1963). In a simple causal chain, the correlations between temporally adjacent variables are the path coefficients (this is an immediate consequence of the definition of path coefficient). Using these three correlations as reported by Hodge *et al.*, we may infer that the correlation between NORC (1963) and Smith is (.990) (.982) = .972; between NORC (1963) and Counts is (.990) (.982) (.968) = .942; and between NORC (1947) and Counts is (.982) (.968) = .951. The observed values of these correlations (with differences from the inferred values in parentheses) are r_{YS} = .971 (−.001), r_{YC} = .934 (−.008), and r_{XC} = .955 (.004). Acceptance of this causal chain model is consistent with the conclusion of Hodge *et al.* that the amount of change in the relative positions of occupations in a prestige hierarchy is a direct function of elapsed time.

Although the discrepancies between inferred and observed correlations seem trivial, it is worth noting that acceptance of the estimates shown in Fig. 4.3a, along with the assumption of a simple causal chain, requires us to postulate a complex pattern of correlations (most of them negligible in size) among the residuals or errors. This pattern is shown in Fig. 4.3b. In obtaining this solution, we assume that each residual is uncorrelated with the immediately preceding variable in the chain but not necessarily with variables two or more links behind it. In the present example, then, the crucial assumptions are that $r_{VS} = r_{WX} = 0$. We can then, using equation (4.5) or the verbal algorithm, write the number of equations required to solve for the quantities to be entered on the diagram (for convenience, lowercase letters designate paths):

$$r_{YX} = a = .990$$

$$r_{XS} = b = .982$$

$$r_{SC} = c = .968$$

$$r_{YY} = 1 = a^2 + w^2$$

$$r_{XX} = 1 = b^2 + v^2$$

$$r_{SS} = 1 = c^2 + u^2 \tag{4.10}$$

$$r_{XC} = .955 = bc + vr_{VC}$$

$$r_{YC} = .934 = abc + avr_{VC} + wr_{CW}$$

$$r_{YS} = .971 = ab + cwr_{CW} + uwr_{UW}$$

$$r_{VS} = 0 = ur_{UV} + cr_{CV}$$

$$r_{WX} = 0 = vr_{VW} + br_{SW}$$

where $r_{SW} = cr_{CW} + ur_{UW}$.

In general, if we are considering a k-variable causal chain, we shall have to estimate $k - 1$ residual paths, $(k - 1)(k - 2)/2$ correlations between residuals, $k - 1$ paths for the links in the chain, and $k - 2$ correlations between the initial variable and residuals 2, 3, . . . , k in the chain. This is a total of $(k^2 + 3k - 6)/2$ quantities to be estimated. We shall have at our disposal $k(k - 1)/2$ equations expressing known correlations in terms of paths, $k - 1$ equations of complete determination (for all variables in the chain except the initial one), and $k - 2$ equations in which the correlation of a residual with the immediately preceding variable in the chain is set equal to zero. This amounts to $(k^2 + 3k - 6)/2$ equations, exactly the number required for a solution. The solution may, of course, include meaningless results (e.g., $r > 1.0$), or results that strain one's credulity. In this event, the chain hypothesis had best be abandoned or the estimated paths modified.

In the present illustration, the results are plausible enough. Both the Counts and the Smith studies differed from the two NORC studies and from each other in their techniques of rating and sampling. A further complication is that the studies used different lists of occupations, and the observed correlations are based on differing numbers of occupations. There is ample opportunity, therefore, for correlations of errors to turn up in a variety of patterns, even though the chain hypothesis may be basically sound. We should observe, too, that the residual factors here include not only extrinsic disturbances but also real though temporary fluctuations in prestige, if there be such.

What should one say, substantively, on the basis of such an analysis of the prestige ratings? Certainly, the temporal ordering of the variables is unambiguous. But whether one wants to assert that an aspect of social structure (prestige hierarchy) at one date "causes" its counterpart at a later date is perhaps questionable. The data suggest there is a high order of persistence over time, coupled with a detectable, if rather glacial, drift in the structure. The calculation of numerical values for the model hardly resolves the question of ultimate "reasons" for either the pattern of persistence or the tempo of change. These are, instead, questions raised by the model in a clear way for further discussion and, perhaps, investigation.

THE SYNTHETIC COHORT AS A PATTERN OF INTERPRETATION

Although, as the example from Turner indicates, it is often difficult in sociological analysis to find unequivocal bases for causal ordering, there is one happy exception to this awkward state of affairs. In the life cycles of individuals and families, certain events and decisions commonly, if not universally, precede others. Despite the well-known fallibility of retrospective data, the investigator is not at the mercy of respondents' recall in deciding to accept the completion of schooling as an antecedent to the pursuit of an occupational career (exceptions granted) or in assuming that marriage precedes divorce. Some observations, moreover, may be made and recorded in temporal sequence, so that the status observed at the termination of a period of observation may logically be taken to depend on the initial status (among other things). Path analysis may well prove to be most useful to sociologists studying actual historical processes from records and reports of the experience of real cohorts whose experiences are traced over time, such as a student population followed by the investigator through the first stages of postgraduation achievement.[2]

_____

[2] For example, Eckland [13].

The final example, however, concerns not real cohorts but the usefulness of a hypothetical synthesis of data from several cohorts. As demographers have learned, synthetic cohort analysis incurs some specific hazards [28]; yet the technique has proved invaluable for heuristic purposes. Pending the execution of full-blown longitudinal studies on real cohorts, the synthetic cohort is, at least, a way of making explicit one's hypotheses about the sequential determination of experiences cumulating over the life cycle.[3]

In a study of the social mobility of a sample of Chicago white men with nonfarm backgrounds surveyed in 1951, Duncan and Hodge [12] used data on father's occupational status, respondent's educational attainment, and respondent's occupational status in 1940 and 1950 for four cohorts: men 25 – 34, 35 – 44, 45 – 54, and 55 – 64 years old on the survey date. Their main results, somewhat awkwardly presented in the source publication, are compactly summarized by the first four diagrams in Figure 4.4. (The superfluous squared term in their equations has been eliminated in the present calculations. The amount of curvilinearity was found to be trivial, and curvilinear relations cannot be fitted directly into a causal chain by the procedure employed here.)

These data involve partial records of the occupational careers of the four cohorts and thus depict only segments of a continuous life history. In the original analysis, it was possible to gain some insights from the interperiod and intercohort comparisons on which that analysis was focused. Here, attention is given to a different use of the same information. Suppose we thought of the four sets of data as pertaining to a single cohort, studied at four successive points in time, at decade intervals. Then, all the data should fit into a single causal or processual sequence.

It is obvious that one cannot achieve perfect consistency on this point of view. The initial correlation, r_{UX}, varies among cohorts, for example. Moreover, age-constant intercohort comparisons of the other correlations (the Y's with X and U) suggest that some variations result from genuine differences between the conditions of 1940 and 1950. But if one is willing to suppress this information for the sake of a necessarily hypothetical synthesis, it is possible to put all the data together in a single model of occupational careers as influenced by socioeconomic origins.

The four correlations r_{UX} were averaged. The remaining correlations for adjacent cohorts were likewise averaged; for example, r_{1U} based on 1950 data for men 25 – 34 years old was averaged with r_{1U}, based on 1940 data for men 35 – 44 in 1951, and so on. Only r_{4X}, r_{4U}, and the three intertemporal correlations, r_{21}, r_{32}, and r_{43}, had to be based on the experience of just one cohort. (In deriving this compromise, one does, of course, lose the temporal specificity of the data by smoothing out apparently real historical fluctuations.) When the correlations had been averaged, the results shown in the

---

[3] See, for example, [15, p. 53 (n. 6) and Table 13].

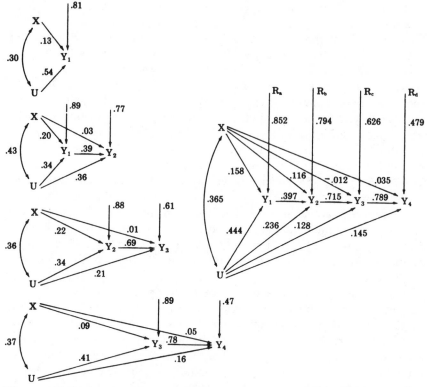

Figure 4.4. Respondent's occupational status (Y) at successive ages, in relation to father's occupational status (X) and respondent's educational attainment (U). Occupational status at age $25-35 = Y_1$; at $35-44 = Y_2$; at $45-54 = Y_3$; at $55-64 = Y_4$. (Source: Duncan and Hodge [12] and unpublished calculations kindly supplied by the authors.)

"composite" model on the right of Figure 4.4 were obtained. The estimates of path coefficients here are simply the partial regression coefficients, in standard form, of Y_1 on X and U; Y_2 on Y_1, X, and U; Y_3 on Y_2, X, and U; and Y_4 on Y_3, X, and U.

The results for the synthetic cohort make explicit the following interpretations:

1. The background factors, father's education (X) and respondent's education (U), have an important direct impact during early stages of a cohort's life cycle; after age $35-44$, their direct effects become small or negligible, although they exert indirect effects via preceding achieved statuses (Y_1 and Y_2).

2. Careers tend to stabilize after age 35 – 44, as indicated by the sharp rise in the path coefficients representing persistence of status over a decade (compare p_{21} with p_{32} and p_{43}) and by the decreasing magnitudes of the residual paths from $R_a, \ldots, R_d$.

3. During the life cycle, many circumstances essentially independent of background factors affect occupational mobility, so that achievement in the later stages of the career becomes more and more dependent upon intervening contingencies while continuing to reflect the indirect influence of conditions determinate at the outset. Thus, for example, r_{4c} —the correlation of occupational status at age 55 – 64 with residual for age 45 – 54 — may be computed as $(.789)\,(.636) = .494$, and the residual path to Y_4 itself is $p_{4d} = .479$. These are comparable in size with the correlations $r_{4X} = .301$ and $r_{4U} = .525$. The residuals are, by definition, uncorrelated with X and U and represent, therefore, the influence of factors quite unrelated to social origins and schooling. The prevailing impression that the United States enjoys a rather "loose" stratification system is thus quantified by a model such as this one.

4. While the data include observed interannual correlations of occupational statuses separated by a decade $(r_{43}, r_{32},$ and $r_{21})$, the synthetic cohort model also implies such correlations for statuses separated by two or three decades. These may be computed from the following formulas based on equation (4.5):

$$r_{42} = p_{4X}r_{2X} + p_{43}r_{32} + p_{4u}r_{2u}$$

$$r_{31} = p_{3X}r_{1X} + p_{32}r_{21} + p_{3u}r_{1u}$$

$$r_{41} = p_{4X}r_{1X} + p_{43}r_{31} + p_{4u}r_{1u}$$

inserting the value of r_{31} obtained from the second equation into the third. The observed and implied correlations are assembled in Table 4.2. The latter represent, in effect, hypotheses to be checked whenever data spanning twenty or thirty years of the occupational experience of a cohort become available. In the meantime, they stand as reasonable estimates, should any one have use for such estimates. If forthcoming evidence casts doubt on these estimates, the model will, of course, be called into question. It is no small virtue of a model that it is capable of being rejected on the basis of evidence.

This last example, since it rests on an explicit fiction — that of a synthetic cohort — perhaps makes clearer than previous examples the point that the role of path analysis is to *render an interpretation* and not merely to provide a format for presenting conventional calculations. In all the examples, the

Table 4.2. Observed and implied (*) correlations for synthetic cohort model of occupational achievement

| | Variable | | |
Variable (age and occupational status)	Y_2	Y_3	Y_4
25–34 (Y_1)	.552	.455*	.443*
35–44 (Y_2)	—	.772	.690*
45–54 (Y_3)	—	—	.866
55–64 (Y_4)	—	—	—

Source: Duncan and Hodge [12] and calculations from model in Figure 4.4.

intention has been to adhere to the purpose of path analysis as Wright formulated it: "The method of path coefficients is not intended to accomplish the impossible task of deducing causal relations from the values of the correlation coefficients [35, p. 193]."

> The method depends on the combination of knowledge of the degrees of correlation among the variables in a system with such knowledge as may be possessed of the causal relations. In cases in which the causal relations are uncertain, the method can be used to find the logical consequences of any particular hypothesis in regard to them [32, p. 557].

> Path analysis is an extension of the usual verbal interpretation of statistics not of the statistics themselves. It is usually easy to give a plausible interpretation of any significant statistic taken by itself. The purpose of path analysis is to determine whether a proposed set of interpretations is consistent throughout [40, p. 444].

NEGLECTED TOPICS

This chapter, for lack of space and especially for lack of "convincing" examples, could not treat several potentially important applications of path analysis: (1) Models incorporating feedback were explicitly excluded. Whether our present techniques of social measurement are adequate to the development of such models is perhaps questionable. (2) The problem of two-wave, two-variable panel analysis, recently discussed by Pelz and Andrews [22], might well be formulated in terms of path coefficients. The present writer, however, has made little progress in attempts to clarify the panel problem by means of path analysis. (3) The pressing problem of the disposition of measurement errors [3,5] may perhaps be advanced toward solution by explicit representation in path diagrams. The well-known "correction for attenuation," where measurement errors are assumed to be uncorrelated, is easily derived on this approach [35,39]. It seems possible that under very special conditions a solution may also be obtained on certain assumptions about correlated errors. (4) Wright has shown [40] how certain ecological models of the interaction of populations can be stated in terms of path coefficients. The inverse method of using path analysis for studies of multiple time series [40] merits consideration by sociologists. (5) Where the investigation involves unmeasured variables, path analysis may be helpful in deciding what deductions, if any, can be made from the observed data. Such unmeasured variables may, in principle, be observable; in this case, path analysis may lead to hypotheses for testing on some future occasion when measurements can be made. If the unmeasured variable is a theoretical construct, its explicit introduction into a path diagram [1] may well point up the nature of rival hypotheses. Ideally, what are sometimes called "validity coefficients" should appear explicitly in the causal model so that the latter accounts for both the "true causes" under study and the ways in which "indicator variables" are thought to represent "underlying variables." A particular case is that of factor analysis. As Wright's work demonstrates [37], a factor analysis is prone to yield meaningless results unless its execution is controlled by explicit assumptions that reflect the theoretical structure of the problem. An indoctrination in path analysis makes one skeptical of the claim that "modern factor analysis" allows us to leave all the work to the computer.

PATH ANALYSIS: SOCIOLOGICAL EXAMPLES (ADDENDA)*

1. Perhaps the most serious defect of the chapter is that it glosses over the distinction between sample and population or between unobservable parameters and the coefficients computed from sample observations (which may, under certain assumptions, serve as estimates of parameters). This defect is manifest not only in the rather casual attitude taken toward the problem of statistical inference throughout the chapter, but more particularly in the confused discussion of correlations involving residuals in connection with the chain model (Figure 4.3). It should have been made clear that non-zero correlations involving residuals computed from sample data may be compatible with a model specifying zero correlations of disturbances (among themselves and with predetermined variables) in the theoretical population. Hence, the correlations shown in Fig. 4.3b may merely reflect sampling variation about zero parameter values. If so it is not necessary to "postulate a . . . pattern of correlations . . . among . . . the errors," but merely to acknowledge that such correlations involving residuals will inevitably appear in any sample data, even if the structural model is

---

* Among other useful comments on the 1966 paper, I should like to acknowledge especially those of Arthur S. Goldberger.

entirely correct as a representation of the process occurring in the population, providing that the model is overidentified, as is the simple chain model.

2. The statement on p. 70 that the solution (for correlations involving residuals) may include the result $r > 1.0$ is incorrect for the model under discussion. Implied correlations in excess of unity may, indeed, turn up in connection with other kinds of models, such as those including unmeasured variables; but this possibility is not relevant to the case at hand.

3. I now believe that the concept of *indirect effect* presented on p. 62 is not very useful and that further discussion[4] of this particular concept was ill-advised. As defined on p. 62, the indirect effect of X_j on X_i is given by $(r_{ij} - p_{ij})$. But, of course, r_{ij} includes not only the indirect effects of X_j on X_i via compound paths leading from X_j to X_i, but also the effects of common causes (if any) of X_j and X_i or the effects of variables correlated with X_j (if X_j is an exogenous variable in the system). In place of the "single summary measure of indirect effect" suggested in the chapter, I would now recommend a calculation that may be illustrated with Figure 4.1. First, consider the correlation of one dependent variable with a subsequent dependent variable, for example, r_{35}.

Correlation:	r_{35}
equals	
Direct effect:	p_{53}
plus	
Indirect effect:	$p_{54} p_{43}$
plus	
Correlation due to common or correlated causes:	$p_{52} r_{23} + p_{51} r_{13} + p_{54} (p_{42} r_{23} + p_{41} r_{13})$

This, of course, is equivalent to either of the expansions of r_{35} given by equations (4.2) and (4.6). Second, consider the correlation of an exogenous variable with a dependent variable, for example, r_{15}.

Correlation:	r_{15}
equals	
Direct effect:	p_{51}
plus	

_____
[4] Land [17, p. 23].

Indirect effect: $p_{54} (p_{41} + p_{43}p_{31}) + p_{53}p_{31}$

   plus

Effect shared with other exogenous variable(s)

$$r_{12}[p_{52} + p_{54} (p_{42} + p_{43}p_{32}) + p_{53}p_{32}]$$

The term in square brackets represents the combined direct and indirect effects of the other exogenous variable in the system.

4. The mention of "neglected topics" on p. 74 is now obsolete. Several papers in the recent sociological literature have provided examples of models incorporating unmeasured variables, measurement errors, and reciprocal causation ("feedback"); and the treatment of two-wave, two-variable panel data has been explicated from the viewpoint of path analysis [11].

5. I am in general agreement with Blalock's comment on the chapter [4]. I would perhaps stress two points more strongly, however. First, in problems involving unmeasured variables or measurement error it may not be practicable to work with (raw-score) path regressions in preference to (standardized) path coefficients. Second, the choice between the two conventions is of no great moment with respect to the formulation, testing, and estimation of models, since the coefficients may be rescaled at will. An exception to this statement may be the kind of model that includes an explicit specification concerning error variance (say, that it is constant over time); in such a case, use of the standardized form may create unnecessary mathematical difficulties. Blalock is correct in observing that, for purposes of interpretation across populations, the raw-score regression form may be more useful in many cases; this is illustrated by a study of the process of stratification in the white and nonwhite populations [10].

ACKNOWLEDGMENTS

Prepared in connection with a project on "Socioeconomic Background and Occupational Achievement," supported by contract OE-5-85-072 with the U.S. Office of Education. Useful suggestions were made by H. M. Blalock, Jr., Beverly Duncan, Robert W. Hodge, Hal H. Winsborough, and Sewall Wright, but none of them is responsible for the use made of his suggestions or for any errors in the chapter.

REFERENCES

[1] Blalock, Hubert M., Jr. "Making Causal Inferences for Unmeasured Variables from Correlations among Indicators." *American Journal of Sociology* 69 (1963): 53 – 62.

[2] Blalock, Hubert M., Jr. *Causal Inferences in Nonexperimental Research.* Chapel Hill: University of North Carolina Press, 1964.

[3] Blalock, Hubert M., Jr. "Some Implications of Random Measurement Error for Causal Inferences." *American Journal of Sociology* 71 (1965): 37–47.

[4] Blalock, Hubert M., Jr. "Path Coefficients versus Regression Coefficients." *American Journal of Sociology* 72 (1967): 675–676.

[5] Bogue, Donald J., and Murphy, Edmund M. "The Effect of Classification Errors upon Statistical Inference: A Case Analysis with Census Data." *Demography* 1 (1964): 42–55.

[6] Boudon, Raymond. "A Method of Linear Causal Analysis: Dependence Analysis." *American Sociological Review* 30 (1965): 365–374.

[7] Campbell, Eleanor D., Turner, Malcolm E., and Wright, Mary Francis. *A Handbook of Path Regression Analysis*, Part I: *Estimators for Simple Completely Identified Systems*, with editorial collaboration of Charles D. Stevens. Preliminary ed. Richmond: Medical College of Virginia, Department of Biophysics and Biometry, 1960.

[8] Costner, Herbert L., and Leik, Robert K. "Deductions from 'Axiomatic Theory.'" *American Sociological Review* 29 (1964): 819–835.

[9] Davidson, F. A. *et al.* "Factors Influencing the Upstream Migration of the Pink Salmon (*Oncorhynchus gorbuscha*)." *Ecology* 24 (1943): 149–168.

[10] Duncan, Otis Dudley. "Inheritance of Poverty or Inheritance of Race?" In *On Understanding Poverty*, edited by Daniel P. Moynihan. New York: Basic Books, 1969.

[11] Duncan, Otis Dudley. "Some Linear Models for Two-Wave, Two-Variable Panel Analysis." *Psychological Bulletin* 72 (1969): 177–182.

[12] Duncan, Otis Dudley, and Hodge, Robert W. "Educational and Occupational Mobility." *American Journal of Sociology* 68 (1963): 629–644.

[13] Eckland, Bruce K. "Academic Ability, Higher Education, and Occupational Mobility." *American Sociological Review* 30 (1965): 735–746.

[14] Hodge, Robert W., Siegel, Paul M., and Rossi, Peter H. "Occupational Prestige in the United States, 1925–1963." *American Journal of Sociology* 70 (1964): 286–302.

[15] Jaffe, A. J., and Carleton, R. O. *Occupational Mobility in the United States: 1930–1960.* New York: King's Crown Press, 1954.

[16] Kempthorne, Oscar. *An Introduction to Genetic Statistics.* New York: John Wiley & Sons, 1957.

[17] Land, Kenneth C. "Principles of Path Analysis." In *Sociological Methodology 1969,* edited by Edgar F. Borgatta and George W. Bohrnstedt. San Francisco: Jossey-Bass, 1969.

[18] Le Roy, Henri Louis. *Statistische Methoden der Populationsgenetik.* Basel: Birkhäuser, 1960.

[19] Li, C. C. *Population Genetics.* Chicago: University of Chicago Press, 1955.

[20] Li, C. C. "The Concept of Path Coefficient and Its Impact on Population Genetics." *Biometrics* 12 (1956): 190–210.

[21] Moran, P. A. P. "Path Coefficients Reconsidered." *Australian Journal of Statistics* 3 (1961): 87–93

[22] Pelz, Donald C., and Andrews, Frank M. "Detecting Causal Priorities in Panel Study Data." *American Sociological Review* 29 (1964): 836–854.

[23] Simon, Herbert A. *Models of Man.* New York: John Wiley & Sons, 1957.

[24] Tukey, J. W. "Causation, Regression, and Path Analysis." In *Statistics and Mathematics in Biology*, edited by O. Kempthorne *et al.* Ames: Iowa State College Press, 1954.

[25] Turner, Ralph H. *The Social Context of Ambition.* San Francisco: Chandler Publishing Co., 1964.

[26] Turner, Malcolm E., and Stevens, Charles D. "The Regression Analysis of Causal Paths." *Biometrics* 15 (1959): 236–258.

[27] Walker, Helen M., and Lev, Joseph. *Statistical Inference.* New York: Holt, Rinehart & Winston, 1953.

[28] Whelpton, P. K. "Reproductive Rates Adjusted for Age, Parity, Fecundity, and Marriage." *Journal of the American Statistical Association* 41 (1946): 501–516.

[29] Wilber, George L. "Growth of Metropolitan Areas in the South." *Social Forces* 42 (1964): 491.

[30] Winsborough, Hal H. "City Growth and City Structure." *Journal of Regional Science* 4 (1962): 35–49.

[31] Wold, Herman, and Juréen, Lars. *Demand Analysis.* New York: John Wiley & Sons, 1953.

[32] Wright, Sewall. "Correlation and Causation." *Journal of Agricultural Research* 20 (1921): 557–585.

[33] Wright, Sewall. *Corn and Hog Correlations.* U.S. Department of Agriculture Bulletin 1300. Washington: Government Printing Office, 1925.

[34] Wright, Sewall. "Statistical Methods in Biology." *Journal of the American Statistical Association* 26 (1931, suppl.): 155–163.

[35] Wright, Sewall. "The Method of Path Coefficients." *Annals of Mathematical Statistics* 5 (1934): 161–215.

[36] Wright, Sewall. "The Genetical Structure of Populations." *Annals of Eugenics* 15 (1951): 323–354.

[37] Wright, Sewall. "The Interpretation of Multivariate Systems." In *Statistics and Mathematics in Biology,* edited by O. Kempthorne *et al.* Ames: Iowa State College Press, 1954.

[38] Wright, Sewall. "The Genetics of Vital Characters of the Guinea Pig." *Journal of Cellular and Comparative Physiology* 56 (1960, suppl. 1): 123–151.

[39] Wright, Sewall. "Path Coefficients and Path Regressions: Alternative or Complementary Concepts?" *Biometrics* 16 (1960): 189–202.

[40] Wright, Sewall. "The Treatment of Reciprocal Interaction, with or without Lag, in Path Analysis." *Biometrics* 16 (1960): 423–445.

Causal Inferences, Closed Populations, and Measures of Association*

H. M. Blalock, Jr.

5

Two of the most important traditions of quantitative research in sociology and social psychology are those of survey research and laboratory or field experiments. In the former, the explicit objective is usually that of generalizing to some specific population, whereas in the latter it is more often that of stating relationships among variables. These two objectives are not thought to be incompatible in any fundamental sense, but nevertheless we lack a clear understanding of their interrelationship.

One of the most frequent objections to laboratory experiments turns on the question of generalizability, or what Campbell and Stanley refer to as "external validity."[1] In essence, this question seems to reduce to at least two related problems: (1) that of representativeness or typicality, and (2) the possibility of interaction effects that vary with experimental conditions. In the first case, the concern would seem to be with central tendency and dispersion of single variables, that is, whether the means and standard deviations of variables in the experimental situation are sufficiently close to those of some larger population. The second involves the question of possi-

* Reprinted by permission of the publisher from the *American Political Science Review*, 61: 130–136. Copyright 1967, The American Political Science Association.
[1] See especially Campbell and Stanley [4, pp. 171–246]. See also Kish [7].

ble disturbing influences introduced into the experimental setting that produce nonadditive effects when combined with either the experimental variable or the premeasurement.[2] These same variables may of course be operative in larger populations. But presumably they take on different numerical values, with the result that one would infer different relationships between major independent and dependent variables in the two kinds of research settings.

As a sociologist who is only superficially acquainted with trends and developments in quantitative behavioral political science, it is my impression that, with the exception of simulation studies, the overwhelming emphasis has been on stating generalizations appropriate to specific populations rather than stating general laws of political behavior. Consistent with this has been the frequent use of correlation coefficients as measures of degree of relationship, as contrasted with the use of unstandardized regression coefficients as measures of the *nature* and form of relationships appropriate for general structural equations representing causal laws. One reason for this emphasis on particular populations is perhaps that the kinds of populations dealt with by political scientists are often of more inherent descriptive or practical importance than many of the (usually smaller) populations sampled by sociologists. I shall return to this question in discussing the Miller–Stokes study of constituency influence.

The question of whether or not one is generalizing only to specific populations or attempting to state scientific laws has created considerable confusion in the sociological literature, and I hope that political scientists will be in a position to benefit from this confusion so as to by-pass controversies such as those concerning whether or not one should make tests of statistical significance when data for the entire population are available. The purpose of the present chapter is to bring this general issue into sharper focus. In doing so, I take a position with which many political scientists will undoubtedly disagree.

Rather than taking as the ultimate objective the goal of generalizing to specific populations, I would maintain that it is preferable to attempt to state general laws that interrelate variables in terms of hypothetical "if-then" statements. These could be of the form, "If X changes by one unit under conditions A, B, and C, then Y should change by b_{yx} units." In effect, then, I would consider generalizations to populations as means rather than ends. But given the limitations imposed by most data collection techniques, it is often necessary to carry out studies on specific populations at a single point in time, or at most several points in time. If so, then what must one assume about these populations? Since no real populations will be completely isolated or "closed," what kinds of assumptions concerning less than com-

---

[2] For a systematic discussion of the handling of various interaction effects in experimental designs see Ross and Smith [10].

pletely closed populations can realistically be made? And what bearing does this have on choice of measures of association? Let me first consider the question of the closure of populations.

CLOSED POPULATIONS AND CLOSED THEORETICAL SYSTEMS

A completely closed theoretical system would be one in which no variables have been omitted and which (if mathematically formulated) would imply perfect mathematical functions with no stochastic or error terms. Obviously, such completely closed systems are impossible to find in the social sciences. Furthermore, our approximations to this ideal are not even close. The error terms one finds empirically are usually too large to be considered negligible. Therefore, one is faced with the necessity of making assumptions about these errors and why they occur. Such assumptions will become quite complex whenever one is dealing with theoretical systems involving reciprocal causation, time lags, interaction terms, and the like.

Let us therefore confine our attention to simple recursive equations of the form:

$$X_1 = e_1$$

$$X_2 = b_{21}X_1 + e_2$$

$$X_3 = b_{31}X_1 + b_{32}X_2 + e_3$$

$$X_4 = b_{41}X_1 + b_{42}X_2 + b_{43}X_3 + e_4$$

One can show in this case that unbiased estimates of the regression coeffi-cients can be obtained by assuming that the error terms in each equation are uncorrelated with each other and also with all of the *independent* variables that appear in their respective equations.[3] Thus, e_2 is assumed uncorrelated with X_1 and also the remaining error terms e_i. Similarly, e_3 is unrelated to X_1 and X_2, and so forth.

What do these assumptions mean in terms of the behavior of outside variables not explicitly contained in the model? In brief, if one assumes that outside variables have a direct effect on *only one* of the explicit variables, then the assumptions can be met. Notice that an implicit variable might have an *indirect* effect on some variable through one of the remaining X_i without violating the assumptions. But if an implicit factor *directly* affects two or more explicit variables, then it will ordinarily be correlated with one

[3] See Wold and Juréen [15, Chapter 2].

of the independent variables in its equation, and the assumptions will not be met. If this is the case, least-squares estimates will be biased, and one's inferences will be incorrect. Such a variable should be explicitly included in the system. At some point, one must stop and make the simplifying assumption that all remaining implicit factors operate (in a major way) on only one explicit variable.

Analogously, a completely closed population is one that is subject to no outside influences, including immigration or emigration. A completely closed theoretical system that implied a stable equilibrium could be tested in such a closed population by collecting data at a single point in time, provided it was assumed that stability had actually been reached.[4] But given that such completely closed populations can never be found empirically, the major problem seems to be that of specifying a satisfactory analogue to the imperfectly closed theoretical system that allows for error terms in the equations.

It seems to me that in order to state and test theoretical generalizations based on population data, one must assume that the lack of closure of the population does not violate the preceding assumptions regarding error terms.[5] That is, disturbances that systematically affect one variable should not affect the others in a major way. For example, migration factors that directly affect X_2 should not also affect X_1, nor should they affect X_3 or X_4 except through the operation of X_2. This means that migration processes (or other disturbances emanating from outside populations) should not affect relationships among the variables, though they may very well affect measures of central tendency or dispersion for single variables.

For example, individuals obviously migrate for economic reasons. Certain cities will have higher income or occupational levels than others, or may be more homogeneous economically than others. Individuals will of course also migrate for other reasons as well—to join relatives or for cultural or recreational purposes. In many instances, however, it will be plausible to assume that these additional reasons are "idiosyncratic" and that, when aggregated, do not systematically distort relationships among variables. If, however, large numbers of persons move for combinations of reasons, then one's inferences regarding relationships among variables may very well be

[4] There will of course be numerous dynamic models all of which predict the same stability conditions. Ideally, theories should be formulated in dynamic terms in such a way that time enters in in an essential way (as in difference equations). The approach of "comparative statics" can then be used to study equilibrium conditions. For a very readable discussion of this question see Baumol [1]. See also Samuelson [11] and Simon [12, Chapters 6–8].

[5] Of course, somewhat less restrictive assumptions concerning the error terms might be used. Whatever set of assumptions are used, however, it would seem necessary to assume that the population is "closed" in the sense of meeting these assumptions.

misleading. For example, suppose that high-income liberals move into communities containing large numbers of low-income conservatives (e.g., rural college communities). A study relating income to conservatism could lead to erroneous conclusions if based solely on this community.

I assume that many objections to "atypical" populations are based on this kind of concern about selective migration. Detroit is certainly not typical of metropolitan centers with respect to income distribution. Whether or not it is considered objectionable, from the standpoint of generalizations to other populations, however, would seem to depend on whether or not Detroit is peculiar with respect to combinations of characteristics of interest to the investigator. Ideally, one should define the boundaries of his population in such a way that such selective migration (by several variables) is reduced to a minimum. Thus, he might prefer to use an entire metropolitan area, rather than the central city, arguing that although migration *within* the metropolitan area may be selective, this will not be the case *between* such larger units. Had he used central cities, he might have found this assumption much less plausible. For example, high-income blacks might be more likely to remain within the city, as might also be true for certain extremely high-status families (e.g., the "Proper Bostonians").

AN EXAMPLE: CONSTITUENCY INFLUENCE

Before proceeding with a rather general discussion of standardized and unstandardized coefficients, I should like to introduce as a concrete example the study of constituency influence in Congress by Miller and Stokes [9]. The authors give correlational data based on a sample of 116 congressional districts and then interpret these data in terms of a causal model as indicated in Fig. 5.1. The direction of influence between the representative's attitudes and his *perceptions* of the constituency's attitudes is left unspecified, though the authors discuss the implications of the two limiting models in which the one or the other of the two arrows is erased. Cnudde and McCrone suggest some possible revisions of this model, based on the magnitudes of the

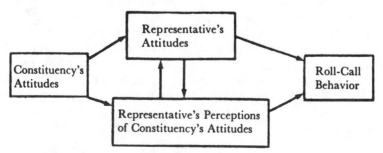

Figure 5.1. A causal model of constituency influence.

coefficients.[6] In particular, they suggest that the data are compatible with a model in which (1) the arrow from constituency's attitude to representative's attitude is erased, and (2) it is assumed that the representative's perception affects his own attitudes, rather than vice versa. There is no need, here, to concern ourselves with this particular substantive issue.

The basic problem with which Miller and Stokes deal is that of measuring the relative importance of the two major paths from constituency's attitude to roll-call behavior. In the case of civil rights, they conclude that the path via the representative's perceptions is more important than that via the representative's own attitudes. The measures they display in the paper are correlation coefficients, but they also make use of path coefficients (which will be discussed below). The conclusion reached is that in the case of civil rights roll-call behavior, and the least favorable assumptions regarding the importance of the representative's perceptions, the path via perceptions accounts for more than twice as much of the variance as does the path involving the representatives' own attitudes.

This is an excellent example of a study in which there is inherent interest in generalizing to a single population, since there is only one U.S. House of Representatives. Once inferences have been made from the sample of 116 districts to this total population, one might then take the position that there is no point in attempting to formulate more general "laws" of constituency influence. But there are a number of respects in which the data are not completely general. First, only a sample of political issues could be studied. Second, the study is obviously time-bound, and it might be desirable to compare results (for these same issues) over a period of time in order to ascertain whether or not the basic processes have remained unaltered. Third, one might wish to compare these results with those of legislative bodies in other countries. Had the Miller–Stokes data pertained to legislative behavior within a single state, then one would obviously be interested in comparing results across states. For example, one might determine whether the coefficients for Southern and Northern states were similar with respect to civil rights issues. If different with respect to civil rights, they might be similar with respect to other issue areas. These types of problems require one to formulate propositions more abstractly than in terms of specific populations and periods of time.

The question of the possible lack of closure of a population may not be so relevant for this type of substantive problem as would be the case in voting-behavior studies. Legislative bodies are closed at least in the sense that, with minor exceptions, persons are elected for specified periods of time and do not migrate into and out of such "populations." In developing measures of constituency attitudes, one likewise need not be too concerned about lack of closure, since presumably representatives are concerned about the makeup

---

[6] Cnudde and McCrone [5].

of *present* constituencies and ordinarily assume that the distribution of attitudes will not be modified in any major way by migration. Each constituency, however, will be influenced by neighboring districts, and it will undoubtedly be necessary to make simplifying assumptions about how this influence process affects the closure of these smaller units.

With this example in mind, let us return to the question of one's choice of appropriate measures. I shall then comment briefly on why I believe that unstandardized measures would be more suitable for certain purposes for which the Miller–Stokes type of data might be used.

STANDARDIZED VERSUS UNSTANDARDIZED REGRESSION COEFFICIENTS

The method of "path coefficients" or "dependence coefficients" has been recommended as an important methodological tool for measuring the relative contributions to associations between pairs of variables in multivariate analyses. Raymond Boudon [3] implies that sociologists might have failed to use such coefficients not only because of a certain confusion over the meaning of regression coefficients in simultaneous equations but also because of the identification problems that may arise.[7] It is important, however, to be well aware of the major differences between standardized measures, such as correlation and path coefficients, and unstandardized regression coefficients. The former seem most appropriate for describing relationships in particular populations; the latter for comparing populations or stating general laws.[8]

It is instructive to examine the relationship between correlation and regression coefficients by conceiving of the numerical value of a correlation coefficient as a dependent variable, being a function of (1) the causal law connecting two (or more) variables, and (2) the relative amounts of variation that happen to exist in any particular population or that may be induced in experimental manipulations. Making the assumptions necessary for least squares, we may write

$$r_{xy} = b_{yx}\left(\frac{s_x}{s_y}\right)$$

Let us assume that the sample size is sufficiently large that we can ignore sampling error. Similar expressions can be written in the case of three or more variables; for example:

[7] See also Wright [16] and Duncan [6].
[8] This position is basically similar to that taken by Tukey [13, Chapter 3].

$$r_{xy \cdot w} = b_{yx \cdot w} \left(\frac{s_x}{s_y} \frac{\sqrt{1 - r_{xw}^2}}{\sqrt{1 - r_{yw}^2}} \right)$$

I shall confine the discussion to the simple two-variable case, as the extension is straightforward.

The coefficient b_{yx} (or $b_{yx \cdot w}$) represents the change in Y produced by a unit change in X and is appropriate for use in a general statement of a causal law.[9] Such a law is expressed in the hypothetical "if-then" form. There is no assertion that X has changed, or will change, by a given amount. But in order to apply or test such a law, one must deal with specific populations or manipulations in particular experiments. The factor s_x / s_y, or its extensions in the multivariate case, involves *actual* variations in X and Y. Assuming that there are no measurement errors in either variable, we would presume that variation in X (e.g., constituency attitudes) is produced by factors not explicitly considered, and that variation in Y is jointly affected by X plus additional factors left out of the theoretical system. Least-squares procedures will give unbiased estimates of the true regression coefficients in causal laws only if certain assumptions are met concerning the effect of variables left out.[10] Ordinarily, however, one is in no theoretical position to specify *a priori* the *amount* of actual variation produced by omitted factors in real populations. In the case of populations that are not completely closed, the existence of migration makes such assumptions even less plausible.

Thus, the amount of variation in X relative to variation in Y produced by factors not considered may be taken as "accidental" from the point of view of one's theory. Of course one may take X as itself being determined by other variables, in which case variation in X may also be partly explained. But we are here considering X as "exogenous," and it is of course necessary to have some such exogenous variables in one's theoretical system. These are the "givens" that the theorist makes no effort to explain. In terms of the

[9] The numerical value of b_{yx} is of course also affected by one's choice of units of measurement. Unlike the expression s_x / s_y, however, these units of measurement are interrelated by purely *a priori* or definitional operations (e.g., 100 pennies = one dollar). It is true, as McGinnis notes, that one can transform r_{xy} into b_{yx} by multiplying by a simple scalar quantity. It does not follow, however, that correlation and regression coefficients are essentially interchangeable as McGinnis implies. For this scalar quantity, s_y / s_x is a function of standard deviations peculiar to each population. A reader who is given only the correlation coefficient is therefore likely to be misled in interpreting results of comparative studies. For a further discussion of this point, see Blalock [2, Chapter 4]. See also McGinnis [8].

[10] In particular, one must assume that the error terms in each equation have zero means and are uncorrelated with each other and with any of the independent variables that appear in their respective equations.

coefficients, he cannot account for the numerical values of the variance in exogenous variables. In this sense, they are taken as accidental, and are unique to each population even where the same causal laws are operative on all populations.

Let me illustrate with a simple example. Suppose one finds stronger correlations between constituency's and representative's attitudes in the North than is true in the South. It is quite conceivable that the same laws are operative, giving the same values of b_{yx} in each region. Yet there may be more variation in constituency's attitudes in the North, and if extraneous factors operated to the same extent in both regions, this would account for the larger correlation. Fortunately, the amount of variation in X can be measured and the regions compared in this respect. But uncontrolled and unknown disturbing influences cannot, and one would have no way of determining whether or not these also varied more in the North than in the South. It would therefore be more meaningful to compare the slope estimates than the respective correlations.

The same basic issues concerning standardization arise in the case of more complex causal models. As Miller and Stokes have noted, the method of path coefficients provides a simple and useful way of representing a total correlation between any two variables as a function of the causal paths that connect them. Consider, for example, the following model:

$$X_1 \longrightarrow X_2$$
$$X_4 \longleftarrow X_3$$

Letting p_{ij} represent the path coefficient from X_j to X_i, we can write down expressions for each of the r_{ij}. For example:

$$r_{23} = p_{21}p_{31}$$

and

$$r_{14} = p_{21}p_{42} + p_{31}p_{43}$$

These expressions can either be derived algebraically, as indicated in Boudon's paper, or they can be obtained directly by following certain rules or "algorithms" for tracing paths. The latter method has more intuitive appeal, and is easier to apply in simple models, but may lead one astray in more complex situations. The general rule is that one can trace paths by proceeding backward and then forward, or forward only, but it is not legitimate to move forward and *then* backward. Thus, in the case of the paths between X_2 and X_3, one finds a path through X_1 (going back against the direction of the arrow, and then forward) but *not* through their common effect X_4 (which would require going forward and then back).

One can then attribute a certain proportion of the total correlation to each component path, thereby obtaining a measure of the relative contribution of each variable to this correlation. But why take the *correlation* between two variables as something to be explained? According to my previous argument, this should be a function of variation in exogenous variables peculiar to particular populations. Breaking up a correlation coefficient into component parts would seem to require that these parts, themselves, be peculiar to the population. This is, in fact, the case, as can be seen from the definition of a path coefficient.

A path coefficient, p_{ij}, is defined as the ratio of two quantities. The standard deviation in the dependent variable, X_i, is in the denominator. The numerator is essentially an adjusted standard deviation, being the standard deviation in X_i that would result if X_j retained the same amount of variation, but if all other causes of X_i (direct and indirect) remained constant. In symbols:

$$p_{ij} = \frac{s_{i \cdot j}}{s_i}$$

where $s_{i \cdot j}$ represents the standard deviation in X_i that can be attributed to X_j, with the other variables held constant. It is possible, therefore, that a path coefficient can take on a value greater than unity, though this is not likely in most realistic examples.

Notice that this definition of path coefficients, from which the simple algebraic relationships with correlations can be derived, involves a combination of the hypothetical and the real. On the one hand, the *actual* variations in both X_j and X_i are accepted as givens. On the other, we are asked to imagine what would happen if the remaining variables were held constant. This mixture of real and hypothetical poses some interesting questions. How, for example, would one retain the same variation in X_3, while holding constant X_1, which is one of its causes? To do so would require one to manipulate some *other* cause of X_3 not included in the model. It is of course possible to imagine experiments in which this could be accomplished. The real question, it seems, is: "Why combine these hypothetical manipulations with the actual variations peculiar to a given population?" Put another way, if one were interested in controlling some of the variables, why would he insist that the independent variable under study retain the same amount of variation? This might make sense if the so-called independent variables were not themselves causally interrelated, but the rationale seems to be less clearcut in more complex situations.[11]

Unstandardized regression coefficients provide direct answers to the kind of hypothetical question that I believe to be more appropriate. For example, one could ask what would be the expected change in X_4 produced by a unit change in X_1, given that X_1 affects both X_2 and X_3, and assuming that outside factors produce only random variation. Suppose, for example, that a unit change in X_1 will increase X_2 by three units, and decrease X_3 by four units. Suppose, also, that unit increases in X_2 and X_3 will increase X_4 by two and four units, respectively. Then an increase of one unit in X_1 should increase Z_4 by $3(2) = 6$ units via X_2 and change X_4 by $(-4)(4) = -16$ units via X_3. The total expected change would then be ten units in the negative direction.

Notice that nothing is being said here about actual changes in a given population. The population data are used to *estimate* the regression coefficients, but the formulation is purely hypothetical. I have used a simple numerical example for illustrative purposes, but the general procedure is quite straightforward *provided* one assumes one-way causation and no feedback, plus the usual least-squares assumptions regarding error terms. Turner and Stevens provide both an algebraic procedure and an algorithm for tracing paths similar to that given by Wright for the standardized path coefficients.[12]

Returning to the Miller–Stokes data, the reason why the path via the representative's perception was found to explain more than twice as much of the variance as the other major path was that the correlation between constituency's attitude and perception was .63, whereas that between constituency's and representatives' attitudes was only .39. The correlations between the two intervening variables and roll-call behavior were almost identical (being .82 and .77, respectively). Let us rule out the possibility of sampling error and assume that these correlations adequately describe the magnitudes of the relationships for this particular population, at this given time (1958), for civil rights issues.

---

[11] If interest is in generalizing to a population, it might make more sense to use a measure that does not involve any hypothetical manipulations. One may of course break down $R^2_{4\cdot123}$ into the components r^2_{14}, $r^2_{24\cdot1} (1 - r^2_{14})$, and $r^2_{34\cdot12} (1 - R^2_{4\cdot12})$. Since x_1, x_2, and x_3 are intercorrelated, these components cannot be directly associated with these variables. However, the exogenous causes of the X_i are assumed orthogonal in the case of simple least squares, and therefore one may — if he wishes — link the above components of $R^2_{4\cdot123}$ with the respective error terms, given these assumptions about the causal ordering of x_1, x_2, and x_3. This of course raises the question of why one would want to associate components with exogenous variables that have not been included in the theoretical system.

[12] Turner and Stevens [14].

Suppose, now, that additional data were available for other populations, time periods, or issues. To be specific, suppose one wished to compare the Miller–Stokes results with those for several different periods of time, posing the question as to whether or not basic changes in influence processes were occurring. As noted previously, it is entirely possible that the basic laws, as measured by slopes, would remain unchanged, whereas the relative magnitudes of the correlations could be altered. In the case of constituency attitudes, which are the starting points for both paths, an increase in the variance would be expected to increase *both* correlations with the two intervening variables. But it is quite possible that outside or exogenous factors affecting the representative's attitudes might produce a smaller variance in subsequent periods, though they might continue to create essentially random disturbances. If so, the correlation between constituency's attitude and representative's attitude would increase, perhaps to a level comparable to that between constituency's attitude and representative's perception. With reduced variation in representative's attitude, the correlation between this variable and roll-call behavior might also decrease, again with no basic changes in the slope coefficients.

The general point, of course, is that both the simpler correlation coefficients and more complex path coefficients are functions of the nature of the underlying processes *and* the relative magnitudes of disturbance terms. As long as one is dealing with a single set of data — as is true with respect to the Miller–Stokes example — the use of path coefficients will not be misleading. When one is interested in making comparisons, however, he should become sensitized to the differences between the two types of measures.

CONCLUSIONS

I suspect that one reason why both standardized and unstandardized regression coefficients have not found favor with sociologists and political scientists is a general reluctance to work with interval–scale assumptions and an understandable resistance to committing oneself to specific causal models. I hope that we will begin to move more and more in these directions. In doing so, however, we must keep clearly in mind the distinction between working with causal laws and unstandardized coefficients and attempting to measure relative importance of variables in specific populations. We must recognize that relative importance cannot be evaluated in the abstract: The contribution of each factor to the total variation in a dependent variable is a function of how much the various independent variables happen to vary in that given population. Since I am arguing against the advisability of generalizing to populations, at least as an ultimate objective, I must also argue that we would not be interested in these standardized measures except for descriptive or practical purposes.

Tukey [13] has taken essentially the same position and has recommended working with unstandardized coefficients. Wright [16] has pointed out that, like the ordinary correlation coefficients, standardized path coefficients have much simpler properties than the unstandardized measures. I would agree with Tukey, however, who suggests that the price of simplicity may be too high. It would seem preferable to make use of unstandardized coefficients that have some chance of being invariant from one population to the next, rather than measures that have admittedly simpler descriptive properties.

ACKNOWLEDGMENT

I am indebted to the National Science Foundation for support of this research.

REFERENCES

[1] Baumol, William J. *Economic Dynamics*. New York: The Macmillan Company, 1959.
[2] Blalock, H. M. *Causal Inferences in Nonexperimental Research*. Chapel Hill: University of North Carolina Press, 1964.
[3] Boudon, Raymond "A Method of Linear Causal Analysis: Dependence Analysis." *American Sociological Review* 30 (1965): 365–374.
[4] Campbell, Donald T., and Stanley, Julien S. "Experimental and Quasi-experimental Designs for Research on Teaching." In *Handbook of Research on Teaching*, edited by N.L. Gage. Chicago: Rand McNally & Company, 1963.
[5] Cnudde, Charles F., and McCrone, Donald J. "The Linkage between Constituency Attitudes and Congressional Voting Behavior: A Causal Model." *The American Political Science Review* 60 (1966): 66–72.
[6] Duncan, Otis Dudley. "Path Analysis: Sociological Examples." *American Journal of Sociology* 72 (1966): 1–16.
[7] Kish, Leslie. "Some Statistical Problems in Research Design." *American Sociological Review* 24 (1959): 328–338.
[8] McGinnis, Robert. "Review of Causal Inferences in Nonexperimental Research." *Social Forces* 44 (1966): 584–586.
[9] Miller, Warren E., and Stokes, Donald E. "Constituency Influence in Congress." *The American Political Science Review* 57 (1963): 45–56.
[10] Ross, John A., and Smith, Perry. Experimental Designs of the Single-Stimulus, All-or-Nothing Type." *American Sociological Review* 30 (1965): 68–80.
[11] Samuelson, Paul A. *The Foundations of Economic Analysis*. Cambridge: Harvard University Press, 1947.
[12] Simon, Herbert A. *Models of Man*. New York: John Wiley & Sons, 1957.
[13] Tukey, John W. "Causation, Regression, and Path Analysis." In *Statistics and Mathematics in Biology*, edited by Oscar Kempthorne et al. Ames, Iowa: Iowa State College Press, 1954.
[14] Turner, Malcolm E., and Stevens, Charles D. "The Regression Analysis of Causal Paths." *Biometrics* 15 (1969): 236–258.
[15] Wold, Herman, and Juréen, Lars. *Demand Analysis*. New York: John Wiley & Sons, 1953.
[16] Wright, Sewall. "Path Coefficients and Path Regressions: Alternative or Complementary Concepts?" *Biometrics* 16 (1960): 189–202.

SIMULTANEOUS-EQUATION TECHNIQUES

II

Part II is a difficult section involving a good many rather elusive, though fundamental, issues facing all of the social sciences. The literature on simultaneous-equation estimation procedures is becoming extensive but is relatively more technical than materials contained in the remaining sections of this volume. The reader who is totally unfamiliar with nonrecursive systems may therefore wish to begin with the Mason and Halter chapter, which explains rather simply the main essentials of two-stage least squares as applied to the substantive example of a diffusion model. The reader should then be in a position to follow some of the more abstract chapters by Koopmans [6], Strotz and Wold [7], and Fisher and Ando [10]. The latter set of readings might best be studied in conjunction with textbook materials on simultaneous-equation estimation.[1]

Part II begins with a classic chapter by Koopmans that conceptualizes the identification problem in the context of supply and demand models and points to the need for exogenous variables and *a priori* restrictive assumptions in order to reduce the number of unknowns relative to knowns in the general k-equation case. Koopmans is dealing with static models that require equilibrium assumptions. In these models, it is generally impossible to estimate the relative effects of each variable on the others when we allow for reciprocal causation among all "endogenous" or mutually dependent variables. The proposed resolution to the identification problem arising in the case of such static models is to introduce predetermined variables in such a way that at least $k - 1$ variables have been left out of each equation that is to be identified. These variables may be truly exogenous or "independent" of any of the endogenous variables, or they may be lagged values of endogenous variables.

The next chapter, by Strotz and Wold, states the case for recursive models involving lagged endogenous variables.[2] In brief, the authors argue that especially in instances where each equation can be linked with autonomous actors (e.g., suppliers, customers, and retailers), a model is most adequately conceived in terms of stimulus–response situations in which each party is responding (with a lag) to the actions of the others. If so, then simultaneous equations involve "specification errors" in that they fail to capture the time lags involved. In effect, Strotz and Wold argue that adequate models will be recursive in nature, and that the use of lagged endogenous variables permits one to conceptualize reciprocal causation in dynamic terms, where differences in time periods are explicitly taken into consideration. The issues raised in these first two chapters are indeed fundamental ones that have implications for sociology, political science, and psychology. One of the greatest difficulties faced in these latter fields, however, is that of the appro-

[1] See especially Christ [1] and Johnston [2].
[2] Wold's more recent position is a good deal more complex but seems to entail a convergence with Fisher's emphasis on block-recursive models. See Wold [4,5,3].

priateness of lagged variables in instances when the time lags are unknown or not uniform. Furthermore, lagged variables involve one with difficult problems of autocorrelation of the disturbance terms.

The chapters by Fisher [9] and Fisher and Ando [10] deal with additional practical issues that introduce further technical complications. Obviously, no model can be completely realistic. Therefore, it involves specification errors of one kind or another—errors produced by faulty assumptions about omitted variables, about linearity and additivity, and about the lack of measurement errors involved. Assumptions used to identify a system (e.g., that certain coefficients are zero) are never strictly correct, and therefore the question arises as to the seriousness of the errors produced whenever these assumptions are in fact invalid. Fisher and Ando give a nontechnical discussion of an important theorem to the effect that as long as assumptions are *approximately* correct, we can count on only minor distortions in our estimates and that even our inferences for long-run dynamic models can be reasonably safe. It is important to realize, however, that assumptions that are totally unjustified empirically *will* produce misleading results. Here we see the necessity of having a reasonably sound theory prior to accurate estimation.

The chapter by Fisher on instrumental variables explores the problem of making practical decisions as to one's choice among exogenous and lagged endogenous variables in the case of macrolevel models involving large numbers of variables. A careful reading of Fisher's chapter should make it abundantly clear that there will always be unmet assumptions and compromises with reality as well as decisions that must be based on inadequate information.

A point that has been emphasized by Fisher in his chapter on the choice among instrumental variables, as well as elsewhere, is that what have been referred to as "block-recursive" models afford a realistic compromise between the simplicity of the recursive models stressed by Wold and the more general linear systems of Koopmans and others. In a block-recursive system, one allows for reciprocal causation *within* blocks of variables but one-way causation *between* blocks. Thus, certain blocks of variables are assumed to be causally prior to others, with no feedback being permitted from the latter to the former. This kind of assumption is, of course, needed in order to treat certain variables as predetermined in any given system. Thus, if variables in Blocks A and B are assumed to be possible causes of variables in Block C, whereas the possibility of Block C variables affecting Blocks A and B is ruled out, then when one is studying the interrelationships among variables in Block C, he may utilize variables from Blocks A and B as predetermined. These variables may be truly exogenous, or they may be lagged values of variables in Block C, and Fisher discusses the relative merits of both kinds of factors. The notion of "instrumental variables" is used as a generic term to refer to any variables that can to help identify and estimate the parameters in the system.

Perhaps the most fundamental idea that pervades Part II, and that has been implicit in the previous sections, is that of "structural parameters" which are seen as providing the "true" causal structure of the theoretical system. In the case of recursive systems, these parameters were the b_{ij} that could be estimated in a straightforward manner by ordinary least squares. But in the more general case, one is faced with the fact that there will ordinarily be indefinitely many different sets of parameter values that all imply the same empirical data. The identification problem can be conceptualized as that of recovering the true structural parameters from the empirical data, and in the case of static formulations this cannot be accomplished without the aid of untestable *a priori* assumptions. Usually, though not necessarily, these take the form of assumptions that certain of the parameters have zero values. In the special case of recursive systems, for example, one-half of the possible b_{ij} have been set equal to zero so as to produce a triangular slope matrix. Additionally, we also assume that the covariances of all pairs of disturbance terms e_i are zero, and these two kinds of assumptions make identification possible in the case of recursive systems.

In the general k-equation case, it will always be possible to rewrite the system in terms of a mathematically equivalent set of equations referred to as a "reduced form." If we represent the endogenous variables as X_i and the predetermined variables as Z_j, each of the equations for the endogenous X_i can be written as a linear function of the Z_j alone. One may therefore estimate each X_i by using only the Z_j, and in fact these reduced forms are used in the first stage of the two-stage least-squares procedure that is discussed in several chapters in this section. However, it does *not* follow that one can always estimate the true structural parameters from the estimates of the reduced-form parameters. Thus, the identification problem can also be conceived as the problem of moving from the reduced-form equations to the structural equations. Whenever there are too many unknown structural parameters for this to be done, we say that the equation(s) is underidentified. Whenever there are exactly the right number and combination of unknown parameters, there may be a unique solution for the structural parameters in terms of the reduced-form parameters, in which case we say that the equation(s) is exactly identified. Whenever there are fewer unknown structural parameters than necessary for exact identification, the equation(s) will be overidentified, and there will be no *unique* way of estimating the parameters.

In the overidentified case, therefore, questions arise as to the "best" method of estimating the parameters, and there has been extensive discussion of this technical problem in the econometrics literature. In general, so-called "full information" methods that utilize information from the entire set of equations at once tend to be relatively more efficient than two-stage least squares, which deals with the equations sequentially. However, the full-information procedures seem more sensitive to specification errors that arise in instances where theoretical foundations for the model are weak.

It would appear as though two-stage least squares is entirely adequate for less advanced fields such as political science and sociology, given the presence of relatively poor measurement procedures and the very tentative nature of existing theories.

As explained by Mason and Halter [8], the essential idea behind two-stage least squares is that of purifying the endogenous variables that appear as the "independent" variables in any given equation by replacing them by their predicted values based on the estimation of the parameters in the reduced-form equations. Since these predicted values of the endogenous variables are *exact* linear functions of the predetermined variables — which are assumed to be uncorrelated with the disturbances — these purified endogenous variables will also be uncorrelated with the disturbances in each equation. Thus, two-stage least-squares estimators are consistent (having negligible biases in large samples), whereas ordinary least-squares estimators will generally have nonnegligible biases in nonrecursive systems.

Overidentified systems are especially necessary whenever *tests* of a model are being considered. In the case of recursive systems, we saw that each time we erased an arrow between a pair of variables, we added a prediction that a partial slope (or correlation) should be approximately zero except for sampling errors. In fact, a recursive system with no additional b_{ij} set equal to zero turns out to be exactly identified given the assumption that all covariances among disturbance terms are zero. If we set one additional $b_{ij} = 0$, this produces one empirical prediction that will not automatically be satisfied by the data. In effect, we may use all but one of the equations in Simon's procedure to estimate the parameters, considering the remaining equation as "redundant" and therefore usable for testing purposes. The more coefficients assumed equal to zero, the more such excess equations we have for testing purposes. As a general principle, it would seem as though the less sure we are of our theories, the more desirable that they be highly overidentified, so that multiple predictions can be made from the excess equations. This point should be more apparent in Part III, where we deal with identification problems brought about by the presence of unmeasured variables in a causal system.

The final chapter in this section, by Land and Felson [11], is concerned with another very important consideration that arises whenever we are not very certain about our theories yet find it necessary to make a series of somewhat arbitrary assumptions in order to reduce the number of unknowns so as to achieve identification. These authors discuss a number of strategies for evaluating the relative sensitivities of our estimates to differing identifying restrictions, suggesting that methods employed in mathematical programming constitute a rather broad approach that may prove highly practical in many such ambiguous situations. This chapter, plus those of Fisher and Fisher and Ando, indicate that the approaches emphasized in Part II may be applied on a flexible basis in those highly common situations in which one's *a priori* knowledge is less than complete, and where one's theoretical assumptions are only approximately valid.

REFERENCES

[1] Christ, Carl F. *Econometric Models and Methods.* New York: John Wiley, 1966.

[2] Johnston, J. *Econometric Methods.* New York: McGraw-Hill, 1972.

[3] Wold, H. O. A. "Toward a Verdict on Macroeconomic Simultaneous Equations." *Pontificiae Academiae Scientiarum Scripta Varia* 28 (1965):115–185.

[4] Wold, H. O. A. "Mergers of Economics and Philosophy of Science." *Synthese* 20 (1969):427–482.

[5] Wold, H. O. A. "Nonexperimental Statistical Analysis from the General Point of View of Scientific Method." *Bulletin of the International Statistical Institute* 42, Part I (1969):391–424.

Identification Problems in Economic Model Construction*

Tjalling C. Koopmans

6

INTRODUCTION

The construction of dynamic economic models has become an important tool for the analysis of economic fluctuations and for related problems of policy. In these models, macroeconomic variables are thought of as determined by a *complete system of equations*. The meaning of the term *complete* is discussed more fully below. At present, it may suffice to describe a complete system as one in which there are as many equations as endogenous variables, that is, variables whose formation is to be "explained" by the equations. The equations are usually of, at most, four kinds: equations of economic behavior, institutional rules, technological laws of transformation, and identities. We shall use the term *structural equations* to comprise all four types of equations.

Systems of structural equations may be composed entirely on the basis of economic *theory*. By this term, we shall understand the combination of (1) principles of economic behavior derived from general observation — partly introspective, partly through interview or experience — of the motives of

* Reprinted by permission of the author and publisher from *Econometrica* 17: 125–143. Copyright 1949, The Econometric Society.

economic decisions; (2) knowledge of legal and institutional rules restricting individual behavior (tax schedules, price controls, reserve requirements, etc.); (3) technological knowledge; and (4) carefully constructed definitions of variables. Alternatively, a structural equation system may be determined on the dual basis of such "theory" combined with systematically collected statistical data for the relevant variables for a given period and country or other unit. In this chapter, we shall discuss certain problems that arise out of model construction in the second case.

Where statistical data are used as one of the foundation stones on which the equation system is erected, the modern methods of statistical inference are an indispensable instrument. However, without economic "theory" as another foundation stone, it is impossible to make such statistical inference apply directly to the equations of economic behavior that are most relevant to analysis and to policy discussion. Statistical inference unsupported by economic theory applies to whatever statistical regularities and stable relationships can be discerned in the data.[1] Such purely empirical relationships, when discernible, are likely to be due to the presence and persistence of the underlying structural relationships and (if so) could be deduced from a knowledge of the latter. However, the direction of this deduction cannot be reversed — from the empirical to the structural relationships — except possibly with the help of a theory that specifies the form of the structural relationships, the variables that enter into each, and any further details supported by prior observation or deduction therefrom. The more detailed these specifications are made in the model, the greater scope is thereby given to statistical inference from the data to the structural equations. We propose to study the limits to which statistical inference, from the data to the structural equations (other than definitions), is subject, and the manner in which these limits depend on the support received from economic theory.

This problem has attracted recurrent discussion in econometric literature with varying terminology and degree of abstraction. Reference is made to Pigou [16], Schultz [17, especially Chapter II, Section IIIe], Frisch [4,5], Marschak [15, especially Sections IV and V], Haavelmo [6, especially Chapter V]. An attempt to systematize the terminology and to formalize the treatment of the problem has been made over the past few years by various authors connected in one way or another with the Cowles Commission for Research in Economics. Since the purpose of this chapter is expository, I shall draw freely on the work by Koopmans, Rubin, and Leipnik [14], Wald [18], Hurwicz [7,8], Koopmans and Reiersöl [13], without specific acknowledgment in each case. We shall proceed by discussing a sequence of examples, all drawn from econometrics, rather than by a formal logical presentation, which can be found in references [14], [7], and [13].

---

[1] See Koopmans [12].

CONCEPTS AND EXAMPLES

The first example, already frequently discussed, is that of a competitive market for a single commodity, of which the price p and the quantity q are determined through the intersection of two rectilinear schedules, of demand and supply, respectively, with instantaneous response of quantity to price in both cases. For definiteness' sake, we shall think of observations as applying to successive periods in time. We shall further assume that the slope coefficients α and γ of the demand and supply schedules, respectively, are constant through time, but that the levels of the two schedules are subject to not directly observable shifts from an equilibrium level. The structural equations can then be written as:

$$q + \alpha p + \epsilon = u \quad \text{(demand)} \tag{6.1d}$$

$$q + \gamma p + \eta = v \quad \text{(supply)} \tag{6.1s}$$

Concerning the shift variables, u and v, we shall assume that they are random drawings from a stable joint probability distribution with mean values equal to zero:

$$\phi(u, v) \qquad \xi u = 0 \qquad \xi v = 0 \tag{6.2}$$

We shall introduce a few terms that we shall use with corresponding meaning in all examples. The not directly observable shift variables u, v are called *latent variables*, as distinct from the *observed variables*, p, q. We shall further distinguish *structure* and *model*. By a structure we mean the combination of a specific set of structural equations (1) (such as is obtained by giving specific numerical values to α, γ, ϵ, η) and a specific distribution function (2) of the latent variables (for instance a normal distribution with specific, numerically given, variances and covariance). By a model, we mean only a specification of the form of the structural equations (for instance, their linearity and designation of the variables occurring in each equation) and of a class of functions to which the distribution function of the latent variables belongs (for instance, the class of all normal bivariate distributions with zero means). More abstractly, a model can be defined as a set of structures. For a useful analysis, the model will be chosen so as to incorporate relevant *a priori* knowledge or hypotheses as to the economic behavior to be described. For instance, the model here discussed can often be narrowed down by the usual specification of a downward sloping demand curve and an upward sloping supply curve:

$$\alpha > 0 \qquad \gamma < 0 \tag{6.3}$$

Let us assume, for the sake of argument, that the observations are produced by a structure, to be called the "true" structure, which is contained in (permitted by) the model. In order to exclude all questions of sampling variability (which are a matter for later separate inquiry), let us further make the unrealistic assumption that the number of observations produced by this structure can be increased indefinitely. What inferences can be drawn from these observations toward the "true" structure?

A simple reflection shows that in our present example, neither the "true" demand schedule nor the "true" supply schedule can be determined from any number of observations. To put the matter geometrically, let each of the two identical scatter diagrams in Figure 6.1A and 6.1B represent the jointly observed values of p and q. A structure compatible with these observations can be obtained as follows: Select arbitrarily "presumptive" slope coefficients α and γ of the demand and supply schedules. Through each point $S(p, q)$ of the scatter diagrams, draw two straight lines with slopes given by these coefficients. The presumptive demand and supply schedules will intersect the quantity axis at distances $-\epsilon + u$ and $-\eta + v$ from the origin, *provided* the presumptive slope coefficients α and γ are the "true" ones. We shall assume this to be the case in Figure 6.1A. In that case, the values of ϵ and η can be found from the consideration that the averages of u and v in a sufficiently large sample of observations are practically equal to zero.

However, nothing in the situation considered permits us to distinguish the "true" slopes α, γ (as shown in Figure 6.1A) from any other presumptive slopes (as illustrated in Figure 6.1B). Any arbitrary set of slope coefficients α, γ (supplemented by corresponding values ϵ, η, of the intercepts) represents another, statistically just as acceptable, hypothesis concerning the formation of the observed variables.

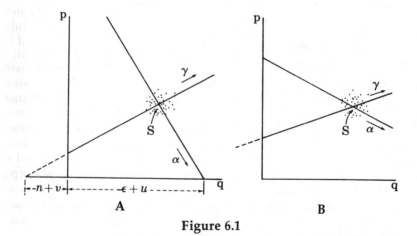

Figure 6.1

Let us formulate the same remark algebraically in preparation for further examples in more dimensions. Let the numerical values of the "true" parameters α, γ, ϵ, η in (6.1) be known to an individual who, taking delight in fraud, multiplies the demand equation (6.1d) by $2/3$, the supply equation (6.1s) by $1/3$, and adds the result to form an equation

$$q + \frac{2\alpha + \gamma}{3}p + \frac{2\epsilon + \eta}{3} = u' \qquad (6.4d)$$

which he proclaims to be the demand equation. This equation is actually different from the "true" demand equation (6.1d) because (6.3) implies $\alpha \neq \gamma$. Similarly, he multiplies the same equations by $2/5$ and $3/5$, respectively, say, to produce an equation

$$q + \frac{2\alpha + 3\gamma}{5}p + \frac{2\epsilon + 3\eta}{5} = v' \qquad (6.4s)$$

different from the "true" supply equation (6.1s) but which he presents as if it were the supply equation. If our prankster takes care to select his multipliers in such a manner as not to violate the sign rules (6.3) imposed by the model, the deceit cannot be discovered by statistical analysis of any number of observations.[2] For the equations (6.4), being derived from (6.1), are satisfied by all data that satisfy the "true" equations (6.1). Moreover, being of the same form as the equations (6.1), the equations (6.4) are equally acceptable *a priori*.

Our second example differs from the first only in that the model specifies a supply equation containing in addition an exogenous variable. To be definite, we shall think of the supply of an agricultural product as affected by the rainfall r during a critical period of crop growth[3] or crop gathering. This variable is called exogenous to our model to express the plausible hypothesis that rainfall r, while affecting the market of the commodity concerned, is not itself affected thereby. Put in mathematical terms, this hypothesis specifies that the disturbances u and v in

[2] The deceit could be discovered if the model were to specify a property (e.g., independence) of the disturbances u and v, which is not shared by $u' = (2u + v)/3$ and $v' = (2u + 3v)/5$. We have not made such a specification.

[3] With respect to this example, the assumption of a linear relationship can be maintained only if we think of a certain limited range of variation in rainfall. Another difficulty with the example is that for most agricultural products, the effect of price on supply is delayed instead of instantaneous, as here assumed. A practically instantaneous effect can, however, be expected in the gathering of wild fruits of nature.

$$q + \alpha p \quad\quad + \epsilon = u \quad \text{(demand)} \quad\quad\quad (6.5\text{d})$$

$$q + \gamma p + \delta r + \eta = v \quad \text{(supply)} \quad\quad\quad (6.5\text{s})$$

are statistically independent[4] of the values assumed by r.

It will be seen at a glance that the supply equation still cannot be determined from a sample of any size. If, starting from "true" structural equations (6.5), we multiply by $-1/2$ and $3/2$, say, and add the results to obtain a pretended supply equation,

$$q + \frac{3\gamma - \alpha}{2}p + \frac{3\delta}{2}r + \frac{3\eta - \epsilon}{2} = v' \quad\quad\quad (6.6\text{s})$$

of the same prescribed form as (6.5s), any data will satisfy this equation (6.6s) as well as they satisfy the two equations (6.5).

A similar reasoning can *not* be applied to the demand equation in the present model. Any attempt to construct another pretended demand equation by a linear combination involving the supply equation (6.5s) would introduce into that pretended demand equation the variable r, which by the hypotheses underlying the model does not belong in it.

It might be thought that if r has the properties of a random variable, its presence in the pretended demand equation might be concealed because its "contribution" cannot be distinguished from the random disturbance in that equation. To be specific, if $4/3$ and $-1/3$ are arbitrarily selected multipliers, the disturbance in the pretended demand equation might be thought to take the form

$$u' = \frac{4u - v}{3} - \frac{\delta}{3}r$$

---

[4] It is immaterial for this definition whether the exogenous variable is regarded as a given function of time—a concept perhaps applicable to a variable set by government policy—or as itself a random variable determined by some other structure involving probability distributions—a concept applicable particularly to weather variables. It should further be noted that we postulate independence between r and (u, v), not between r and (p, q), although we wish to express that r "is not affected by" p and q. The meaning to be given to the latter phrase is that in other equations explaining the formation of r, the variables (p, q) do not enter. Precisely this is implied in the statistical independence of r and (u, v) because (p, q) is, by virtue of (6.5), statistically dependent on (u, v), and any role of (p, q) in the determination of r would therefore create statistical dependence between r and (u, v). On the other hand, the postulated statistical independence between r and (u, v) is entirely compatible with the obvious influence, by virtue of Eq. (6.5), of r on (p, q).

This, however, would violate the specification that r is exogenous and that therefore r and u' are to be statistically independent as well as r and (u, v). The relevance of the exogenous character of r to our present discussion is clearly illustrated by this remark.

Our analysis of the second example suggests (and below we shall cite a theorem establishing proof) that a sufficiently large sample does indeed contain information with regard to the parameters α, ϵ of the demand equation (it being understood that such information is conditional upon the validity of the model). It can already be seen that there must be the following exception to the foregoing statement. If, in fact (although the model does not require it), rainfall has no influence on supply, that is, if in the "true" structure $\delta = 0$, then any number of observations must necessarily be compatible with the model (6.1), and hence does not convey information with regard to either the demand equation or the supply equation.

As a third example we consider a model obtained from the preceding one by the inclusion in the demand equation of consumers' income i as an additional exogenous variable. We assume the exogenous character of consumers' income merely for reasons of exposition, and in full awareness of the fact that actually price and quantity on any market do affect income directly to some extent, while furthermore the disturbances u and v affecting the market under consideration may well be correlated with similar disturbances in several other markets which together have a considerably larger effect on consumers' income.

The structural equations are now

$$q + \alpha p + \beta i \quad + \epsilon = u \quad \text{(demand)} \tag{6.7d}$$

$$q + \gamma p \quad + \delta r + \eta = v \quad \text{(supply)} \tag{6.7s}$$

Since each of the two equations now excludes a variable specified for the other equation, neither of them can be replaced by a different linear combination of the two without altering its form. This suggests, and proof is cited below, that from a sufficiently large sample of observations, the demand equation can be accurately determined provided rainfall actually affects supply ($\delta \neq 0$), and the supply equation can be determined provided consumers' income actually affects demand ($\beta \neq 0$).

The fourth example is designed to show that situations may occur in which some, but not all, parameters of a structural equation can be determined from sufficiently many observations. Let the demand equation contain both this year's income i_0 and last year's income i_{-1}, but let the supply equation not contain any variable absent from the demand equation:

$$q + \alpha p + \beta_0 i_0 + \beta_{-1} i_{-1} + \epsilon = u \tag{6.8d}$$

$$q + \gamma p \qquad\qquad\qquad + \eta = v \tag{6.8s}$$

Now obviously we cannot determine either α or ϵ, because linear combinations of the equations (6.8) can be constructed that have the same form as (6.8d) but other[5] values α' and ϵ' for the coefficients α and ϵ. However, as long as (6.8d) enters with some nonvanishing weight into such a linear combination, the ratio β_{-1}/β_0 is not affected by the substitution of that linear combination for the "true" demand equation. Thus, if the present model is correct, the observations contain information with respect to the relative importance of present and past income to demand, whereas they are silent on the price elasticity of demand.

The fifth example shows that an assumption regarding the joint distribution of the disturbances u and v, where justified, may open the door to a determination of a structural equation that is otherwise indeterminate. Returning to the equation system (6.5) of our second example, we shall now make the model specify in addition that the disturbances u in demand and v in supply are statistically independent. Remembering our previous statement that the demand equation can already be determined without the help of such an assumption, it is clear that in attempting to construct a "pretended" supply equation, no linear combination of the "true" demand and supply equations (6.5), other than the "true" supply equation (6.5s) itself, can be found that preserves the required independence of disturbances in the two equations. Writing λ and $1 - \lambda$ for the multipliers used in forming such a linear combination, the disturbance in the pretended supply equation would be

$$v' = \lambda u + (1 - \lambda)v \tag{6.9}$$

Since u and v are by assumption independent, the disturbance v' of the pretended supply equation is independent of the disturbance u in the demand equation already found determinable, if and only if $\lambda = 0$ (i.e., if the pretended supply equation coincides with the "true" one).

We emphasize again the expository character of the foregoing examples. It has already been indicated that the income variable i is not truly exogenous. By assuming it to be so, we have held down the size of the equation system underlying our discussion, and we may as a result have precluded ourselves from seeing indeterminacies that could come to light only by a

---

[5] As regards ϵ' this is true whenever $\epsilon \neq \eta$. As regards α' it is safeguarded by Eq. (6.3).

study of all relationships participating in the formation of the variables involved. It will therefore be necessary to develop criteria by which indeterminacies of the coefficients of larger equation systems can be detected. Before discussing such criteria for linear systems, we shall formalize a few of the concepts used or to be used.

THE IDENTIFICATION OF STRUCTURAL PARAMETERS

In our discussion, we have used the phrase "a parameter that can be determined from a sufficient number of observations." We shall now define this concept more sharply and give it the name *identifiability* of a parameter. Instead of reasoning, as before, from "a sufficiently large number of observations," we shall base our discussion on a hypothetical knowledge of the probability distribution of the observations, as defined more fully below. It is clear that exact knowledge of this probability distribution cannot be derived from any finite number of observations. Such knowledge is the limit approachable but not attainable by extended observation. By hypothesizing, nevertheless, the full availability of such knowledge, we obtain a clear separation between problems of statistical inference arising from the variability of finite samples and problems of identification in which we explore the limits to which inference even from an infinite number of observations is subject.

A *structure* has been defined as the combination of a distribution of latent variables and a complete set of structural equations. By a *complete set of equations* we mean a set of as many equations as there are endogenous variables. Each endogenous variable may occur with or without time lags and should occur without lag in at least one equation. Also, the set should be such as to permit unique determination of the nonlagged values of the endogenous variables from those of the lagged endogenous, the exogenous, and the latent variables. Finally, by *endogenous variables* we mean observed variables that are not exogenous, that is variables that are not known or assumed to be statistically independent of the latent variables and whose occurrence in one or more equations of the set is necessary on grounds of "theory."

It follows from these definitions that, for any specific set of values of the exogenous variables, the distribution of the latent variables (i.e., one of the two components of a given structure) entails or generates, through the structural equations (i.e., the other component of the given structure), a probability distribution of the endogenous variables. The latter distribution is, of course, conditional upon the specified values of the exogenous variables for each time point of observation. This conditional distribution, regarded again as a function of all specified values of exogenous variables, shall be the hypothetical datum for our discussion of identification problems.

We shall call two structures S and S' (observationally) *equivalent* (or indistinguishable) if the two conditional distributions of endogenous variables generated by S and S' are identical for all possible values of the exogenous variables. We shall call a structure S permitted by the model (uniquely) *identifiable* within that model if there is no other equivalent structure S' contained in the model. Although the proof has not yet been completely indicated, it may be stated in illustration that in our third example almost all structures permitted by the model are identifiable. The only exceptions are those with either $\beta = 0$ or $\delta = 0$ (or both). In the first and second examples, however, no structure is identifiable, although in the second example, we have stated that the demand equation by itself is determinate. To cover such cases, we shall say that a certain parameter θ of a structure S is uniquely *identifiable* within a model, if that parameter has the same value for all structures S' equivalent to S, contained in the model. Finally, a *structural equation* is said to be *identifiable* if all its parameters are identifiable.

This completes the formal definitions with which we shall operate. They can be summarized in the statement that anything is called identifiable, the knowledge of which is implied in the knowledge of the distribution of the endogenous variables, given the model (which is accepted as valid). We now proceed to a discussion of the application of this concept to linear models of the kind illustrated by our examples.

IDENTIFIABILITY CRITERIA IN LINEAR MODELS

In our discussion of these examples, it has been possible to conclude that a certain structural equation is not identifiable whenever we are able to construct a different equation, obtained by linear combination of some or all structural equations, which likewise meets the specifications of the model. In the opposite case, where we could show that no such different linear combination exists, we could not yet conclude definitely that the equation involved is identifiable. Could operations other than linear combination, perhaps, be used to derive equations of the same form?

We shall now cite a theorem that establishes that no such other operations can exist. The theorem relates to models specifying a complete set of structural equations as defined above and in which a given set of endogenous and exogenous variables enters linearly. Any time lags with which these variables may occur are supposed to be integral multiples of the time interval between successive observations. Furthermore, the exogenous variables (considered as different variables whenever they occur with a different time lag) are assumed not to be linearly dependent (i.e., in the functional sense).[6] Finally, although simultaneous disturbances in different structural equations are permitted to be correlated, it is assumed that any disturbances operating in different time units (whether in the same or in different structural equations) are statistically independent.

Suppose the model does not specify anything beyond what has been stated. That is, no restrictions are specified yet that exclude some of the variables from specific equations. Obviously, with respect to such a broad model, not a single structural equation is identifiable. However, a theorem has been proved [14] to the effect that, given a structure S within that model, any structure S' in the model, equivalent to S, can be derived from S by replacing each equation by some linear combination of some or all equations of S.

It will be clear that this theorem remains true if the model is narrowed down by excluding certain variables from certain equations or by other restrictions on the parameters. Thus, whenever in our examples we have concluded that different linear combinations of the same form prescribed for a structural equation did not exist, we have therewith established the identifiability of that equation. More in general, the analysis of the identifiability of a structural equation in a linear model consists in a study of the possibility to produce a different equation of the same prescribed form by linear combination of all equations. If this is shown to be impossible, the equation in question is thereby proved to be identifiable. To find criteria for the identifiability of a structural equation in a linear model is therefore a straightforward mathematical problem to which the solution has been given elsewhere [14]. Here we shall state without proof what the criteria are.

A *necessary condition* for the identifiability of a structural equation within a given linear model is that the number[7] of variables excluded from that equation (more generally: the number of linear restrictions on the parameters of that equation) be at least equal to the number (G, say) of structural equations less one. This is known as the *order condition* of identifiability. A *necessary and sufficient condition* for the identifiability of a structural equation within a linear model, restricted only by the exclusion of certain variables from certain equations, is that we can form at least one nonvanishing determinant of order $G - 1$ out of those coefficients, properly arranged, with which the variables excluded from that structural equation appear in the $G - 1$ other structural equations. This is known as the *rank condition* of identifiability.

---

[6] The criteria of identifiability to be stated would require amended formulation if certain identities involving endogenous variables would be such that each variable occurring in them also occurs, in some equation of the complete set, with a time lag, and if this time lag were the same for all such variables. In this case, a complication arises from linear (functional) dependence among lagged endogenous (and possibly exogenous) variables.

[7] Again, counting lagged variables as separate variables.

The application of these criteria to the foregoing examples is straightforward. In all cases considered, the number of structural equations is $G = 2$. Therefore, any of the equations involved can be identifiable through exclusion of variables only if at least $G - 1 = 1$ variable is excluded from it by the model. If this is so, the equation is identifiable provided at least one of the variables so excluded occurs in the other equation with nonvanishing coefficient (a determinant of order 1 equals the value of its one and only element). For instance, the conclusion already reached at the end of the discussion of our second example is now confirmed: The identifiability of the demand equation (6.5d) is only then safeguarded by the exclusion of the variable r from that equation if $\delta \neq 0$, that is, if that variable not only possibly but actually occurs in the supply equation.

THE STATISTICAL TEST OF *A PRIORI* UNCERTAIN IDENTIFIABILITY

The example just quoted shows that the identifiability of one structural parameter, θ, say, may depend on the value of another structural parameter, η, say. In such situations, which are of frequent occurrence, the identifiability of θ cannot be settled by *a priori* reasoning from the model alone. On the other hand, the identifiability of θ cannot escape all analysis because of possible nonidentifiability of η. As is argued more fully elsewhere [13], since the identifiability of any parameter is a property of the distribution of the observations, it is subject to some suitable statistical test, of which the degree of conclusiveness tends to certainty as the number of observations increases indefinitely. The validity of this important conclusion is not limited to linear models.

In the case of a linear model as described in the previous section the present statement can also be demonstrated explicitly by equivalent reformulation of the rank criterion for identifiability in terms of identifiable parameters only. By the *reduced form* of a complete set of linear structural equations as described in the previous section, we mean the form obtained by solving for each of the *dependent* (i.e., nonlagged endogenous) variables, in terms of the *predetermined* (i.e., exogenous or lagged endogenous) variables, and in terms of transformed disturbances (which are linear functions of the disturbances in the original structural equations). It has been argued more fully elsewhere [14, Section 3.1.6], that the coefficients of the equations of the reduced form are parameters of the joint distribution of the observations and as such are always identifiable.

It may be stated briefly without proof that the following rank criterion for identifiability of a given structural equation, in terms of coefficients of the reduced form, is equivalent to that stated in the previous section: Consider only those equations of the reduced form that solve for dependent variables, specified by the model as occurring in (strictly: as not excluded from) the structural equation in question. Let the number of the equations so obtained be H, where $H \leq G$. Now form the matrix Π^{**} of the coefficients, in these H equations, of those predetermined variables that are excluded by the model from the structural equation involved. A necessary and sufficient condition for the identifiability of that structural equation is that the rank of Π^{**} be equal to $H - 1$. A direct proof of the equivalence of the two identification criteria will be published in due course.

IDENTIFICATION THROUGH DISAGGREGATION AND INTRODUCTION OF SPECIFIC EXPLANATORY VARIABLES

As a further exercise in the application of these criteria, we shall consider a question that has already been the subject of a discussion between Ezekiel [2,3] and Klein [9,10]. The question is whether identifiability of the investment equation can be attained by the subdivision of the investment variable into separate categories of investment. In the discussion referred to, which took place before the concepts and terminology employed in this chapter were developed, questions of identifiability were discussed alongside questions regarding the merit of particular economic assumptions incorporated in the model and questions of the statistical method of estimating parameters that have been recognized as identifiable. In the present context, we shall avoid the latter two groups of problems and concentrate on the formal analysis of identifiability, accepting a certain model as economically valid for purposes of discussion.

As a starting point we shall consider a simple model expressing the crudest elements of Keynesian theory. The variables are, in money amounts,

$$\begin{cases} S & \text{savings} \\ I & \text{investment} \\ Y & \text{income} \\ Y_{-1} & \text{income lagged one year} \end{cases} \tag{6.10}$$

The structural equations are:

$$S - I \qquad\qquad\qquad = 0 \tag{6.11id}$$

$$S \quad - \alpha_1 Y - \alpha_2 Y_{-1} - \alpha_0 = u \tag{6.11S}$$

$$I - \beta_1 Y - \beta_2 Y_{-1} - \beta_0 = v \tag{6.11I}$$

Of these, the first is the well-known savings–investment identity arising from Keynes's definitions of these concepts.[8] The second is a behavior equation of consumers, indicating that the money amount of their savings (income not spent for consumption) is determined by present and past income, subject to a random disturbance u. The third is a behavior equation of entrepreneurs, indicating that the money amount of investment is determined by present and past income, subject to a random disturbance v.

Since the identity (6.11id) is fully given *a priori*, no question of identifiability arises with respect to the first equation. In both the second and third equations, only one variable is excluded that appears in another equation of the model, and no other restrictions on the coefficients are stated.[9] Hence, both of these equations already fail to meet the necessary order criterion of identifiability. This could be expected because the two equations connect the same savings–investment variable with the same two income variables and therefore can not be distinguished statistically.

Ezekiel attempts to obtain identifiability of the structure by a refinement of the model as a result of subdivision of aggregate investment I into the following four components:

$$\begin{cases} I_1 \text{ investment in plant and equipment} \\ I_2 \text{ investment in housing} \\ I_3 \text{ temporary investment: changes in consumers' credit} \\ \quad \text{and in business inventories} \\ I_4 \text{ quasi-investment: net contributions from foreign} \\ \quad \text{trade and the government budget} \end{cases} \quad (6.12a)$$

If each of these components were to be related to the same set of explanatory variables as occurs in (6.11), the disaggregation would be of no help toward identification. Therefore, for each of the four types of investment decisions, Ezekiel introduces a separate explanatory equation, either explicitly or by

[8] These definitions include in investment all increases in inventory, including undesired inventories remaining in the hands of manufacturers or dealers as a result of falling demand. In principle, therefore, the "investment" equation should include a term or terms explaining such inventory changes. The absence of such terms from Eq. (6.11) and from later elaborations thereof may be taken as expressing the "theory" that for annual figures, say, such changes can be regarded as random. Alternatively, investment may be defined so as to exclude undesired inventory changes, and Eq. (6.11id) may be interpreted as an "equilibrium condition," expressing the randomness of such changes by replacing the zero in the right-hand member by a disturbance w. The obvious need for refinement in this crude "theory" does not preclude its use for illustrative purposes.

[9] The normalization requirement that the variables S and I shall have coefficients $+1$ in Eqs. (6.11S) and (6.11I), respectively, does not restrict the relationships involved but merely serves to give a common level to coefficients that otherwise would be subject to arbitrary proportional variation.

implication in his verbal comments. In attempting to formulate these explanations in terms of a complete set of behavior equations, we shall introduce two more variables:

$$\begin{cases} H \text{ semi-independent cyclical component of housing} \\ \quad \text{investment} \\ E \text{ exogenous component of quasi-investment} \end{cases} \quad (6.12b)$$

In addition, linear and quadratic functions of time are introduced as trend terms in some equations by Ezekiel. For purposes of the present discussion, we may as well disregard such trend terms, because they would help toward identification only if they could be excluded *a priori* from some of the equations while being included in others — a position advocated neither by Ezekiel nor by the present author.

With these qualifications, "Ezekiel's model" can be interpreted as follows:

$$S - I_1 - I_2 - I_3 - I_4 \qquad\qquad\qquad\qquad = 0 \quad (6.13\text{id})$$

$$S \qquad\qquad - \alpha_1 Y - \alpha_2 Y_{-1} \qquad - \alpha_0 = u \quad (6.13S)$$

$$I_1 \qquad\qquad - \beta_1 Y - \beta_2 Y_{-1} \qquad - \beta_0 = v_1 \quad (6.13I_1)$$

$$I_2 \qquad - \gamma_1 Y - \gamma_2 Y_{-1} - H \qquad - \gamma_0 = v_2 \quad (6.13I_2)$$

$$I_3 \quad - \delta_1 Y + \delta_1 Y_{-1} \qquad - \delta_0 = v_3 \quad (6.13I_3)$$

$$I_4 - \epsilon_1 Y - \epsilon_2 Y_{-1} \qquad - E - \epsilon_0 = v_4 \quad (6.13I_4)$$

(6.13id) is the savings–investment identity. (6.13S) repeats (6.11S), and (6.13I_1) is modeled after (6.11I). More specific explanations are introduced for the three remaining types of investment decisions.

Housing investment decisions I_2 are explained partly on the basis of income[10] Y, partly on the basis of a "semi-independent housing cycle" H. In Ezekiel's treatment, H is not an independently observed variable but a smooth long cycle fitted to I. We share Klein's objection [9, p. 255] to this procedure but do not think that his proposal to substitute a linear function of time for H does justice to Ezekiel's argument. The latter definitely thinks of H as produced largely by a long-cycle mechanism peculiar to the housing

[10] We have added a term with Y_{-1} because the exclusion of such a term could hardly be made the basis for a claim of identifiability.

market and quotes in support of this view a study by Derksen [1] in which this mechanism is analyzed. Derksen constructs an equation explaining residential construction in terms of the rent level, the rate of change of income, the level of building cost in the recent past, and growth in the number of families; he further explains the rent level in terms of income, the number of families, and the stock of dwelling units (all of these subject to substantial time lags). The stock of dwelling units, in its turn, represents an accumulation of past construction diminished by depreciation or demolition. Again accepting without inquiry the economic assumptions involved in these explanations, the point to be made is that H in $(6.13I_2)$ can be thought to represent specific observable exogenous and *past* endogenous variables.

Temporary investment I_3 is related by Ezekiel to the rate of change in income. Quasi-investment I_4 is related by him partly to income[11] (especially via government revenue, imports), partly to exogenous factors underlying exports and government expenditure where used as an instrument of policy. The variable E in $(6.13I_4)$ is therefore similar to H in that it can be thought to represent observable exogenous or past endogenous variables.

It cannot be said that this interpretation of the variables H and E establishes the completeness of the set of equations (6.13) in the sense defined above. The variable H has been found to depend on the past values of certain indubitably endogenous variables (building cost, rent level) of which the present values do not occur in the equation system (6.13), and which therefore remain unexplained by (6.13). The reader is asked to accept what could be proved explicitly: that incompleteness of this kind does not invalidate the criteria of identifiability indicated.[12]

Let us then apply our criteria of identifiability to the behavior equations in (6.13). In each of these, the number of excluded variables is at least 5, that is, at least the necessary number of identifiability in a model of 6 equations. In order to apply the rank criterion for the identifiability of the savings equation (6.13S), say, we must consider the matrix

$$
\begin{array}{cccccc}
(I_1) & (I_2) & (I_3) & (I_4) & (H) & (E) \\
\end{array}
$$
$$
\begin{bmatrix}
-1 & -1 & -1 & -1 & 0 & 0 \\
1 & 0 & 0 & 0 & 0 & 0 \\
0 & 1 & 0 & 0 & -1 & 0 \\
0 & 0 & 1 & 0 & 0 & 0 \\
0 & 0 & 0 & 1 & 0 & -1
\end{bmatrix}
\tag{6.14}
$$

[11] We have again added a term with Y_{-1} on grounds similar to those stated with respect to $(6.13I_2)$.

[12] Provided, as indicated in footnote 6, there is no linear functional relationship between the exogenous and lagged endogenous variables occurring in Eq. (6.13).

There are several ways in which a nonvanishing determinant of order 5 can be selected from this matrix. One particular way is to take the columns labeled I_1, I_2, I_3, H, E. It follows that if the present model is valid, the savings equation is indeed identifiable.

It is easily seen that the same conclusion applies to the equations explaining investment decisions of the types I_1 and I_3. Let us now inspect the rank criterion matrix for the identifiability of (6.13I_2):

$$
\begin{array}{c}
\begin{array}{ccccc} (S) & (I_1) & (I_3) & (I_4) & (E) \end{array} \\
\begin{bmatrix}
1 & -1 & -1 & -1 & 0 \\
1 & 0 & 0 & 0 & 0 \\
0 & 1 & 0 & 0 & 0 \\
0 & 0 & 1 & 0 & 0 \\
0 & 0 & 0 & 1 & -1
\end{bmatrix}
\end{array}
\tag{6.15}
$$

Again the determinant value of this square matrix of order 5 is different from zero. Hence the housing equation is identifiable. A similar analysis leads to the same conclusion regarding the equation (6.13I_4) for quasi-investment.

It may be emphasized again that identifiability was attained not through the mere subdivision of total investment but as a result of the introduction of specific explanatory variables applicable to some but not all components of investment.[13] Whenever such specific variables are available in sufficient number and variety of occurrence, on good grounds of economic theory as defined above, the door has been opened in principle to statistical inference regarding behavior parameters — inference conditional upon the assumptions derived from "theory."

How wide the door has been opened, that is, how much accuracy of estimation can be attained from given data, is of course a matter depending

---

[13]In fact, more specific detail was introduced than the minimum necessary to produce identifiability. Starting again from (6.11), identifiability can already be obtained if it is possible to break off from investment I some observable exogenous component, like public works expenditure P (supposing that to be exogenous for the sake of argument). Writing $Q = I - P$ for the remainder of investment, (6.11) is then modified to read

$$
\begin{cases}
S - Q - P & = 0 \\
S & - \alpha_1 Y - \alpha_2 Y_{-1} - \alpha_0 = u \\
Q & - \beta_1 Y - \beta_2 Y_{-1} - \beta_0 = v
\end{cases}
\tag{6.11a}
$$

of which each equation meets our criteria of identifiability. The intent of this remark is largely formal, because Eq. (6.11a) is not so defensible a "theory" as Eq. (6.13).

on many circumstances and to be explored separately by the appropriate procedures of statistical inference.[14] In the present case, the extent to which the exclusion of H and/or E from certain equations contributes to the reliability of estimates of their parameters depends very much on whether or not there are pronounced differences in the time-paths of the three *predetermined variables* Y_{-1}, H, E, that is, the variables determined either exogenously or in earlier time units. These time-paths represent in a way the basic patterns of movement in the economic model considered, such that the time-paths of all other variables are linear combinations of these three paths, modified by disturbances. If the three basic paths are sufficiently distinct, conditions are favorable for estimation of identifiable parameters. If there is considerable similarity between any two of them, or even if there is only a considerable multiple correlation between the three, conditions are adverse.

IMPLICATIONS OF THE CHOICE OF THE MODEL

It has already been stressed repeatedly that any statistical inference regarding identifiable parameters of economic behavior is conditional upon the validity of the model. This throws great weight on a correct choice of the model. We shall not attempt to make more than a few tentative remarks about the considerations governing this choice.[15]

It is an important question to what extent certain aspects of a model of the kind considered above are themselves subject to statistical test. For instance, in the model (6.13), we have specified linearity of each equation, independence of disturbances in successive time units, time lags that are an integral multiple of the chosen unit of time, as well as exclusions of specific variables from specific equations. It is often possible to subject one particular aspect or set of specifications of the model to a statistical test that is conditional upon the validity of the remaining specifications. This is, for instance, the case

---

[14] We are not concerned here with an evaluation of the particular estimation procedures applied by Ezekiel.

[15] In an earlier article [11] I have attempted, in a somewhat different terminology, to discuss that problem. That article needs rewriting in the light of subsequent developments in econometrics. It unnecessarily clings to the view that each structural equation represents a causal process in which one single dependent variable is determined by the action upon it of all other variables in the equation. Moreover, use of the concept of *identifiability* will contribute to sharper formulation and treatment of the problem of the choice of a model. However, the most serious defect of the article, in my view, cannot yet be corrected. It arises from the fact that we do not yet have a satisfactory statistical theory of choice among several alternative hypotheses.

with respect to the exclusion of any variable from any equation whenever the equation involved is identifiable even without that exclusion. However, at least *four* difficulties arise that point to the need for further fundamental research on the principles of statistical inference.

In the first place, on a given basis of maintained hypotheses (not subjected to test), there may be several alternative hypotheses to be tested. For instance, if there are two variables whose exclusion, either jointly or individually, from a given equation is not essential to its identifiability, it is possible to test separately (*a*) the exclusion of the first variable, or (*b*) of the second variable, or (*c*) of both variables simultaneously, as against (*d*) the exclusion of neither variable. However, instead of three separate tests, of (*a*) against (*d*), (*b*) against (*d*), and (*c*) against (*d*), we need a procedure permitting selection of one of the four alternatives (*a*), (*b*), (*c*), (*d*). An extension of current theory with regard to the testing of hypotheses, which is concerned mainly with choices between two alternatives, is therefore needed.

Second, if certain specifications of a model can be tested given all other specifications, it is usually possible in many different ways to choose the set of "other" specifications that is not subjected to test. It may not be possible to choose the minimum set of untested specifications in any way so that strong *a priori* confidence in the untested specifications exists. Even in such a case, it may nevertheless happen that for any choice of the set of untested specifications, the additional specifications that are confirmed by test also inspire some degree of *a priori* confidence. In such a case, the model as a whole is more firmly established than any selected minimum set of untested specifications. However, current theory of statistical inference provides no means of giving quantitative expression to such partial and indirect confirmation of anticipation by observation.

Third, if the choice of the model is influenced by the same data from which the structural parameters are estimated, the estimated sampling variances of these estimated parameters do not have that direct relation to the reliability of the estimated parameters that they would have if the estimation were based on a model of which the validity is given *a priori* with certainty.

Finally, the research worker who constructs a model does not really believe that reality is exactly described by a "true" structure contained in the model. Linearity, discrete time lags, are obviously only approximations. At best, the model builder hopes to construct a model that contains a structure that approximates reality to a degree sufficient for the practical purposes of the investigation. The tests of current statistical theory are formulated as an (uncertain) choice, from two or more sets of structures (single or composite hypotheses), of that one which contains the "true" structure. Instead, we need to choose the simplest possible set — in some sense — that contains a structure sufficiently approximative — in some sense — to economic reality.

FOR WHAT PURPOSES IS
IDENTIFICATION NECESSARY?

The question should finally be considered why it is at all desirable to postulate a structure behind the probability distribution of the variables and thus to become involved in the sometimes difficult problems of identifiability. If we regard as the main objective of scientific inquiry to make prediction possible and its reliability ascertainable, why do we need more than a knowledge of the probability distribution of the variables to permit prediction of one variable on the basis of known (or hypothetical) simultaneous or earlier values of other variables?

The answer to this question is implicit in Haavelmo's discussion of the degree of permanence of economic laws [6, see p. 30] and has been formulated explicitly by Hurwicz [8]. Knowledge of the probability distribution is in fact sufficient whenever there is no change in the structural parameters between the period of observation from which such knowledge is derived and the period to which the prediction applies. However, in many practical situations it is required to predict the values of one or more economic variables either under changes in structure that come about independently of the economist's advice or under hypothetical changes in structural parameters that can be brought about through policy based in part on the prediction made. In the first case, knowledge may, and in the second case it is likely to, be available as to the effect of such structural change on the parameters. An example of the first case is a well-established change in consumers' preferences. An example of the second case is a change in the average level or in the progression of income tax rates.

In such cases, the "new" distribution of the variables on the basis of which predictions are to be constructed can only be derived from the "old" distribution prevailing before the structural change, if the known structural change can be applied to identifiable structural parameters, that is, parameters of which knowledge is implied in a knowledge of the "old" distribution combined with the *a priori* considerations that have entered into the model.

ACKNOWLEDGMENT

I am indebted to present and former Cowles Commission staff members and to my students for valuable critical comments regarding contents and presentation of this article. An earlier version of this paper was presented before the Chicago Meeting of the Econometric Society in December 1947. This article was reprinted with the addition of a sixth example in Section 2, as Chapter II of *Studies in Econometric Method,* edited by William C. Hood and Tjalling C. Koopmans, John Wiley, 1953.

REFERENCES

[1] Derksen, J. B. D. "Long Cycles in Residential Building: An Explanation." *Econometrica* 8 (1940): 97–116.

[2] Ezekiel, M. "Saving, Consumption and Investment." *American Economic Review* 32 (1942): 22–49; (1942): 272–307.

[3] Ezekiel, M. "The Statistical Determination of the Investment Schedule." *Econometrica* (1944): 89–90.

[4] Frisch, R. *Pitfalls in the Statistical Construction of Demand and Supply Curves.* Veröffentlichungen der Frankfurter Gesellschaft für Konjunkturforschung, Neue Folge, Heft 5, Leipzig, 1933.

[5] Frisch, R. "Statistical versus Theoretical Relations in Economic Macrodynamics." Mimeographed document prepared for a League of Nations conference concerning Tinbergen's work, 1938.

[6] Haavelmo, T. "The Probability Approach in Econometrics." *Econometrica* 12 (1944), Supplement; also Cowles Commission Paper, New Series, No. 4.

[7] Hurwicz, L. "Generalization of the Concept of Identification," in *Statistical Inference in Dynamic Economic Models.* Cowles Commission Monograph 10, New York, John Wiley and Sons (forthcoming).

[8] Hurwicz, "Prediction and Least-Squares," in *Statistical Inference in Dynamic Economic Models.* Cowles Commission Monograph 10. New York: John Wiley & Sons (forthcoming).

[9] Klein, L. "Pitfalls in the Statistical Determination of the Investment Schedule." *Econometrica* 11 (1943): 246–258.

[10] Klein, L. "The Statistical Determination of the Investment Schedule: A Reply." *Econometrica* 12 (1944): 91–92.

[11] Koopmans, T. C. "The Logic of Econometric Business Cycle Research." *Journal of Political Economy* 49 (1941): 157–181.

[12] Koopmans, T. C. "Measurement without Theory." *The Review of Economic Statistics* 29, no. 3 (1947): 161–172; also Cowles Commission Paper, New Series, No. 25.

[13] Koopmans, T. C., and Reiersöl, O. "Identification as a Problem in Inference," to be published.

[14] Koopmans, T. C., Rubin, H., and Leipnik, R. B. "Measuring the Equation Systems of Dynamic Economics." In *Statistical Inference in Dynamic Economic Models.* Cowles Commission Monograph 10. New York: John Wiley & Sons (forthcoming).

[15] Marschak, J., "Economic Interdependence and Statistical Analysis." In *Studies in Mathematical Economics and Econometrics,* in memory of Henry Schultz. Chicago: The University of Chicago Press, 1942. pp. 135–150.

[16] Pigou, A. C., "A Method of Determining the Numerical Values of Elasticities of Demand." *Economic Journal* 20 (1910): 636–640. Reprinted as Appendix II in *Economics of Welfare.*

[17] Schultz, Henry. *Theory and Measurement of Demand.* Chicago: The University of Chicago Press, 1938.

[18] Wald, A. "Note on the Identification of Economic Relations." In *Statistical Inference in Dynamic Economic Models.* Cowles Commission Monograph 10. New York: John Wiley & Sons (forthcoming).

Recursive versus Nonrecursive Systems: An Attempt at Synthesis*

Robert H. Strotz

H. O. A. Wold

7

Over the past 15 years, there has been an extended discussion of the meaning and applicability of nonrecursive as distinct from recursive systems in econometrics, and throughout this discussion there has been a marked divergence of views as to the merits of the two types of models. It is not the purpose of this chapter to extend that controversy further but rather to attempt a constructive statement of the relationship between the two approaches and the circumstances under which each is applicable.

We assume that the reader is generally familiar with the past discussion,[1] and that it will suffice here simply to recall that a recursive, or causal-chain, system has the formal property that the coefficient matrix of the nonlagged endogenous variables is triangular (upon suitable ordering of rows and columns), whereas a nonrecursive, or interdependent, system is one for which this is not the case. While the triangularity of the coefficient matrix is a formal property of recursive models, the essential property is that each relation is provided a causal interpretation in the sense of a stimulus–response relationship. The question of whether and in what sense nonre-

* Reprinted by permission of the authors and publisher from *Econometrica* 28 417–427. Copyright 1960, The Econometric Society.
[1] See references appended at end. An extensive bibliography is included in [8].

cursive systems allow a causal interpretation is the main theme of this chapter. Much controversy can, in our opinion, be resolved once there is agreement on some initial points of principle.

1. The first thing to consider when constructing an economic model is its purpose, that is, how it is to be applied in dealing with economic facts. We want to distinguish in this connection between descriptive and explanatory models. A descriptive model simply sets forth a set of relationships that have "bound together" different variables in situations in which they have previously been observed. More generally, these relationships may be described in probability terms, certain terms in these relationships representing the "disturbances" that in fact occurred. One can in this way describe given observations as a random drawing from a joint conditional probability distribution. Methodologically, the estimation of such a distribution is an exercise in n-dimensional "curve fitting." A descriptive model is thus cognate to the notion of a vector *function* such as (in the linear case)

$$Ax' = u' \tag{7.1}$$

where A is a (not necessarily square) matrix of constants, x' is a column vector of the variables in question, and u' is a vector of zeros in the exact case or of stochastic variables in the case of a probability model. Whatever the validity of such a specification, the validity of any other model obtained by applying any linear transformation is the same. If *a priori* restrictions are imposed upon the sort of distribution that is to be used for this descriptive model, this may, of course, circumscribe the acceptable transformations.

Explanatory models, by contrast, are causal. This means that each relation (equation) in the model states something about "directions of influence" among the variables. (But see section, "Vector Causality" below.) In the case of explanatory models, then, the theorist asserts more than functional relationships among the variables; he also invests those relationships with a special interpretation, that is, with a causal interpretation. But what is a "causal interpretation" to mean?

2. No one has monopoly rights in defining *causality*. The term is in common parlance, and the only meaningful challenge is that of providing an explication of it. No explication need be unique, and some may prefer never to use the word at all. For us, however, the word in common scientific and statistical–inference usage has the following general meaning.[2] z is a cause of y if, by hypothesis, it is or "would be" possible by *controlling z* indirectly to control y, at least stochastically. But it may or may not be possible by controlling y indirectly to control z. A causal relation is therefore in essence asymmetric, in that in any instance of its realization it is asymmet-

---

[2] Wold has elaborated his views in [6,7].

ric. Only in special cases may it be reversible and symmetric in a causal sense. These are the cases in which sometimes a controlled change in z may cause a change in y and at other times a controlled change in y may cause a change in z, but y and z cannot both be subjected to simultaneous controlled changes independently of one another without the causal relationship between them being violated.

The asymmetry of causation in any instance of its realization has the following probability counterpart. It may make sense to talk about the probability distribution of y as being *causally conditional* on z but not make sense to talk about the probability distribution of z as being *causally conditional* on y. This asymmetry is classical in statistical theory. It appears in the difference between a sample statistic and a population parameter. We speak of the probability that the sample frequency of successes will be 0.5 conditional upon the population frequency being 0.4. We do not speak of the probability that the population frequency is 0.4 conditional upon the sample frequency being 0.5.[3] Thus, if we wish to estimate (by the maximum likelihood method) a population parameter knowing a sample of observations, we write the likelihood function as the conditional probability distribution of the *sample*.

Suppose we were to estimate by the maximum likelihood method the nth value of a causal variable (population parameter) $z(n)$ on the basis of the nth value of a resultant variable (a sample observation) $y(n)$ by use of a regression fitted to $n - 1$ previous observations in all of which z has been causal. We should use the regression of y on z—not of z on y— over the previous observations, although this point is occasionally misunderstood.[4] Causality as used here is an essential notion in the statistical inference of population parameters by the maximum likelihood method. We must hypothesize how the sample observations are *generated* (i.e., caused) in order to proceed.

The concept of *causality* presented here is intended to be that of the everyday usage in the laboratory and emphasizes mainly the notion of control. Now, others may present a different explication of causality. Other versions may involve strange and seemingly unnatural notions, two of which are of particular interest to us. The first (a) involves accepting simultaneously the two statements: (1) "y has the value 100 because of (by cause of) z having the value 50" and (2) "z has the value 50 because of (by cause of) y

---

[3] "Probability" is used here in the "relative frequency," not in the "degree of belief" sense.

[4] For what we regard as the correct treatment, see Mood, [2, Sec. 13.4]. For the contrary view see Waugh [5]. A qualification is needed: One must not have any constraining *a priori* knowledge about the possible values of z, or, if z is itself a random variable, about its probability distribution. The model we have in mind is given by $y = a + \beta z + u$, with u and z statistically independent. Otherwise, the likelihood function is $f[(1), \ldots, y(n)|z(1), \ldots z(n)] \cdot g[z(n)]$, rather than the f function alone, where f and g are probability density functions.

having the value of 100." The second (b) involves accepting simultaneously the two statements: (1) "z causes y in accordance with the function $y = f(z)$" and (2) "z causes y in accordance with the function $y = g(z) \neq f(z)$." Usage (a) we shall describe as a "causal circle" and discuss in the section "Causal Circles." Usage (b) we describe as "bicausality" and discuss in the section "Bicausality."

Whether such notions of causality seem weird or not, whether or not they conform to usage in the scientific workshop, there is nothing to prevent their use in theory construction. Argument for the interpretation of causality presented by us is not an argument for "strait-jacketing" the freedom of the theorist and econometrician to use other interpretations. But we would (and will) argue that the notion of causality in economics is the one we have presented. Examples are: "Income is the cause of consumer expenditure (the consumption function)," "Price is a cause of quantity demanded," and "Price is a cause of quantity supplied." Most of the problems in assessing this claim arise in equilibrium models.

THE CAUSAL INTERPRETATION OF
A RECURSIVE SYSTEM

It was not our purpose in the previous section to provide a precise definition of *causation*. The term enters our discussion essentially as a "primitive," and what efforts we have made at definition have been ostensive: We have pointed to the familiar usage of the word in the laboratory. With reference to this primitive meaning of "causation," however, we wish to define the concept of *the causal interpretability of a parameter*. This is what will occupy us next.

Suppose a recursive system is written in the form

$$y_1 + \beta_{12}y_2 + \cdots + \beta_{1g}y_g + \beta_{1,g+1}y_{g+1} + \cdots + \beta_{1G}y_G + \sum_k \gamma_{1k}z_k = u_1$$

$$y_2 + \cdots + \beta_{2g}y_g + \beta_{2,g+1}y_{g+1} + \cdots + \beta_{2G}y_G + \sum_k \gamma_{2k}z_k = u_2$$

$$\cdot$$
$$\cdot$$
$$\cdot$$

$$y_g + \beta_{g,g+1}y_{g+1} + \cdots + \beta_{gG}y_G + \sum_k \gamma_{gk}z_k = u_g$$

$$\cdot$$
$$\cdot$$
$$\cdot$$

$$y_G + \sum_k \gamma_{Gk}z_k = u_G$$

$$(7.2)$$

where the y's are causally dependent variables, the z's predetermined variables, and the u's stochastic variables statistically independent of the z's. Each u_g is assumed, moreover, to be statistically independent of y_{g+1}, y_{g+2}, . . . , y_G.[5] In each equation, the y variable with unit coefficient is regarded as the resultant variable, and the other y variables and the z's are regarded as causal variables. We now consider the possibility that we gain direct control over y_g, that is, we can manipulate y_g by use of variables other than the z's appearing in the model. In this case, we now need a new model. It can be obtained, however, by a single change in the old one. We merely strike out the gth equation and reclassify y_g as an exogenous (predetermined) variable rather than as a dependent variable. The coefficients of y_g in the $G - 1$ equations of the new model will be the same as they were before, namely, β_{1g}, . . . , $\beta_{g-1,g}$, and zeros. It is in this sense that each nonunit coefficient in a recursive system has a causal interpretation. It describes the influence of the variable whose coefficient it is on the resultant variable, irrespective of whether the causal variable is dependent or exogenous in the system. Such a parameter has causal interpretability.

THE CAUSAL INTERPRETATION OF NONRECURSIVE SYSTEMS

We now turn our attention to nonrecursive systems. What is the possibility of causal interpretation in these systems?

No Causal Interpretation

It may be that no causal interpretation of a nonrecursive system is intended. The relations in the system may be asserted only to define the joint probability distribution of the dependent variables conditional upon the predetermined variables. The coefficients to be estimated are then simply parameters in the joint conditional distribution of y given z. With nothing further claimed, there is no objection to such a model or to efforts to estimate the parameters of the distribution.

Vector Causality

It may be asserted for the *nonrecursive* model

[5] Although Wold has imposed this specification in his definition of a causal chain system, Strotz feels it may be too restrictive and would classify causal chain systems under two headings: those that are causal chains in their stochastic form (Wold's case) and those that are causal chains only in their exact part. In the latter case, the covariance matrix of the u_g's need not be diagonal. For further discussion of this, see Strotz, *infra*, p. 430, fn. 6.

$$\beta y' + \Gamma z' = u' \tag{7.3}$$

that the vector z *causes* the vector y. Causality in this sense goes beyond the definition of "causality" given in Wold [6,7]. It may readily be accepted, however, as an abstract terminological extension of the more usual notion of causation and may be employed in the everyday sense of the statement, "The food supply causes the fish population." An example may be useful. Suppose z is a vector whose elements are the amounts of various fish feeds (different insects, weeds, etc.) available in a given lake, and that y is a vector whose elements are the numbers of fish of various species in the lake. The reduced form $y' = -\beta^{-1} \Gamma z' + \beta^{-1} u'$ would tell us specifically how the number of fish of any species depends on the availabilities of different feeds. The coefficient of any z is the partial derivative of a species population with respect to a food supply. It is to be noted, however, that the reduced form tells us nothing about the interactions among the various fish populations — it does not tell us the extent to which one species of fish feeds on another species. Those are causal relations among the y's.[6]

Suppose, in another situation, we continuously restock the lake with species g, increasing y_g by any desired amount. How will this affect the values of the other y's? If the system were recursive and we had estimates of the elements of β, we would simply strike the gth equation out of the model and regard y_g, the number of fish of species g, as exogenous — as a food supply or, when appearing with a negative coefficient, as a poison. It will be the purpose of the sections "Causal Circles" and "Bicausality" to determine whether if the model is not recursive the problem can be dealt with in this same way. The nonrecursive model does, in any case, enable us to predict the effects on the y's of controlled variations in the z's.

Simon [3] in developing a sense of causality for econometric models, has used this notion of vector causality. He defines causal relations among *subsets* of dependent variables by using a model recursive in these subsets. Partition y' into three subsets, that is, into three column vectors, y_1', y_2', and y_3', so that $y' = (y_1, y_2, y_3)'$ and partition β conformally. Consider the system (7.3) in which β may be written as

$$\beta = \begin{pmatrix} \beta_{11} & \beta_{12} & \beta_{13} \\ 0 & \beta_{22} & \beta_{23} \\ 0 & 0 & \beta_{33} \end{pmatrix} \tag{7.4}$$

consisting of nine submatrices. Then y_3' is caused by z'; y_2' is caused by y_3' and

---

[6] Indeed, even though $\partial y_g / \partial z_k > 0$, this does not imply that fish species g consumes food supply k. It may be that species g consumes species h which consumes food supply k.

z'; and y_1' is caused by y_2', y_3', and z'. In the previous sentence, the word "caused" is used in the sense of vector causation, and β is "block triangular." No causal relations among the variables *within* a subset are defined. Press this logic further. If each subset consists of but a single endogenous variable and a causal sequence is established among subsets, β is triangular and the system is recursive.

If, by way of contrast with vector causality, each effect variable is given as an explicit function of only variables that are its causes, we may speak of *explicit causality*.

Causal Circles, Mutual Causation, and Equilibrium Conditions

By a "causal circle" we shall mean a system such as

$$p(t) = \alpha_1 + \beta_1 q(t) + \gamma_1 z_1(t) + u_1(t) \tag{7.5a}$$

$$q(t) = \alpha_2 + \beta_2 p(t) + \gamma_2 z_2(t) + u_2(t) \tag{7.5b}$$

where $z_1(t)$ and $z_2(t)$ are exogenous and for which the following two statements are asserted: (1) In equation (7.5a), $q(t)$ is a cause of $p(t)$; (2) in equation (7.5b), $p(t)$ is a cause of $q(t)$. Causation is here used in a sense not allowed by the operative meaning that causation has in an experimental laboratory. To accept a causal circle is, in the laboratory meaning of the word "cause," to suppose that the value of one variable is determined by the value of another variable whose value cannot be determined until that of the first has been determined. To assume that the values of the two variables determine each other makes sense only in an equilibrium system, and such a system provides no explanation of how the equilibrium comes about (of change or of causal connections among the endogenous variables of the system).[7]

The familiar illustration of the three balls in the bowl mutually *causing* one another's location, which has been advanced as an example of mutual causation,[8] might be considered in this connection. For the steady state (equilibrium), there are certain mutual conditions that must be satisfied; but

---

[7] If, for example, an entrepreneur wishing to produce a given amount of product at minimum cost decides simultaneously on how much of each of two factors of production, x_1 and x_2, to employ, it might be said that his decision as to x_1 causes his decision as to x_2 and his decision as to x_2 causes his decision as to x_1. We do not believe this conforms to the "laboratory" meaning of causation, and we therefore reject this usage. On this, see [1].

[8] This illustration, due to Marshall, has been referred to recently by Stone [4] in this connection. Marshall spoke of mutual "determination" rather than "causation," and it is not clear whether these words are to be regarded as synonyms.

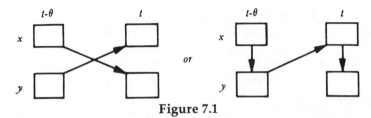

Figure 7.1

if the balls are displaced and then roll toward equilibrium, they are either in mutual contact and roll (or slide) as a single mass or the position of each can depend on the positions of the others only when the latter are lagged in time. Indeed, mutual causation in a dynamic system can have meaning only as a limit form of the arrow schemes shown in Figure 7.1 where the time lag θ is reduced toward zero.[9]

There is, however, a sense in which the coefficients of an equilibrium system may be given a causal interpretation, even though the relations in an equilibrium system may not themselves be causal relations. Let us take first an imaginary case in ecology: the balanced aquarium. Suppose there are two species of fish in an aquarium, big, b, and small, s. Their populations are y_b and y_s. The big fish feed on the small ones and on weed type a, available in quantity x_a. The small fish feed only on weed type c, available in quantity x_c. It takes time for the big fish to catch the small ones. The model is linear and stochastic, thus:

$$y_b(t) = \alpha_1 + \beta_1 y_s(t - \theta) + \gamma_1 x_a(t) + u_1(t) \qquad (7.6a)$$

$$y_s(t) = \alpha_2 + \beta_2 y_b(t - \theta) + \delta_2 x_c(t) + u_2(t) \qquad (7.6b)$$

Suppose now that every time we observe the aquarium it is in equilibrium. Moreover, we wish to estimate two numbers — are they β_1 and β_2? — so that we can answer these questions:

If we controlled the population of the big (small) fish, y_b (resp., y_s), by adding them to the aquarium or taking them out, and thereby held y_b (resp., y_s) at some arbitrary level, what would the expected value of the population of small (resp., big) fish, y_s (resp., y_b), be at the new equilibrium level, conditional upon the values of x_a and x_c?

Suppose we next formulated the model:

$$y_b(t) = \alpha_1 + \beta_1 y_s(t) + \gamma_1 x_a(t) + u_1(t) \qquad (7.7a)$$

[9] In equilibrium when $x(t) = x(t - \theta)$ and $y(t) = y(t - \theta)$, these structures may be collapsed into one of apparent "mutual causation," but what are simultaneous equilibrium conditions ought not to be confused with causal relations.

$$y_s(t) = \alpha_2 + \beta_2 y_b(t) + \delta_2 x_c(t) + u_2(t) \tag{7.7b}$$

To control the population of big fish is to wipe out or invalidate relation (7.6a) and to regard $y_b(t - \theta)$ as exogenous. To control the population of small fish is to wipe out (7.6b) and to regard $y_s(t - \theta)$ as exogenous. In the absence of such intervention, we certainly cannot say that $y_s(t)$ *causes* $y_b(t)$ and that $y_b(t)$ *causes* $y_s(t)$, that is, we cannot use model (7.7) as a *causal* model telling us what happens through time in the uncontrolled aquarium. Nevertheless, the values of β_1 and β_2 appearing in model (7.7) do tell us what a second fish population will be conditional upon our specifying (controlling, manipulating) a first fish population — provided that the aquarium is then brought back to an equilibrium situation. Equilibrium models do tell us (enable us to predict) something about equilibrium values under *control*.

A comparable example in economics might well be the cobweb model. The model is:

$$p(t) = \alpha - \beta q_h(t) + \epsilon z_1(t) + u_1(t) \quad \text{(demand)} \tag{7.8a}$$

$$q_h(t) = \gamma + \delta p(t - 1) + \eta z_2(t) + u_2(t) \quad \text{(supply)} \tag{7.8b}$$

where $p(t)$ is price at time t, q_h is quantity harvested, z_1 and z_2 are exogenous variables, and u_1 and u_2 are stochastic shocks. This system is recursive, but if observed in equilibrium may, by use of the equilibrium condition

$$p(t - 1) = p(t)$$

be written as

$$p(t) = \alpha - \beta q_h(t) + \epsilon z_1(t) + u_1(t) \quad \text{(demand)} \tag{7.9a}$$

$$q_h(t) = \gamma + \delta p(t) + \eta z_2(t) + u_2(t) \quad \text{(supply)} \tag{7.9b}$$

and will be subject to the same causal interpretation as given in the previous example. The question of how β and δ are best to be estimated is left open.

Bicausality

Suppose we confront a demand–supply model of the following sort

$$q(t) = \alpha_{10} + \alpha_{11} p(t) + \alpha_{13} z_1(t) + u_1(t) \quad \text{(demand)} \tag{7.10a}$$

$$q(t) = \alpha_{20} + \alpha_{21} p(t) + \alpha_{24} z_2(t) + u_2(t) \quad \text{(supply)} \qquad (7.10b)$$

where $q(t)$ is quantity, $p(t)$ is price, $z_1(t)$ and $z_2(t)$ are exogenous variables, and $u_1(t)$ and $u_2(t)$ are stochastic shocks.

Now suppose this system is given the following causal interpretation: $p(t)$ causes $q(t)$ in accordance with equation (7.10a) and $p(t)$ also causes $q(t)$ in accordance with equation (7.10b). This notion of causality is certainly out of accord with the usual laboratory or control notion that we find so natural. Those who write such systems do not, however, really mean what they write, but introduce an ellipsis that is familiar to economists. What is meant is that

$$q_d(t) = \alpha_{10} + \alpha_{11} p(t) + \alpha_{13} z_1(t) + u_1(t) \qquad (7.11a)$$

$$q_s(t) = \alpha_{20} + \alpha_{21} p(t) + \alpha_{24} z_2(t) + u_2(t) \qquad (7.11b)$$

$$q_d(t) = q_s(t) \qquad (7.11c)$$

where $q_d(t)$ is quantity demanded and $q_s(t)$ is quantity supplied. This is an equilibrium model (nonrecursive), and (7.11c) is not an *identity* but an *equality* that is assumed to hold in fact over the observations.[10]

A somewhat far-fetched example, but in a surer context, is the following. Suppose there are two crops, d and s, whose yield is measured in bushels, and that (over the relevant range) the yield of one, q_d, is a positive linear function of rainfall, p, while the yield of the other, q_s, is a negative linear function of rainfall, and that these functions intersect within the relevant range. An amount z_1 of fertilizer 1 is applied to crop d and an amount z_2 of fertilizer 2 is applied to s. Suppose for each of N years we conduct an experiment for each crop, applying different amounts of fertilizer. Rainfall is uncontrolled. Imagine now the amazing result that each year of the experiment, Nature chooses a rainfall that makes the two yields equal. We may then represent the experiment by system (7.11) and reduce this to system (7.10) simply by dropping the subscripts on q and keeping track of which equation is for which crop.

Now, while it would be remarkable for Nature to choose rainfall so as to give us this strange result, it may be not so remarkable for the market to

---

[10] Especially for one not familiar with the economist's ellipsis, difficulty may result from the careless use of symbols in this reduction. Strictly speaking, the q in (7.10a and b) should be either a q_d or a q_s, or an additional equation such as $q = \min (q_d, q_s)$ or $q = q_d$ should be added to model (7.10). Otherwise, identity (and not simply equality) of quantities demanded and supplied is technically implied.

choose price so that quantity demanded equals quantity supplied, at least approximately. This is because, while rainfall is independent of past crop yields, price may well depend on past quantities demanded and supplied, and, if the system is not subject to violent change, the price adjustment relation may work with great efficiency. Whether it will or not is, of course, an empirical question: The answer may vary from market to market and time to time; but this is a matter of realism that need not concern us here. When the equality holds the theoretical system (7.11) may be represented by system (7.10), although if (7.11c) holds only approximately, system (7.10) is one with errors in the variables.

The causal system that underlies the equilibrium model (7.11) is then one in which (7.11c) is replaced by some function such as

$$p(t) = f[q_d(t - \theta), q_s(t - \theta), p(t - \theta), z_1(t), z_2(t), z_1(t - \theta),$$
$$z_2(t - \theta)] + u_3(t) \tag{7.11d}$$

and the causal (and, in this case, dynamic) model is recursive. If $p(t)$ were now to be subject to direct control, (7.11d) must be abandoned and $p(t)$ must be regarded as exogenous. Equations (7.11a) and (7.11b) would then answer questions regarding the causal effect of controlled variation in $p(t)$ on $q_d(t)$ and $q_s(t)$, and α_{11} and α_{12} would be causally interpretable coefficients. They are, moreover, the same coefficients that enter the "bicausal" system (7.10).

What is to be concluded from all this is that equilibrium systems may appear to entail "causal circles" or "bicausality," but that this is not what is intended. The causal interpretation of a coefficient in either of these types of equilibrium models is to be found in the underlying dynamic model which, *if the laboratory notion of causality is to be sustained,* will be recursive in character.[11]

Two major questions remain: (1) Can a stochastic shock model of the sort commonly considered—that is, equation (1)—ever be assumed to be always in equilibrium whenever observed?[12] (2) If not, must it not be said that either there are measurement errors introduced by the assumption of equilibrium (for example, ought not (7.11c) be regarded as an *approximate* equality) or there is a specification error? If so, this raises questions as to appropriate estimation procedure and as to the properties of estimates that ignore these model qualifications.

---

[11] Incidentally, differential equation systems are regarded as recursive with respect to infinitesimal time intervals. See [8].

[12] For (7.11a, b, and c) to hold exactly, (7.11d) would need to include $u_1(t)$ and $u_2(t)$ as arguments and exclude $u_3(t)$.

REFERENCES

[1] Bentzel, R., and Hansen, B. "On Recursiveness and Interdependency in Economic Models." *Review of Economic Studies* 22:153–168.

[2] Mood, A. *Introduction to the Theory of Statistics.* New York: McGraw-Hill, 1950.

[3] Simon, Herbert A. "Causal Ordering and Identifiability." In *Studies in Econometric Method,* edited by Wm. C. Hood and Tjalling C. Koopmans. New York: Wiley & Sons, 1953.

[4] Stone, Richard. "Discussion on Professor Wold's Paper. *Journal of the Royal Statistical Society,* Series A, 119, I (1956) 51.

[5] Waugh, Frederick. "Choice of the Dependent Variable in Regression." *Journal of the American Statistical Association* 38, no. 22 (1943) 210–214.

[6] Wold, H. O. A. "Causality and Econometrics. *Econometrica* 22:114 ff.

[7] Wold, H. O. A. "On the Definition and Meaning of Causal Concepts." Manuscript, submitted to *Philosophy of Science.*

[8] Wold, H. O. A. "Ends and Means in Econometric Model Building. Basic Considerations Reviewed." In *Probability and Statistics,* edited by U. Grenander (The Harald Cramer volume). Stockholm: Almqvist & Wiksell, 1959.

[9] Wold, H. O. A., and Juréen, L. *Demand Analysis.* Stockholm: Almqvist & Wiksell, 1952.

The Application of a System of Simultaneous Equations to an Innovation Diffusion Model*

Robert Mason

Albert N. Halter

8

Systems of interdependent simultaneous equations have been used extensively to describe social behavior in the field of economics or, more specifically, in econometrics. A system of equations in the social sciences is usually considered a nonexperimental model, that is, a model for which data came from a nonexperimental setting such as time-series or cross-sectional studies. Two problems that arise when using a system of simultaneous equations in a nonexperimental setting are (1) identification and (2) estimation. The problem of identification has been given extensive treatment in the econometric literature and has recently been introduced into the literature of sociology by Blalock [5] and Boudon [6].

A logical next step is to consider the estimation problem in the context of building behavioral theory. Since econometrics has pioneered in the development of estimation procedures for determining coefficients in systems of simultaneous equations, a reasonable starting place for other behavioral sciences is to consider these developments and their applicability to their own problems. The purpose of this chapter is to show the results of applying the Theil–Basmann method (two-stage least squares) to a system of simul-

* Reprinted from *Social Forces* 47 (1968): 182–195, by permission of the authors and the publisher. Copyright 1968, The University of North Carolina Press.

taneous equations involving both sociological and economic variables. In this chapter, first, the identification problem is reviewed. Then the logic of the two-stage estimation procedure is presented. Finally, a specific model concerning the diffusion of technical innovations is formulated and tested.

REVIEW OF IDENTIFICATION PROBLEM

Blalock [5] has shown how the identification problem arises in attempting to specify a system of equations (model) to describe status inconsistency. One of the models had the following three equations:

$$Y = b_1X_1 + b_2X_2 + b_3W + e_y$$

$$W = b_4Z + e_w$$

$$Z = f(X_1 - kX_2)$$

where Y is votes for liberal candidates; X_1 is income; X_2 is education; W represents the strain factor; Z stands for status inconsistency; b_1, b_2, b_3, and b_4 are coefficients to be estimated from some data; and k is a known parameter. The error terms e_y and e_w are assumed to account for variables that are excluded from the equations and are assumed uncorrelated with one another. In order for the coefficients in a given equation to be called identifiable, Blalock states that "the number of variables excluded from this equation must be at least equal to one less than the number of equations."[1] Applying this counting rule to each of the equations in the above system, we find: (1) the first equation has only the variable Z excluded, thus it would need another variable such as a V in the second equation to identify it; (2) the second equation is identified since it has two variables, X_1 and X_2, excluded, which is equal to one less than the number of equations; (3) the third equation is an exact mathematical relation without an error term, and hence the identification question is not relevant. The first equation is said to be underidentified and the second, exactly identified.

Coefficients in individual equations in a system can be estimated by ordinary least-squares regression techniques, provided the equation is just identified. If a model contains one or more underidentified equations, then certain alterations must be made before the coefficients can be estimated. Blalock has noted that by bringing into the second equation another exogenous variable (i.e., one that is not to be explained within the system), one that would not logically appear in the first equation, the equation for Y can

---

[1] An equivalent form of this counting rule can be expressed as "the number of endogenous variables appearing in any given equation cannot be greater than one more than the number of exogenous variables left out of this equation."

be identified. This adding of a variable would not affect the identification of the second equation.

Another case of identification is where an equation is overidentified.[2] For example, if the first equation contained another variable such as age, X_3, which did not appear in the equation for W, then the second equation would be overidentified. By the counting rule, the second equation would have three variables excluded, which is equal to the number of equations. The case of overidentification would seem to be the most prevalent case in behavioral science theory construction; that is, one is more likely to think of more variables to include in an explanation than to attempt to rely on a parsimonious interpretation.

Underidentification is a more serious hurdle than overidentification. Alterations must be made in the model to remove the former. One does not need to remove the latter before estimation of the coefficients, but use of ordinary least-squares regression is not recommended.[3] Work in statistical techniques compatible with the overidentification problem has advanced rapidly since Haavelmo's contribution in 1943 [10]. The Cowles Commission has sponsored further research in the development of limited-information–maximum-likelihood techniques for estimating the coefficients of a single equation.[4] Recently, a new technique has been developed, attributed to both Theil [27] and Basmann [2], which is closely related to the limited-information approach but is computationally more simple. This technique is frequently called the Theil–Basmann or two-stage least-squares technique of estimation. Although the complete mathematical and statistical justification for this technique is beyond the scope of this chapter, the technique would seem of sufficient importance that the steps in its application should be described and applied to a specific model, viz., the diffusion of technical innovations.

TWO-STAGE LEAST-SQUARES ESTIMATION

First, assume the following system of equations with endogenous and exogenous variables and error terms.[5] (The meaning of the variables will be discussed in the next section.)

---

[2] Identified means that the equation is either exactly identified or overidentified. Not identified means the equation is underidentified.

[3] Least-squares regression applied to an overidentified equation will lead to estimates of the coefficients that are biased and inconsistent. In addition, such estimates are less efficient than those obtained by alternative procedures. See Valavanis, [29].

[4] See, for example, Koopmans [15]. and Hood and Koopmans [11].

[5] Error terms can cover a multitude of sins. These include omission of possible variables or equations, imperfect specification of relationships, and errors of measurement. For application of the two-stage procedure, it is assumed that error is due to omission of variables and not to errors of measurement. For a discussion of assumptions concerning the error term, see Valavanis [29, pp. 5–6, 9–18].

$$Y_1 = c_1 + a_{11}Y_2 + a_{12}Y_4 + b_{11}X_2 + b_{12}X_4 + b_{13}X_5 + e_1 \qquad (8.1)$$

$$Y_2 = c_2 + a_{21}Y_1 + a_{22}Y_4 + b_{21}X_1 + b_{22}X_2 + b_{23}X_7 + e_2 \qquad (8.2)$$

$$Y_3 = c_3 + a_{31}Y_1 + a_{32}Y_2 + b_{31}X_6 + e_3 \qquad (8.3)$$

$$Y_4 = c_4 + a_{41}Y_3 + b_{41}X_1 - b_{42}X_3 + b_{43}X_4 + e_4 \qquad (8.4)$$

This system is first transformed into reduced form equations:

$$Y_1 = c_1' + z_{11}X_1 + z_{12}X_2 - z_{13}X_3 + z_{14}X_4 + z_{15}X_5 + z_{16}X_6 + z_{17}X_7 + v_1 \qquad (8.5)$$

$$Y_2 = c_2' + z_{21}X_1 + z_{22}X_2 - z_{23}X_3 + z_{24}X_4 + z_{25}X_5 + z_{26}X_6 + z_{27}X_7 + v_2 \qquad (8.6)$$

$$Y_3 = c_3' + z_{31}X_1 + z_{32}X_2 - z_{33}X_3 + z_{34}X_4 + z_{35}X_5 + z_{36}X_6 + z_{37}X_7 + v_3 \qquad (8.7)$$

$$Y_4 = c_4' + z_{41}X_1 + z_{42}X_2 - z_{43}X_3 + z_{44}X_4 + z_{45}X_5 + z_{46}X_6 + z_{47}X_7 + v_4 \qquad (8.8)$$

where each endogenous variable (Y_i) is expressed as a function of all the exogenous variables ($X_1, \ldots, X_7$). Since each equation contains only one endogenous variable, the coefficients in each of these equations can be estimated by ordinary least-squares regression.[6] This is the first stage of the two-stage least-squares procedure.

Having estimated the coefficients in the reduced-form equations, the next step is to obtain estimated values for the endogenous variables Y_1, Y_2, Y_3, and Y_4, using these coefficients and the observed values for the exogenous variables, $X_1, \ldots, X_7$. Using these new values of the endogenous variables, $\hat{Y}_1, \hat{Y}_2, \hat{Y}_3$, and $\hat{Y}_4$, and the same observed values of the exogenous variables on the right-hand side, one then estimates the coefficients in the original system [equations (8.1), (8.2), (8.3), and (8.4)] by ordinary least-squares regression. That is, each equation is estimated separately in the following set of equations:

$$Y_1 = c_1 + a_{11}\hat{Y}_2 + a_{12}\hat{Y}_4 + b_{11}X_2 + b_{12}X_4 + b_{13}X_5 + e_1 \qquad (8.9)$$

[6] For a simple presentation of the two-stage least-squares procedure in matrix notation, see Wallace and Judge [30].

$$Y_2 = c_2 + a_{21}\hat{Y}_1 + a_{22}\hat{Y}_4 + b_{21}X_1 + b_{22}X_2 + b_{23}X_7 + e_2 \qquad (8.10)$$

$$Y_3 = c_3 + a_{31}\hat{Y}_1 + a_{32}\hat{Y}_2 + b_{31}X_6 + e_3 \qquad (8.11)$$

$$Y_4 = c_4 + a_{41}\hat{Y}_3 + b_{41}X_1 - b_{42}X_3 + b_{43}X_4 + e_4 \qquad (8.12)$$

In order to distinguish these estimates from those obtained from the reduced-form equations — (8.5), (8.6), (8.7), and (8.8) — they will be indicated as $\hat{\hat{Y}}_1$, $\hat{\hat{Y}}_2$, $\hat{\hat{Y}}_3$, and $\hat{\hat{Y}}_4$ and referred to as structural equations elsewhere in this chapter. The coefficients estimated by this procedure are still biased but are consistent and more efficient than if ordinary least squares had been applied. Significance of coefficients is usually tested by the usual methods, including Student's t. Reservations must be attached to interpretations of multiple correlation coefficients of the structural equations.[7]

SPECIFICATION OF THE DIFFUSION MODEL

To illustrate the application of the two-stage technique, a diffusion model is specified. The model is composed of both sociological and economic variables that have a bearing on the diffusion of technical innovations. The model is thus not only a vehicle for illustrating the estimation procedure but also illustrates how a system of interdependent equations can be justified and tested. Once the coefficients in the model are estimated, the important task of interpretation and prediction can be undertaken.

When one begins to construct a model there usually are more explanatory or exogenous variables available than there are equations. The problem comes in selecting not only the variables to include in the interdependent system (endogenous variables) but also those that will serve as explanatory or predictor (exogenous) variables. The diffusion model was developed in part from already existing theory and in part from established empirical relationships.

As a starting point, the "accounting scheme" of Katz, Levin, and Hamilton [14] was employed. With their approach, a set of component elements was formulated as key variables in the diffusion process. Their process was characterized as (1) acceptance, (2) over time, (3) of some specific item — an idea or practice, (4) by an adopting unit — individual, group, etc., linked (5) to specific channels of communication, (6) to a social structure, and (7) to a given system of values or culture. As they point out, there is a growing interest in exploring the effects of social structures in which adopting units are linked. They also note that the field will be advanced if effects of other channels of communication, including the mass media, on these structures

---

[7] Basmann [3].

can be determined. We would like to add here that effects of economic variables—such as level of production—also can impinge on a social structure just as the adoption or acceptance of an innovation can have implications for production. It might prove fruitful to consider such an economic variable in the model under consideration.

Specification of Endogenous Variables

The following variables will be considered: (1) adoption, (2) social structure, and (3) production.

1. ADOPTION. Considering the Katz et al. scheme, it is stipulated at the outset that acceptance or the "sustained use" of an innovation, by individuals in a single-occupation category, represents the definition of the adoption variable. There is a considerable body of evidence, particularly in agricultural economics, [8] that suggests the hypothesis that when other attributes are taken into account, the relationship between adoption of agricultural innovations and production is positive. The adoption of innovations derived from research findings of the biological sciences also suggests a basis for increased production.[9]

2. SOCIAL STRUCTURE. Many diffusion researchers classify individuals in terms of social influence and prestige, and there are both conceptual and empirical bases for postulating that an individual's position in one hierarchy is related to his position in the other but is not identical with it. Merton [23, pp. 387–420] states that an individual's position in a local interpersonal influence-structure may be related to his position in other hierarchies and, conversely, his position in the class, power, or prestige hierarchies contributes to the potential for interpersonal influence but does not determine the extent to which it occurs. Influence is considered persuasive and is supported by the giving or withholding of attitudinal sanctions of approval or disapproval. Assuming that an individual has access to some channel of communication, he still must possess other attributes in order to exert influence. Research on persuasion effects has clearly established that the attributes of the apparent source of a message contribute to the effectiveness of the message. The messages of an individual with prestige, for example, tend to carry weight, and the possession of prestige can be conceived as a base for

---

[8] Schultz [26, pp. 110–144].

[9] This is not to say that farmers are better off from an income standpoint when they adopt. Increased output, if the new technology is widely adopted, will depress prices by a greater percentage than the increase in production—assuming a free market, that is, no outside force sets the price at a fixed level. Despite this, however, an individual farmer's failure to adopt an innovation will result in even lower income, since he acquires no production advantage and receives all the price disadvantages.

influence.[10] Prestige is the valuation of a particular position that an individual occupies within a social system. Status refers to the ordinal position in a hierarchy of rankings. Thus, an interdependent relationship between an individual's influence and prestige is postulated — one serves as a basis for the occurrence of the other and vice versa.

But the effects of influence and prestige, while interdependent, also may be related to the adoption of innovations as well. That is, those high in influence or prestige tend to adopt innovations more than those low in these attributes, assuming community norms support innovation. Individuals with these attributes often possess the characteristics that contribute to innovation: viz., greater formal education; greater use of the channels of mass communication, particularly the printed word; more social contacts outside their communities; and higher incomes with commensurate styles of life.[11]

3. PRODUCTION. Production is defined as an entrepreneurial function. Among individuals who exercise the entrepreneurial role are single-proprietorships, such as small retail firms and individual farmers. Thus, it is assumed that all decisions concerning production are centered in the individual, including decisions involving risk. Production is measured in terms of output, total product, or output per unit of some fixed input.

Control of economic resources has been emphasized as a basis of influence in society, as Cartwright [7, pp. 1–47] has pointed out in an excellent summary on the matter. In short, an individual gains the ability to exercise influence by occupying positions that control economic resources and, presumably, the production from these resources as well. Considering only individuals in a single-occupation category who perform an entrepreneurial function, it seems reasonable to expect that level of production will be positively associated with level of social influence. High-producing farmers, for example, living in a community in which production is commonly valued, should be expected to wield more influence among farmer–peers than low-producing farmers. The possession of control of valued resources (land, equipment, for example) is postulated to represent a basis for influence. Those individuals who manage these resources so that their efforts are visible and valued — such as high producers — should also be influential.

Moreover, from a functional point of view, production also may be a basis for prestige. Even when one is concerned with only individuals in a single-occupation category, the prestige of one position may be higher than that of another to the extent that the individual occupying the position has

---

[10] Empirical support for a positive relationship between influence and prestige variables has been reported by Lionberger and Coughenour and by Rogers. See Lionberger and Coughenour [19]; Lionberger [16]; and Rogers [24].

[11] Again, Lionberger and Coughenour and Rogers have offered empirical support concerning the positive relationship between influence, prestige, and adoption variables. See Lionberger and Coughenour [19] and Rogers [24, pp. 289–292].

been perceived by others as having made a larger contribution to the function (i.e., production in this case).[12]

This concludes the specification of endogenous variables in the interdependent system. The empirical relationships expected to be found among these variables are summarized as follows:

$$Y_1 = f(Y_2, Y_4) \tag{8.13}$$

$$Y_2 = g(Y_1, Y_4) \tag{8.14}$$

$$Y_3 = h(Y_1, Y_2) \tag{8.15}$$

$$Y_4 = l(Y_3) \tag{8.16}$$

where Y_1 is an individual's social-influence score, Y_2 is an individual's prestige score, Y_3 is an individual's innovation–adoption score, and Y_4 is an individual's level of production.

Specification of Exogenous Variables

Explanatory or exogenous (predictor) variables are assumed to be outside the interdependent system. They represent crucial operational requirements for analyzing the system empirically, as has already been noted. Furthermore, the specification of these variables and their relationship to endogenous variables can, in itself, represent derivations from existing theory or, in the absence of such theory, from known empirical associations. The set of predictors for each endogenous variable will be discussed in the order that these variables were presented in the previous section.

1. ADOPTION. The adoption of innovations by individuals requires information or knowledge about these innovations. This information may come from many sources, but the effects of one type of source have been clearly established as providing a link between the use of this source of information and the adoption of the innovations it is promoting. This source, which Rogers [24, p. 254] calls a "change agent," is a professional who attempts to influence adoption decisions in the direction he feels is desirable. In education, they may be school counselors; in medicine or health fields, they are

---

[12] Mason and Gross [20], who tested a functional education-prestige hypothesis among Massachusetts school superintendents, did not find empirical support and concluded that a highly visible, variable salary is a more important determinant of prestige. However, if level of production is highly visible, the functional hypothesis should receive empirical support.

physicians or public health workers; in agriculture, they are county agents or commercial field men.[13]

2. SOCIAL STRUCTURE. Three variables are postulated as predictors of social influences. One, as suggested in the discussion on production, is the control of economic resources necessary for production, such as land and machinery. In other words, those who exercise such control gain the ability to exercise social influence. A second variable is the utilization of the mass media—particularly the mass media commensurate with the individual's area of influence such as public affairs and health. The development of the "two-step" hypothesis, which states that "ideas often flow from radio and print to opinion leaders and from these to less active sections of the population," has a bearing here.[14] Assuming that the terms *influential* and *opinion leader* represent identical attributes, one would predict that influentials expose themselves more to pertinent forms of mass media than those less influential.

The third variable is level of education. Education effects are assumed to provide a person with greater knowledge and skills. The knowledge acquired, of whatever quality, can only expand a person's total world of reality. Each increment of schooling adds to the individual's competence through the acquisition of knowledge and incidental skills. Moreover, the advancement into college or university training is taken to reflect the potential of better performance compared to those who are screened from higher education. This acquisition of competence through formal training is postulated as a contributor to social influence, and one would predict a positive relationship between level of education and level of influence.[15]

Three variables are postulated as predictors of an individual's prestige. One of these is, again, level of education. Education effects are pervasive and may be positively related to prestige. Formal recognition of the acquisition of knowledge and skills has consequences that may have a greater effect than the learning of content only. Each additional diploma or degree leads to an inculcation of the appropriate ideology or pattern of tastes, attitudes, and values, as Sanford [25, p. 34] has noted. Such skills provide the basis for appropriate behaviors associated with a status position. It is assumed that each position carries with it a set of appropriate learned behaviors. Given this, level of education is postulated as one mechanism for the learning of these behaviors, particularly for those of higher status.

[13] Whatever the profession, there is considerable empirical support, as Rogers has summarized, to suggest that the use made of the information provided by these professionals is positively associated with the adoption of innovations they are promoting. See Rogers [24, pp. 254–267].

[14] See for example, Lazarsfeld, Berelson, and Gaudet [16]; Katz and Lazarsfeld [13]; Menzel and Katz [22]; and Katz [12].

[15] Katz [12, p. 74] has summarized empirical support for such a relationship.

Utilization of the mass media is a second variable that provides a basis for prestige. As Lazarsfeld and Merton [17, pp. 95–118] have pointed out, one function of the mass media is to confer recognition on people which, in turn, may affect their status. The social standing of a person is enhanced when he commands favorable attention in the mass media. Such recognition testifies that he has "arrived," that he is important enough to be singled out from the mass for public notice. This publicity does not go unnoticed among those who are in positions of high status, and these individuals monitor the appropriate mass media for the cues of status conferral upon themselves, their peers, and others.

A final base for prestige is type of residence. The type of dwelling, according to Barber [1, pp. 144–146] is a likely symbol for social class position in all societies. This is because the home is the place where many important consumption, socialization, educational, or social activities are conducted, that is, where one is more likely to associate intimately and equally with one's peers. Since it is assumed that a status position carries with it a set of appropriate, learned behaviors, the house where these behaviors occur—its style, size, and surroundings—may indicate the status of the owner as well.

3. PRODUCTION. Three exogenous variables are hypothesized as predictors of production. One of these is type of residence. Assuming that production is a primary source for income and wealth, high-producing individuals are likely to afford the type of residence commensurate with their economic position. Another base for production, as noted earlier, is the control of resources. The control of land, for example, is positively associated with production, according to a recent agricultural census.[16] Moreover, age of the operator has been repeatedly shown to be negatively associated with production by the same source. Thus, the three exogenous variables in the production equation are type of residence, control of land, and age of operator.

Combined with the relationships postulated among the endogenous variables, the complete system can be described by the following relations:

$$Y_1 = f(Y_2, Y_4; X_2, X_4, X_5) \tag{8.17}$$

$$Y_2 = g(Y_1, Y_4; X_1, X_2, X_7) \tag{8.18}$$

$$Y_3 = h(Y_1, Y_2; X_6) \tag{8.19}$$

$$Y_4 = l(Y_3; X_1, X_3, X_4) \tag{8.20}$$

[16] U.S., Bureau of the Census [28]. See also Davis and Mumford [8, pp. 52–54].

where Y_1 to Y_4 refer to variables in relations (8.13) to (8.16), and X_1 is an individual's type of residence, X_2 is an individual's educational level, X_3 is an individual's age, X_4 is an individual's level of control over production resources, X_5 is an individual's level of exposure to mass media appropriate for the exercise of influence, X_6 is an individual's level of use of technological information sources, and X_7 is an individual's level of exposure to mass media appropriate for recognition.

Using the counting rule described earlier, all equations in the system postulated are overidentified. The next step is to secure the appropriate operational measures for these variables and to obtain sufficient data to test the model empirically.[17]

OPERATIONAL MEASURES AND SOURCES OF DATA

An agricultural setting afforded an opportunity to empirically test the hypothesized system. Such a setting contains (1) individuals, (2) in a social structure, (3) who adopt innovations that conceivably lead to (4) increased production. The agricultural setting in this case was among full-time farmers in an Oregon community in which virtually all raised rye grass seed as their primary source of income.

A personal interview survey and examination of records available at the county assessor's and other governmental offices were the principal means of securing data. With the survey, each respondent was interviewed twice. The first interview provided data for identifying full-time farmers, measuring interpersonal influence, media exposure, and demographic characteristics. The second interview provided data on sources of information used. A total of 195 schedules was completed for the first interview, 154 for the second—a 79% completion rate. The operational measures employed are described in the order in which the variables were presented earlier. Measurement of endogenous variables will be discussed first, then measurement of exogenous variables.

Measurement of Endogenous Variables

1. ADOPTION. The agricultural area selected suffered from soil drainage and fertility problems, and innovations were selected that, in part, enabled a

---

[17] As used here, "test" refers to the procedure by which variables in the equations meet some predetermined level of statistical significance and in which the signs of the coefficients are compared to those hypothesized. Furthermore, to accept a model subsequent to a test, a necessary condition is that any variable excluded for reasons of lack of significance or incorrect sign must result in an identified system of equations.

farmer to cope with these problems. Moreover, the innovations selected required the involvement of some governmental agency, and unequivocal adoption data were available from these agencies. The innovations were: use of a soil test, installation of tile drainage, completion of land leveling, and joining a community drainage project. Adoption data were subjected to a scalogram analysis, and items met the minimum requirements for a scale.[18] Scale scores represented a measure of the adoption variable.

2. SOCIAL STRUCTURE. Two variables are involved as aspects of the community social structure—social influence and prestige. Three methods were employed to develop items to measure social influence: (1) sociometric choice, (2) self-detection, and (3) rating by a judge.[19]

Scores from each type of measure were subjected to a scalogram analysis.[20] Nondichotomized scale scores were employed as the measure of social influence.

The prestige rating technique described by Ellis [9, p. 324–337] was employed, in part, to develop an operational measure of prestige. This required procedures for (1) selecting ratees and raters,[21] (2) the rating technique,[22] and (3) combining rating scores into a prestige measure.[23]

---

[18] Coefficient of Reproducibility, .95; Coefficient of Scalability (items), .71; Coefficient of Scalability (individuals), .60.

[19] The sociometric item employed was, "May I ask the name of three or four people in this community whom you consider to be your best friends?" A modification of the items developed by Berelson et al. was used as a method for self-detection of influence. (See [4, pp. 109–115].) Two types of items were employed: one set covering farming matters, the other covering activities in the political realm. Scores were summed for each set, and each sum represented an item for subsequent development of the influence measure. The local county extension agent, who had known the respondents for more than 15 years, ranked each according to a definition of influence: the relative ability of a respondent to affect, through persuasion, the opinions or behavior of other farmers in the study area.

[20] Coefficient of Reproducibility, .94; Coefficient of Scalability (items), .82; Coefficient of Scalability (individuals), .78; Spearman–Brown split-half (odd–even) reliability coefficient, .84.

[21] All individuals—heads of households—sampled in the study area (195 in total) were selected as ratees. Eight raters were selected that had the following characteristics in common: (a) were beyond the age of adolescence but not senile; (b) possessed the characteristic cultural values of the community; (c) had been long-time residents of the area; (d) occupied strategic positions in the community that afforded each a wide range of interaction with other residents; and (e) were willing to cooperate.

[22] Names of ratees were typed on eight sets of cards and the cards in each set randomized, one set for each rater. The interviewer first placed seven blank sheets of white paper side by side before each rater in separate and independent judging sessions. Cards on which the words "High Social Standing in the Community" or "Low Social Standing in the Community" were printed served as verbal stimuli that anchored each end of the 7-point scale. The rater was instructed to judge each individual according to his social standing in the community. The cards with each

3. PRODUCTION. Total pounds of cleaned rye grass seed produced by each individual was the measure of production employed. These data were obtained from the Oregon Department of Agriculture, which had gathered the information from warehouse receipts for the production year the personal interview survey was completed. The Department was conducting a referendum among grass seed growers concerning the establishment of a growers' commission which required production information.

Measurement of Exogenous Variables

1. ADOPTION. One exogenous variable was postulated as a predictor for adoption of innovations. This variable was the level of use of technical information sources. Two such sources were available for three of the innovations promoted in the area (soil test, land leveling, tile drainage). These were the county extension agent and soil scientists at Oregon State University. In addition, a third source was available for the innovation of joining a community drainage project, the local Soil Conservation Service technician. This source was added to the other two (county agent, State College scientist) for this practice. Each item contained a source, an innovation, and a set of response alternatives and had the same basic construction.[24] Response scores were summed across all nine items to form a technical information use measure.

2. SOCIAL STRUCTURE. Three exogenous variables were postulated as

---

ratee's name were given to the rater, one at a time, and he placed the card face down on one of the seven sheets of white paper. If a rater was unable to make a rating, the interviewer put down an "unable to rate" card and instructed the rater to place the ratee's card face down on this new card. After judging had been completed, the interviewer asked the rater to divide the "unable to rate" deck on the basis of difficulty to rate or lack of knowledge about the ratee. In all cases, the raters said they were unable to rate because they didn't know the ratee. Following the judging session, each rater was asked, through a series of standardized open-ended items, to describe the criteria employed for his rating.

[23] Cards from each of the several piles for each rater were scored from one to seven, with those in the extreme positive category receiving a score of "7." Scores were arranged in a rater by ratee matrix, converted to standard scores so that scores across raters were comparable. Scores for each ratee were summed and averaged. This score represented the prestige value for each ratee. Where a rater had been unable to rate, a dummy score, the ratee's average, was used. Analysis of variance of the standard scores in the rater by ratee matrix was completed as a preliminary step for estimating the reliability of the prestige measure, following the procedure McNemar has described. (McNemar [21, pp. 296–301].) The reliability estimate, $1-1/F$, equaled 0.76 in this study. One degree of freedom was subtracted from the error term for each dummy value employed in the rater by ratee matrix.

[24] For example, "How much have you talked to someone at State College about a community drainage project—a lot, quite a bit, a little, or not at all?" Ascending or descending order of responses was determined randomly.

predictors of social influence; three of prestige as well. For social influence, these variables were the control of economic resources necessary for production, use of certain forms of the mass media, and level of education.

Control of land, necessary for production, was selected as the measure of resource control. For this study, the number of cropland acres owned per individual, available from the local tax assessor, was the operational measure employed for control of land.[25]

An agricultural setting suggests that mass media exposure appropriate to that sphere provides an adequate subject matter for examining the mass media information-seeking behavior of influentials in the study area. New ideas about farming, for example, come from magazines, and respondents could subscribe to at least four national and regional magazines in this field. The number of subscriptions to these magazines by each individual was employed as the measure of mass media exposure.

Level of education was measured by the response to a single item employed in the personal interview survey.

Two exogenous variables, in addition to level of education, were postulated as predictors of prestige. These were type of residence and exposure to mass media appropriate for recognition.

All respondents resided in single-family dwelling units. This provided a comparable base that would not be found in an urban setting, that is, an admixture of single- and multiple-family dwelling units, and permitted the use of the market value of the house as a measure of type of residence for the purpose of this study. Such market value data were obtained from the county tax assessor.[26]

Weekly newspapers provided the mass media channel appropriate for recognition in the study area. Four such newspapers, with widely overlapping circulations, were available, and the number of subscriptions to these newspapers, by respondents, was the operational measure of mass media exposure for recognition purposes.

[25] The selection of cropland ownership is straightforward. The grass seed industry in Oregon mechanized rapidly following World War II. With mechanization came the demand for more land, commensurate for the operation of an economic farm unit. The pattern of part ownership soon developed in the study area in which some individuals who already owned a block of land rented rather than purchased additional property. Those unable to mechanize and therefore unable to function as an economic farm unit, found it to their advantage to rent and to derive income from rent and from off-farm employment. Thus, a highly positive relationship has developed between ownership and renting. Because of this high association, ownership of cropland in the area is considered a valid measure of control of land.

[26] Oregon law requires that county assessors must employ as appraisers only certified professionals who have passed appropriate state civil service commission examinations. These appraisers, in the judgment of professionals on the faculty of Oregon State University, can be expected to estimate the market value of a house within 5% of its true cash value.

3. PRODUCTION. One exogenous variable, in addition to two already described (control of resources and type of residence), was postulated as a predictor of production. This was the individual's age and was measured by response to a single item employed in the personal interview survey.

Preliminary Analysis

Scatter diagrams were plotted between each exogenous and endogenous variable prior to the analysis by two-stage least squares. The purpose of such diagrams was to ascertain visually the order of polynomial of fitting the data. From this inspection, all reduced-form equations were first written to include both the linear and quadratic forms of each exogenous variable. The quadratic term was significant and was utilized in the reduced-form and the structural equations for three of these variables. These were level of education, age, and mass media exposure for recognition. All other variables were linear.

RESULTS AND DISCUSSION

Multiple correlation coefficients and zero-order correlation coefficients between each endogenous and exogenous variable are given in Table 8.1. One can note that the correlations between the exogenous and endogenous variables are as hypothesized in relations (8.17), (8.18), (8.19), and

Table 8.1. Multiple correlation coefficients and zero-order correlations between each exogenous and each endogenous variable*

	Endogenous variables			
Exogenous variables	Social influence (Y_1)	Prestige (Y_2)	Innovation adoption (Y_3)	Production (Y_4)
Market value of house (X_1)	.389	.196	.223	.341
Education (X_2)	.470	.364	.333	.185
Age (X_3)	−.238	−.181	−.165	−.223
Number of cropland acres owned (X_4)	.188	.021	.277	.649
Number of farm magazine subscriptions (X_5)	.188	.073	.183	.112
Use of technical information sources (X_6)	.455	.319	.482	.367
Number of weekly newspaper subscriptions (X_7)	.128	.239	.108	.039
Multiple correlation coefficient	.611	.463	.541	.719

* $p_{.05} = .159$; $p_{.01} = .208$

(8.20). For example; Education (X_2), Number of cropland acres owned (X_4), and Number of farm magazine subscriptions (X_5) are all significantly related to Social influence (Y_1). Some exogenous variables, however, are significantly related to endogenous variables other than the ones predicted. As an example, Use of technical information sources (X_6) is significantly related to the other endogenous variables (Y_1, Y_2, Y_4) as well as to the predicted variable, Innovation adoption (Y_3). These correlations should not be misinterpreted, since the test of the model is made on its structural form and not on its reduced form.[27] The structural form of the model is shown in equations (8.21), (8.22), (8.23), and (8.24) with the estimated coefficients as given by the two-stage technique. (Numbers under the coefficients are Student's t values: (*) denotes $.05 < p < .10$; (**) $.01 < p < .05$; and (***) $p < .01$, one-tailed test.)

$$\hat{Y}_1 = -4.239 + (.315)\hat{Y}_2 + (.514 \times 10^{-5})\hat{Y}_4 + (.063)X_2 - $$
$$(2.803)^{***} \quad (1.894)^{**} \qquad\qquad (1.376)^{*}$$
$$(.950 \times 10^{-3})X_4 + (.409)X_5 \quad (8.21)$$
$$(-.712) \qquad\qquad (1.478)^{*}$$

$$\hat{Y}_2 = 31.423 + \quad (1.558)\hat{Y}_1 - (.321 \times 10^{-5})\hat{Y}_4 - (.877 \times 10^{-4})X_1 + $$
$$(2.568)^{***} \quad (-.867) \qquad\qquad (-.584)$$
$$(.060)X_2 + (.467)X_7 \quad (8.22)$$
$$(.487) \qquad (2.037)^{**}$$

$$\hat{Y}_3 = -.588 + (.185)\hat{Y}_1 - (.023)\hat{Y}_2 + (.047)X_6 \qquad\qquad (8.23)$$
$$(2.63)^{***} \quad (-.596) \quad (2.423)^{***}$$

$$\hat{Y}_4 = 103705.400 + (92887.630)\hat{Y}_3 + (5.696)X_1 - $$
$$(2.431)^{***} \qquad\qquad (1.739)^{**}$$
$$(1714.761)X_3 + (371.178)X_4 \quad (8.24)$$
$$(-2.748)^{***} \qquad (7.910)^{***}$$

The postulated and empirically supported relationships of the model are shown in Figure 8.1.

In equation (8.21), Prestige ($\hat{Y}_2$), Production ($\hat{Y}_4$), Education (X_2), and Number of farm magazine subscriptions (X_5) are all significantly related to Social influence ($\hat{Y}_1$).[28] Number of cropland acres owned (X_4) is not related,

---

[27] Certain coefficients are assumed to be zero in the structural form of the model, which implies that exogenous variables with zero coefficients are hypothesized to have no influence on the endogenous variables.

[28] It should be noted that X_2, X_3, and X_7 represent the square of the measure of level of education, age, and number of weekly newspaper subscriptions, respectively.

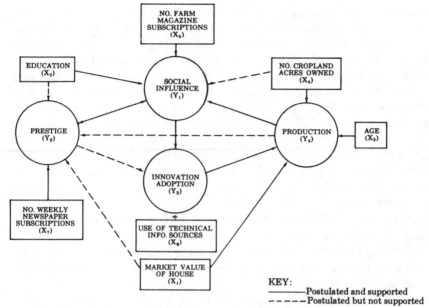

Figure 8.1. Diagram of the innovation diffusion model.

suggesting that the basis for influence in this case is production, not control of resources. Such control is an important predictor of production [equation (8.24)], which in turn is a significant predictor of influence. Furthermore, the hypothesis concerning exposure to mass media by influentials is supported, as are the hypotheses about prestige and education effects.

In equation (8.22), both Social influence ($\hat{Y}_1$) and Number of weekly newspaper subscriptions (X_7) are significantly related to Prestige ($\hat{Y}_2$), while Production ($\hat{Y}_4$), Market value of house (X_1), and Education (X_2) are not. Education does not have a direct effect on prestige, although education is a predictor of influence and influence is in turn, a predictor of prestige. The same may be said for production effects. This cannot be said, however, for the variable of market value of the house. While this exogenous variable is a predictor of production [see equation (8.24)], production is not significantly related to prestige. An examination of responses to the open-ended questions given by the raters concerning the criteria they employed in judging the social standing of respondents revealed that none dealt with level of agricultural production or type of residence in any way.[29] Assuming that the

[29] Rather, respondents were judged on the perceived quality of their interpersonal relations with others in the community. Attributes such as "willingness to work with others for the betterment of the community," "generous with their time and money," "honesty and integrity in their dealings with neighbors," "ability to cooperate with others," and "active in civic affairs," were cited.

pursuit of high production levels requires considerable outlays of scarce resources, such as the individual's time and money, an alternative hypothesis is suggested that those high in production may not necessarily be high in prestige (i.e., that production is not a basis for prestige). Moreover, while type of residence may be a predictor of production (for reasons cited earlier), it would not necessarily predict prestige when house type is measured solely in terms of market value. There appears to be nothing in the notion of prestige as operationalized by the raters (the quality of interpersonal relations), which suggests that a relationship to market value of a house would be found.[30] Exposure to mass media appropriate for recognition as a basis for prestige is straightforward and is supported by the data. Similarly, the interdependence between influence and prestige is straightforward and is supported empirically, too.

In equation (8.23) Social influence ($\hat{Y}_1$) and Use of technical information sources (X_6) are related to Innovation adoption ($\hat{Y}_3$) whereas Prestige ($\hat{Y}_2$) is not. Prestige may be construed not to have a direct effect on innovation adoption. It serves, however, as a base for influence, which in turn is related to adoption.

In equation (8.24), Innovation adoption ($\hat{Y}_3$), Market value of house (X_1), Age (X_3), and Number of cropland acres owned (X_4) are all significantly related to Production ($\hat{Y}_4$) in the predicted direction.

On the basis of the t values, one could reformulate the model by excluding the variables with low t values. The equations in that case would still be overidentified. However, to make the most meaningful test of the new model would require acquisition of a new set of data.

Some Implications of the Model

An established proposition concerning the behavior of influentials is that they are more likely than those less influential to expose themselves to mass communications, especially to those channels and messages likely to deal with topics in their sphere or spheres of influence. As well, this selective exposure contributes to the fulfillment of the influential role. While the findings of this study do not contradict this proposition, they do suggest that the information influentials seek is used for more than the enhancement of their opinion leadership role. Influentials in this study, compared to those less influential, adopt significant innovations that contribute, in turn, to increased production. Decisions leading to the adoption of innovations may

---

[30] Type of residence, however, may be a basis for influence. The variable of Market value of house (X_1) correlated highest with social influence, and, when added to equation (8.21), was significant at the .07 level ($t = 1.875$), two-tailed test. Significance levels of other terms in the equation ($\hat{Y}_2$, $\hat{Y}_4$, X_2, X_4, and X_5) were not increased.

require considerable information from many sources, including the mass media. While influentials may (or may not) pass on information to those who are less influential, there is considerable evidence that suggests that this information is used for the pursuit of purely economic goals. Thus, influentials appear to play a key role in the agricultural production process, a role distinct — but not independent — from their prestige and other attributes of social position. The fact that they are more productive, compared to those less influential, enhances their influence position and undoubtedly has implications for their capital position as well.

The importance of the influential can be noted when the model is used to predict the consequences of changing the level of a variable. For example, if an increase in agricultural production is a desired goal, how does one go about increasing production in an area comparable to one in which this study was conducted? That is, which variables must be manipulated? An examination of the important variables in equation (8.24) suggests that individuals who adopt innovations and who also control resources are likely to be high producers. Those who adopt innovations, according to equation (8.23), are influentials who utilize technological sources of information. And influentials [equation (8.21)] are those who are well educated, who expose themselves to relevant mass media channels, and who are prestigious in their community. A starting point for increasing production would appear to be knowledge of the influentials.

CONCLUSIONS

The purpose of this chapter was to describe the logic of, and to utilize, the Theil–Basmann two-stage estimation procedure for systems of simultaneous equations. Using behavioral data, the methodology appears to have promise for explaining or accounting for the effects of interdependent systems involving sociological and economic variables.

Specifically, variables relating to the diffusion of technical innovations were found to represent an interdependent system — a system in which the relationships among endogenous variables were examined and the effects of a set of exogenous variables were tested. The methodology is recommended for analysis of other kinds of interdependent systems in behavioral areas that heretofore have yielded less with less powerful techniques.

ACKNOWLEDGMENT

The authors wish to thank Dr. Norman McKown, Director of Institutional Research at Oregon State University, for his critical reading and comments on earlier drafts of this chapter. Published as Technical Paper 2459, Oregon Agricultural Experiment Station.

REFERENCES

[1] Barber, Bernard. *Social Stratification.* New York: Harcourt, Brace, & Co., 1957.

[2] Basmann, R. L. "A Generalized Classical Method of Linear Estimation of Coefficients in a Structural Equation." *Econometrica* 25 (1957): 77–84.

[3] Basmann, R. L. "Letter to the Editor." *Econometrica* 30 (1962) 824–826.

[4] Berelson, Bernard R., Lazarsfeld, Paul F., and McPhee, William N. *Voting, A Study of Opinion Formation in a Presidential Campaign.* Chicago: University of Chicago Press, 1954.

[5] Blalock, Hubert M. Jr. "The Identification Problem and Theory Building: The Case of Status Inconsistency." *American Sociological Review* 31 (1966): 52–61.

[6] Boudon, Raymond. "A Method of Linear Causal Analysis: Dependence Analysis." *American Sociological Review* 30 (1965): 365–374.

[7] Cartwright, Dorwin. "Influence, Leadership, Control." In *Handbook of Organizations,* edited by James C. March. Chicago: Rand McNally & Co., 1965.

[8] Davis, G. B., and Mumford, D. Curtis. *Farm Organization and Financial Progress in the Willamette Valley.* Corvallis, Oregon: Agricultural Experiment Station Bulletin 444, 1947.

[9] Ellis, Robert A. "The Prestige-Rating Technique in Community Stratification Research." In *Human Organization Research,* edited by R. N. Adams and J. J. Preiss. Homewood, Ill.: The Dorsey Press, 1960.

[10] Haavelmo, T. "The Statistical Implications of a System of Simultaneous Equations." *Econometrica* 11 (1943): 1–12.

[11] Hood, W. C., and Koopmans, T. C., eds. *Studies in Econometric Method:* New York: John Wiley & Sons, 1953.

[12] Katz, Elihu. "The Two-Step Flow of Communication: An Up-to-Date Report on an Hypothesis." *Public Opinion Quarterly* 21 (1957): 61–78.

[13] Katz, Elihu, and Lazarsfeld, Paul. *Personal Influence.* Glencoe, Ill.: The Free Press, 1955.

[14] Katz, Elihu, Levin, Martin L., and Hamilton, Herbert. "Traditions of Research on the Diffusion of Innovation." *American Sociological Review* 28 (1963): 237–252.

[15] Koopmans, T. C., ed. *Statistical Inference in Dynamic Economic Models.* New York: John Wiley & Sons, 1950.

[16] Lazarsfeld, Paul, Berelson, Bernard, and Gaudet, Hazel. *The People's Choice.* New York: Columbia University Press, 1944.

[17] Lazarsfeld, Paul F., and Merton, Robert K. "Mass Communication, Popular Taste, and Organized Social Action." In *The Communication of Ideas,* edited by Lyman Bryson. New York: Harper & Bros., 1948.

[18] Lionberger, Herbert F. "Community Prestige and The Choices of Sources of Farm Information." *Public Opinion Quarterly* 23 (1959): 111–118.

[19] Lionberger, Herbert F., and Coughenour, C. Milton. *Social Structure and Diffusion of Farm Information.* Columbia, Mo.: Agricultural Experiment Station Research Bulletin 631, 1957.

[20] Mason, Ward S., and Gross, Neal. "Intra-occupational Prestige Differentiation: The School Superintendency." *American Sociological Review* 20 (1955): 326–331.

[21] McNemar, Quinn. *Psychological Statistics.* New York: John Wiley & Sons, 1962.

[22] Menzel, Herbert, and Katz, Elihu. "Social Relations and Innovation in the Medical Profession: The Epidemiology of a New Drug." *Public Opinion Quarterly* 19 (1955): 337–352.

[23] Merton, Robert K. "Patterns of Influence: Local and Cosmopolitan Influentials." In *Social Theory and Social Structure*. Glencoe, Ill.: The Free Press, 1957.

[24] Rogers, Everett M. *Diffusion of Innovations*. New York: The Free Press, 1962.

[25] Sanford, Nevitt, ed. *The American College*. New York: John Wiley & Sons, 1962.

[26] Schultz, Theodore W. *Transforming Traditional Agriculture*. New Haven: Yale University Press, 1964.

[27] Theil, H. *Estimation and Simultaneous Correlation in Complete Equation Systems:* The Hague: Central Plan Bureau, 1953.

[28] U.S., Bureau of the Census, *U.S. Census of Agriculture: Oregon,* vol. 1, part 47 (Washington, D.C.: Government Printing Office, 1964).

[29] Valavanis, Stefan. *Econometrics, An Introduction in Maximum Likelihood Methods*. New York: McGraw-Hill Book Co., 1959.

[30] Wallace, T. D., and Judge, G. G. *Discussion of the Theil–Basmann Method for Estimating Equations in a Simultaneous System*. Stillwater, Okla.: Agricultural Experiment Station, Processed Series P-301, 1958.

The Choice of Instrumental Variables in the Estimation of Economy-Wide* Econometric Models

Franklin M. Fisher

9

INTRODUCTION

This chapter is concerned with an important class of problems encountered in the estimation of economy-wide econometric models. The essential characteristics of such models for our purposes are three. They are dynamic, including lagged endogenous variables as essential parts of the system. They are large and nearly self-contained, so that they include relatively few truly exogenous variables. Finally, they are essentially interdependent in that their dynamic structure is indecomposable.

Because an economy-wide model tends to be large, it is frequently impossible to estimate its reduced form by unrestricted least squares, since the number of exogenous and lagged endogenous variables is greater than the number of available observations. In addition, as will be brought out below, in the presence of serial correlation in the residuals, such estimation will not be consistent given that there are lagged endogenous variables. It is therefore usually necessary to use some form of instrumental variables method.

While there are a number of essentially equivalent ways of describing such methods, for our purposes we may think of estimation by instrumental

* Reprinted by permission of the author and publisher from the *International Economic Review* 6: 245–274. Copyright 1965, Kansai Economic Federation.

variables as consisting of choosing a list of variables to be treated as instruments and then estimating a given equation by two-stage least squares, treating all noninstrumental variables in the equation as endogenous and using the instruments as though they were the only exogenous variables in the system. In other words, all noninstrumental variables on the right-hand side of the given equation are replaced by values calculated from their regressions on the instruments and the equation then estimated by ordinary least squares.[1] Note that there is no reason why the instrumental variables so used need be the same for every equation. We shall argue below that it is positively desirable that they be different.

We shall be concerned with the choice of instrumental variables to be used in such a procedure. Our discussion is in two parts. First, we consider the choice of an eligible list of instruments, the criterion for eligibility being near zero correlation in the probability limit with the disturbance term from the given equation. In this connection, we pay particular attention to the eligibility of lagged endogenous variables. Since the eligible list will generally be too long for every variable on it to be used, we then consider how one should go about selecting from it the instrumental variables that will actually be employed. The latter discussion is essentially independent of the way in which the eligible list is constructed and is therefore of more general application than the model discussed in selecting an eligible list.

THE MODEL TO BE ESTIMATED

We suppose that the model to be estimated is

$$y_t = Ay_t + By_{t-1} + Cz_t + u_t \tag{9.1}$$

where u_t is an m-component column vector of disturbances; y_t is an m-component column vector of current endogenous variables; z_t is an n-component column vector of exogenous variables (known at least to be uncorrelated in the probability limit with all current and past disturbances); A, B, and C are constant matrices to be estimated; and $(I - A)$ is nonsingular, while A has zeroes everywhere on its principal diagonal. The assumption that there are no terms in $y_{t-\theta}$ for $\theta > 1$ involves no loss of generality in the present discussion, since it can always be accomplished by redefinition of y_t and expansion of the equation system and will be used only for convenience in dealing with the solution of (9.1) regarded as a system of stochastic difference equations.

---

[1] Other k-class estimators such as limited-information, maximum-likelihood also have their analogous instrumental variable estimators; essentially the same treatment would apply to them.

To examine the correlations among the disturbance terms and the current and lagged endogenous variables, we solve the system for y_t, obtaining

$$y_t = (I - A)^{-1}By_{t-1} + (I - A)^{-1}Cz_t + (I - A)^{-1}u_t \qquad (9.2)$$

Denote $(I - A)^{-1}$ by D. Assuming that DB is stable, we have[2]

$$y_t = \sum_{\theta=0}^{\infty} (DB)^{\theta}(DCz_{t-\theta} + Du_{t-\theta}) \qquad (9.3)$$

Denoting the covariance matrix of u_t and $y_{t-\theta}$ by $W(\theta)$ with columns corresponding to elements of u_t and rows corresponding to elements of $y_{t-\theta}$, and that of u_t and $u_{t-\theta}$ by $V(\theta)$ (which is assumed to be independent of t), with columns corresponding to u_t and rows to $u_{t-\theta}$,

$$W(0) = \sum_{\theta=0}^{\infty} (DB)^{\theta}[DV(\theta)] \qquad (9.4)$$

Similarly,

$$W(1) = \sum_{\theta=1}^{\infty} (DB)^{\theta-1}[DV(\theta)] \qquad (9.5)$$

Unless one is willing to assume $V(\theta) = 0$ for all $\theta > 0$ (no serial correlation) or to make other assumptions thereon, and on D and B, lagged endogenous variables will generally be correlated with current disturbances.

NEAR-CONSISTENCY, BLOCK-RECURSIVE SYSTEMS AND THE CHOICE OF ELIGIBLE INSTRUMENTAL VARIABLES

Introduction

In this section, we discuss the choice of an eligible list of predetermined instruments. Until further notice, then, we discuss only whether and under

---

[2] We shall not discuss the assumption of the stability of DB in any detail at this point. If it is not stable, then it suffices to assume that the model begins with nonstochastic initial conditions. Obviously, if stability fails, the question of serial correlation in the disturbances becomes of even greater importance than if stability holds. We shall return to this and shall discuss the question of stability in general in a later section.

what circumstances a given single variable ought to be treated as predetermined.

In general, we desire two things of a variable that is to be treated as predetermined in the estimation of a given equation. First, it should be uncorrelated in the probability limit with the disturbance from that equation; second, it should closely causally influence the variables that appear in that equation and should do so independently of the other predetermined variables.[3] If the first criterion is not satisfied, treating the variable as predetermined results in inconsistency; if the second fails, such treatment does not aid much in estimation — it does not reduce variances. In practice, these requirements may frequently not be consistent, and one has to compromise between them. The closer the causal connection, the higher may be the forbidden correlation. Thus, in one limit, the use of ordinary least squares, which treats all variables on the right-hand side of the equation as predetermined, perfectly satisfies the second but not the first criterion. In the other limit, the use of instrumental variables that do not directly or indirectly causally influence any variable in the model perfectly meets the first requirement but not the second.[4] In general, one is frequently faced with the necessity of weakening the first requirement to one of low rather than of zero correlation and accepting indirect rather than direct causal relations between instruments and included variables. (Such a compromise may result in different instruments for different equations when a limited-information estimator is used; this will be the case below.)

In the present section, we discuss the circumstances under which zero or low inconsistencies can be expected, leaving explicit use of the causal criterion to the next section.

Now, two sets of candidates for treatment as instrumental variables are obviously present. The first of these consists of those variables that one is willing to assume truly exogenous to the entire system and the lagged values thereof; the second consists of the lagged endogenous variables. The dynamic and causal structure of the system may well provide a third set, however, and may cast light on the appropriateness of the use of lagged endogenous variables; to a discussion of this we now turn.

The Theory of Block-Recursive Systems

A generalization of the recursive systems introduced by Wold[5] is provided by what I have elsewhere termed "block-recursive systems".[6] In

---

[3] It should therefore be relatively uncorrelated with the other variables used as instruments so that lack of collinearity is not really a separate criterion. We shall return to this in the next section.

[4] If only such variables are available for use (or if an insufficient number of more interesting ones are), then the equation in question is underidentified and even asymptotic variances are infinite.

[5] See Wold and Juréen [19], Wold [17] and other writings.

[6] Fisher [8].

general, such systems have similar properties to those of recursive systems when the model is thought of as subdivided into sets of current endogenous variables and corresponding equations (which we shall call "sectors") rather than into single endogenous variables and their corresponding equations.

Formally, we ask whether it is possible to partition the vectors of variables and of disturbances and the corresponding matrices (renumbering variables and equations, if necessary) to secure a system with certain properties. In such partitionings, the Ith subvector of a given vector x will be denoted as x^I. Similarly, the submatrix of a given matrix M, which occurs in the Ith row and Jth column of submatrices of that matrix, will be denoted by M^{IJ}. Thus,

$$M = \begin{bmatrix} M^{11} & M^{12} & \ldots & M^{1N} \\ M^{21} & M^{22} & \ldots & M^{2N} \\ \cdot & \cdot & & \\ \cdot & \cdot & & \\ \cdot & \cdot & & \\ M^{N1} & M^{N2} & \ldots & M^{NN} \end{bmatrix} \qquad x = \begin{bmatrix} x^1 \\ x^2 \\ \cdot \\ \cdot \\ \cdot \\ x^N \end{bmatrix} \qquad (9.6)$$

We shall always assume the diagonal blocks, M^{II}, to be square.

If when written in this way, the matrix M has the property that $M^{IJ} = 0$ for all $I = 1, \ldots, N$ and $J > I$, the matrix will be called *block-triangular*. If $M^{IJ} = 0$ for all $I = 1, \ldots, N F \neq I$, the matrix will be called *block-diagonal*.[7]

Now consider the system (9.1). Suppose that there exists a partition of that system (with $N > 1$) such that (BR.1) A is block-triangular; (BR.2) $V(0)$ is block-diagonal; (BR.3) $V(\theta) = 0$ for all $\theta > 0$.[8] In this case, it is easy to show that the *current* endogenous variables of any given sector are uncorrelated in probability with the current disturbances of any higher-numbered sector. Such variables may thus be consistently treated as predetermined instruments in the estimation of the equations of such higher-numbered sectors.

To establish the proposition in question, observe that by (BR.3)

$$W(0) = DV(0) \qquad (9.7)$$

By (BR. 1), however, $D = (I - A)^{-1}$ is block-triangular, while $V(0)$ is block-diagonal by (BR. 2). It follows that their product is block-triangular with the same partitioning. Thus,

---

[7] Block-triangularity and block-diagonality are the respective canonical forms of decomposability and complete decomposability.

[8] (BR. 1) and (BR. 2) are generalizations of the corresponding assumption for recursive systems. See Wold [17, pp. 358–359].

$$W(0)^{IJ} = 0 \qquad\qquad (9.8)$$

for all $I, J = 1, \ldots, N$ and $J > I$, but this is equivalent to the proposition in question.

As can also be done in the special case of recursive systems, assumption (BR. 3) can be replaced by a somewhat different assumption. (BR. 3*) B is block-triangular with the same partitioning as A, as is $V(\theta)$ for all $\theta > 0$. Further, either all B^{II} or all $V(\theta)^{II}(\theta > 0)$ are zero $(I = 1, \ldots, N)$.

To see that this suffices, observe that in this case every term in (9.4) will be block-triangular.

Note, however, that whereas (BR. 1) $-$ (BR. 3) patently suffice to give $W(1) = 0$ and thus show that lagged endogenous variables are uncorrelated with current disturbances, this is not the case when (BR. 3) is replaced by (BR. 3*). As in the similar case for recursive systems, what is implied by (BR. 1), (BR. 2), and (BR. 3*) in this regard is that $W(1)$ is also block-triangular with zero matrices on the principal diagonal so that lagged endogenous variables are uncorrelated with the current disturbances of the same or *higher*-numbered blocks, but not necessarily with those of *lower*-numbered ones.

If A and B are both block-triangular with the same partitioning, then the matrix DB is also block-triangular, and the system of difference equations given by (9.2) is decomposable. In this case, what occurs in higher-numbered sectors *never* influences what occurs in lower-numbered ones, so that there is, in any case, no point in using current or lagged endogenous variables as instruments in lower-numbered sectors. This is an unlikely circumstance to encounter in an economy-wide model in any essential way, but it may occur for partitionings that split off a small group of equations from the rest of the model. If it does not occur, then (BR. 3) is generally necessary for the block-triangularity of $W(0)$. Indeed, unless either (BR. 3) or the first statement of (BR. 3*) holds, no $W(0)^{IJ}$ can generally be expected to be zero if $B \neq 0$.

To see this, observe that (9.4) implies that $W(0)$ cannot generally be expected to have any zero submatrices unless every term in the sum that is not wholly zero has a zero submatrix in the same place. This cannot happen unless every matrix involved is either block-diagonal or block-triangular. Hence, if $V(\theta) \neq 0$ for all $\theta > 0$, all such $V(\theta)$ must at least be block-triangular, as must B.[9]

---

[9] Of course, this does not show that (BR. 3) or (BR. 3*) is necessary, since counterexamples may easily be produced in which different non-zero terms in (9.4) just cancel out. The point is that this cannot be assumed to occur in practice. To put it another way, since such cancellation cannot be known to occur, it clearly occurs only on a set of measure zero in the parameter space. Thus, (BR. 3) or (BR. 3*) is necessary with probability 1.

Block-Recursive Assumptions in Economy-Wide Models

Unfortunately, while block-triangularity of A is not an unreasonable circumstance to expect to encounter in practice the assumptions on the disturbances involved in (BR. 2), and (BR. 3) or (BR. 3*) seem rather unrealistic in economy-wide models.[10] Thus, it does not seem reasonable to assume that the omitted effects that form the disturbances in two different sectors have no common elements; nor does it seem plausible to assume either that there is no serial correlation of disturbances or that the dynamic system involved is decomposable.

Note, however, that these assumptions may be better approximations than in the case of recursive systems. Thus, one may be more willing to assume no correlation between contemporaneous disturbances in two different aggregate sectors than between disturbances in any two single equations. A similar assumption may be even more attractive when the disturbances in question are from different time periods, as will be seen below. Thus, also, the dynamic system may be thought *close* to decomposability when broad sectors are in view and feedbacks within sectors explicitly allowed. If such assumptions are approximately satisfied, then the inconsistencies involved in the use of current and lagged endogenous variables, as predetermined in higher-numbered sectors, will be small.[11]

Nevertheless, the assumption of no correlation between contemporaneous disturbances from different sectors, the assumption of no serial correlation in the disturbances, and the assumption of decomposability of the dynamic system all seem rather strong ones to make. If none of these assumptions is in fact even approximately made, then the use of current endogenous variables as instruments in higher-numbered sectors leads to non-negligible inconsistencies. We shall show, however, that this need not be true of the use of some *lagged* endogenous variables in higher-numbered sectors under fairly plausible assumptions as to the process generating the disturbances. We thus turn to the question of the use of lagged endogenous variables, assuming that A is known to be at least nearly block-triangular.

Reasonable Properties of the Disturbances

The problems that we have been discussing largely turn on the presence of common omitted variables in different equations and on the serial correlation properties of the disturbances. It seems appropriate to proceed by

---

[10] It is encountered in preliminary versions of the SSRC model. C. Holt and D. Steward have developed a computer program for organizing a model in block-triangular form.

[11] See Fisher [8]. The theorems involved are generalizations of the Proximity Theorem of Wold for recursive systems. See Wold and Faxér [18].

setting up an explicit model of the process generating the disturbances in terms of such omitted variables and such serial correlation.

We shall assume that the disturbances to any equation are made up of three sets of effects. The first of these will consist of the effects of elements common to more than one sector — in general, common to all sectors. The second will consist of the effects of elements common to more than one equation in the sector in which the given equation occurs. The third will consist of effects specific to the given equation.

Thus, let the number of equations in the Ith sector be n_I. We write

$$u_t^I = \phi^I e_t + \psi^I v_t^I + w_t^I \qquad (I = 1, \ldots, N) \qquad (9.9)$$

where

$$e_t = \begin{bmatrix} e_{1t} \\ \cdot \\ \cdot \\ \cdot \\ e_{Kt} \end{bmatrix} \qquad (9.10)$$

is a vector of implicit disturbances whose effects are common (in principle) to all equations in the model and ϕ^I is an $n_I \times K$ constant matrix;

$$v_t^I = \begin{bmatrix} v_{1t}^I \\ \cdot \\ \cdot \\ \cdot \\ v_{H_I t}^I \end{bmatrix} \qquad (9.11)$$

is a vector of implicit disturbances, whose effects are common (in principle) to all equations in the Ith sector but not to equations in other sectors; ψ^I is an $n_I \times H_I$ constant matrix; and

$$w_t^I = \begin{bmatrix} w_{1t}^I \\ \cdot \\ \cdot \\ \cdot \\ w_{n_I t}^I \end{bmatrix} \qquad (9.12)$$

is a vector of implicit disturbances, the effect of each of which is specific to a given equation in the Ith sector.

Define

$$\phi = \begin{bmatrix} \phi^1 \\ \cdot \\ \cdot \\ \cdot \\ \phi^N \end{bmatrix} \qquad (9.13)$$

$$v_t = \begin{bmatrix} v_t^1 \\ \cdot \\ \cdot \\ \cdot \\ v_t^N \end{bmatrix} \qquad (9.14)$$

$$\psi = \begin{bmatrix} \psi^1 & 0 & \ldots & 0 \\ 0 & \psi^2 & \ldots & 0 \\ \cdot & \cdot & & \cdot \\ \cdot & \cdot & & \cdot \\ \cdot & \cdot & & \cdot \\ 0 & 0 & \ldots & \psi^N \end{bmatrix} \qquad (9.15)$$

and

$$w_t = \begin{bmatrix} w_t^1 \\ w_t^2 \\ \cdot \\ \cdot \\ \cdot \\ w_t^N \end{bmatrix} \qquad (9.16)$$

Then (9.9) may be rewritten more compactly as

$$u_t = \phi e_t + \psi v_t + w_t \qquad (9.17)$$

We shall refer to the elements of e_t, v_t and w_t as *economy-wide, sector,* and *equation* implicit disturbances, respectively, noting that whether an economy-wide or sector implicit disturbance actually affects a given equation depends on the relevant rows of ϕ and ψ, respectively. (The unqualified term *disturbance* will be reserved for the elements of u_t.)

All elements of e_t, v_t, and w_t are composites of unobservables; it is hardly restrictive to assume that

(A. 1) Every element of e_t, v_t, or w_t is uncorrelated in probability with all present or past values of any *other* element of any of these vectors. The vectors can always be redefined to accomplish this.

We shall assume that each element of each of these implicit disturbance vectors obeys a (different) first-order autoregressive scheme.[12] Thus,

$$e_t = \Lambda_e e_{t-1} + e_t^* \tag{9.18}$$

$$v_t = \Lambda_v v_{t-1} + v_t^* \tag{9.19}$$

$$w_t = \Lambda_w w_{t-1} + w_t^* \tag{9.20}$$

where Λ_e, Λ_v, and Λ_w are diagonal matrices of appropriate dimension and e_t^*, v_t^*, and w_t^* are vectors of nonautocorrelated random variables. Assuming that the variance of each element of e_t, v_t, and w_t is constant through time, the diagonal elements of Λ_e, Λ_v, and Λ_w are first-order autocorrelation coefficients and are thus each less than one in absolute value.

Now let Δ_e, Δ_v, and Δ_w be the diagonal variance–covariance matrices of the elements of e_t, v_t, and w_t, respectively. In view of (A.1) and (9.18)–(9.20), it is easy to show that (9.17) implies

$$V(\theta) = \phi \Lambda_e^\theta \Delta_e \phi' + \psi \Lambda_v^\theta \Delta_v \psi' + \Lambda_w^\theta \Delta_w \qquad (\theta \geqq 0) \tag{9.21}$$

Evidently, $V(\theta)$ will be non-zero unless some other assumptions are imposed. Consider, however, the question of whether $V(\theta)$ will be block-diagonal. Since all the Λ and Δ matrices are diagonal, and since ψ is itself block-diagonal by (9.15), we have

$$V(\theta)^{IJ} = \phi^I \Lambda_e^\theta \Delta_e (\phi^J)' \qquad (\theta \geqq 0; I, J, = 1, \ldots, N; J \neq I) \tag{9.22}$$

Thus, the off-diagonal blocks of $V(\theta)$ depend only on the properties of the economy-wide disturbances.

This result is perhaps worth emphasizing. When applied to $\theta = 0$, it merely states formally what we have said previously, that contemporaneous disturbances from the equations of the model that occur in different sectors cannot be assumed uncorrelated if there are common elements in

---

[12] Autoregressive relations of higher orders could be considered in principle, but this would rather complicate the analysis. We shall thus assume that first-order relationships are sufficiently good approximations. If higher-order relationships are involved, there is no essential change in the qualitative results.

each of them, that is, implicit disturbance elements affecting both sectors. When applied to $\theta > 0$, however, the result is at least slightly less obvious. Here it states that despite the fact that contemporaneous disturbances from different sectors may be highly correlated, and despite the fact that every disturbance may be highly autocorrelated, a given disturbance will *not* be correlated with a lagged disturbance from another sector unless the economy-wide implicit disturbances are themselves autocorrelated. To put it another way, the presence of economy-wide implicit disturbances and the presence of substantial serial correlation do not prevent us from taking $V(\theta)$ as block-diagonal for $\theta > 0$, provided that the serial correlation is entirely confined to the sector and equation implicit disturbances.

Is it then reasonable to assume that the serial correlation is so confined? I think it is reasonable in the context of a carefully constructed economy-wide model. Any such model inevitably omits variables, the effects of which are not confined within sectors. Effects that are highly autocorrelated, however, are effects that are relatively systematic over time. In an inevitably aggregate and approximate economy-wide model, there are likely to be such systematic effects influencing individual equations and even whole sectors. Systematic effects that spread over more than one sector, however, seem substantially less likely to occur, especially when we recall that the limits of a sector in our sense are likely to be rather wide.[13] Variables that give rise to such effects are not likely to be omitted variables whose influence lies in the disturbance terms. Rather, they are likely to be explicitly included in the model, if at all possible. If not, if they relate to the occurrence of a war, for example, and are thus hard to specify explicitly, the time periods in which they are most important are likely to be omitted from the analysis. In short, systematic behavior of the disturbances is an indication of incomplete specification. Such incompleteness is much less likely to occur as regards effects that are widespread than as regards effects that are relatively narrowly confined, especially since the former are less likely to be made up of many small effects.[14] (Recall that an economy-wide implicit disturbance is one that affects more than one sector *directly*, not simply one whose effects are transmitted through the dynamic causal structure of the explicit model.) It thus does not seem unreasonable to assume that

$$\Lambda_e = 0 \tag{9.23}$$

[13] As they are in the SSRC model.

[14] A similar argument obviously implies that sector implicit disturbances are less likely to be serially correlated than are equation implicit disturbances. The analysis of the effects of this on $V(\theta)$ and the subsequent discussion is left to the reader. The assumption of no serial correlation in the sector implicit disturbances seems considerably more dangerous than that being discussed in the text.

and therefore

$$V(\theta)^{IJ} = 0 \qquad (\theta > 0; I, J = 1, \ldots, N; J \neq I) \qquad (9.24)$$

as good approximations.

Implications for the Use of Lagged Endogenous Variables

Of course, assuming (9.24) to hold is not sufficient to yield consistency when lagged endogenous variables are treated as predetermined. We have already seen that unless $V(\theta) = 0$, the decomposability of the dynamic system must be assumed in addition to (9.24) to secure such consistency. We argued above, however, that such decomposability was rather unlikely in an interconnected economy, although the fact that (9.24) is likely to hold approximately makes it important to look for *near*-decomposability and thus secure *near*-consistency.[15]

Even if such near-decomposability of the dynamic system does not occur, however, (9.24) has interesting consequences for the treatment of lagged endogenous variables as predetermined.

Consider the expression of $W(1)$ given in equation (9.5). Writing out the first few terms of the sum, we obtain

$$W(1) = DV(1) + DBDV(2) + (DB)^2 DV(3) + \ldots \qquad (9.25)$$

Since D is block-triangular and $V(1)$ block-diagonal by (9.24), the first term in this expansion is also block-triangular. Hence, even if the dynamic system is not decomposable, endogenous variables lagged one period are approximately uncorrelated in the probability limit with disturbances in *higher*-numbered sectors (but not in the same or lower-numbered sectors), to the extent that the right-hand terms in (9.25) other than the first can be ignored.

In what sense is it legitimate, then, to assume that such terms can, in fact, be ignored? Assume that the matrix DB is similar to a diagonal matrix, so that there exists a nonsingular matrix P such that

$$DB = PHP^{-1} \qquad (9.26)$$

where H is diagonal and has for diagonal elements the latent roots of

---

[15] Near-decomposability of a dynamic system has a number of interesting consequences in addition to this. See Ando, Fisher, and Simon [3], especially Ando and Fisher [2].

DB.[16] Let

$$\Lambda = \begin{bmatrix} \Lambda_e & 0 & 0 \\ 0 & \Lambda_v & 0 \\ 0 & 0 & \Lambda_w \end{bmatrix} \qquad (9.27)$$

$$\Delta = \begin{bmatrix} \Delta_e & 0 & 0 \\ 0 & \Delta_v & 0 \\ 0 & 0 & \Delta_w \end{bmatrix} \qquad (9.28)$$

$$Q = [\phi : \varphi : I] \qquad (9.29)$$

Then every such term can be written as

$$(DB)^{\theta-1}DV(\theta) = PH^{\theta-1}P^{-1}DQ\Lambda^{\theta}\Delta Q' \qquad (\theta > 1) \qquad (9.30)$$

We know that every diagonal element of the diagonal matrix, Λ, is less than unity in absolute value (indeed, we are assuming that some of the diagonal elements are zero). Moreover, *if we are prepared to maintain the stability assumption on DB* that was slipped in some time ago, every diagonal element of the diagonal matrix H will also be less than unity in absolute value. It follows that every element of every term in the expansion of $W(1)$ other than the first is composed of a sum of terms, each of which involves at least the product of a factor less than unity and the square of another such factor. There is clearly a reasonable sense in which one may be prepared to take such terms as negligible, at least when compared with the non-zero elements of the first term in the expansion for $W(1)$ which involve only the diagonal elements of Λ to the first power. If one is willing to do this, then one is saying that the use of endogenous variables lagged one period as instruments in higher-numbered sectors involves only negligible inconsistency, at least as compared with the use of the same variables as instruments in their own or lower-numbered sectors.

There may be considerable difficulties in accepting such a judgment, however. In the first place, it is well to be aware that there are two different statements involved. It is one thing to say that the effects in question are negligible compared to others and quite another to say that they are negligible in a more absolute sense. If one accepts the stability assumption, then there certainly is a value of θ beyond which further terms in the expansion of $W(1)$ are negligible by any given standard. These may not be all terms

---

[16] The assumption involved is, of course, very weak and is made for ease of exposition.

after the first, however; we shall discuss the case in which there are non-negligible terms after the first below.

Second (a minor but one worth observing), even our conclusion about *relative* importance *need* not hold although other assumptions are granted. While it is true that as θ becomes large the right-hand side of (9.30) approaches zero, such approach need not be monotonic. To put it another way, every element of the matrix involved is a sum of terms. Each such term involves a diagonal element Λ to the θ and a diagonal element of H to the $\theta - 1$. If all such diagonal elements are less than unity in absolute value, then the absolute value of each separate term approaches zero monotonically as θ increases; this need not be true of the *sum* of those terms, however, and it is easy to construct counter-examples. Nevertheless, there is a sense in which it seems appropriate to assume the terms in the expansion for $W(1)$ to be negligible for θ greater than some value, perhaps for $\theta > 1$.

All this, however, has leaned a bit heavily on the stability of DB. If that matrix has a latent root greater than unity in absolute value, then part of the reason for assuming that the right-hand side of (9.30) is negligible even for high values of θ has disappeared. Of course, the diagonal elements of Λ are known to be less than unity in absolute value, so that the infinite sum involved in $W(1)$ may still converge. However, such convergence is likely to be slow in an unstable case and may not occur at all, so that the effects of serial correlation are even more serious than in the stable case. Clearly, the stability assumption requires additional discussion at this point.

The usual reason for assuming stability of the dynamic model being estimated is one of convenience or of lack of knowledge of other cases. Since the unstable case tends to lead to unbounded moment matrices, the usual proofs of consistency of the limited-information estimators tend to break down in that circumstance. Indeed, maximum-likelihood estimators are presently known to be consistent only in the stable case and in rather special unstable cases.[17] It is therefore customary to assume stability in discussions of this sort. For present purposes, even if limited-information estimators are consistent in unstable cases and even if the Generalized Proximity Theorems, which guarantee small inconsistencies for sufficiently good approximations also hold, the approximations that we are now discussing are relatively unlikely to be good ones in such cases.[18] Even if the existence of $W(1)$ is secured by assuming that the dynamic process (9.1) begins with nonstochastic initial conditions at some finite time in the past (and even this does not suffice for the existence of the probability limit), the effects of serial

---

[17] For example, if *all* latent roots are greater than unity in absolute value. See Anderson [1]. J. D. Sargan has privately informed me that he has constructed a proof of consistency for the general case. The classic paper in this area is that of Mann and Wald [12].

[18] See Fisher [8].

correlation will not die out (or will die out only slowly) as we consider longer and longer lags. The conclusion seems inescapable that if the model is thought to be unstable (and the more so, the more unstable it is), the use of lagged endogenous variables as instruments *anywhere* in an indecomposable dynamic system with serially correlated disturbances is likely to lead to large inconsistencies at least for all but very high lags. The lower the serial correlation and the closer the model to stability, the less dangerous is such use.

Is the stability assumption a realistic one for economy-wide models then? I think it is. Remember that what is at issue is not the ability of the economy to grow, but its ability to grow (or to have explosive cycles) with no help from the exogenous variables and no impulses from the random disturbances. Since the exogenous variables generally include population growth, and since technological change is generally either treated as a disturbance or as an effect that is exogenous in some way, this is by no means a hard assumption to accept. While there are growth and cycle models in economic theory that involve explosive systems, such models generally bound the explosive oscillations or growth by ceilings or floors that would be constant if the exogenous sources of growth were constant.[19] The system *as a whole* in such models is not unstable in the presence of constant exogenous variables and the absence of random shocks.[20] We shall thus continue to make the stability assumption.

Even when the stability assumption is made, however, it may not be the case, as we have seen, that one is willing to take the expression in (9.30) as negligible for all $\theta > 1$. In particular, this will be the case if serial correlation is thought to be very high so that the diagonal elements of Λ are close to unity in absolute value. In such cases, one will not be willing to assume that the use of endogenous variables lagged *one* period as instruments in higher-numbered sectors leads to only negligible inconsistencies. Accordingly, we must generalize our discussion.

Fortunately, this is easy. There clearly does exist a smallest $\theta^* > 0$ such that for all $\theta > \theta^*$ even the diagonal blocks of $V(\theta^*)$ are negligible on any given standard. Consider $W(\theta^*)$, the covariance matrix of the elements of u_t and those of $y_{t-\theta^*}$, with the columns corresponding to elements of u_t and the rows to elements of $y_{t-\theta^*}$. Clearly,

$$
\begin{aligned}
W(\theta^*) &= DV(\theta^*) + \sum_{\theta=\theta^*+1}^{\infty} (DB)^{\theta-\theta^*} DV(\theta) \\
&= DV(\theta^*) + \sum_{\theta=\theta^*+1}^{\infty} FH^{\theta-\theta^*} P^{-1}DQ\Lambda^{\theta}Q'
\end{aligned}
\tag{9.31}
$$

---

[19] See, for example, Hicks [10] and Harrod [9].
[20] Whether a linear model is a good approximation if such models are realistic is another matter.

Since D is block-triangular and $V(\theta^*)$ block-diagonal by (9.24), the product, $DV(\theta^*)$, is also block-triangular. Considering $W(\theta^*)^{IJ}$ for $J > I$, it is apparent that the covariances of endogenous variables lagged θ^* periods, and current disturbances from *higher-numbered sectors* are made up of only negligible terms. Not only is $V(\theta)$ negligible by assumption for $\theta > \theta^*$, but also every such term involves at least one power of H, which by assumption is diagonal and has diagonal elements less than unity in absolute value.

Note, however, that a similar statement is clearly false as regards the covariances of endogenous variables lagged θ^* periods and current disturbances *from the same or lower-numbered sectors*. Such covariances involve the non-zero diagonal blocks of $V(\theta^*)$ in an essential way. It follows that the order of inconsistency so to speak, involved in using endogenous variables lagged a given number of periods as instruments, is less if such variables are used in higher-numbered sectors than if they are used in the same or lower-numbered sectors. To put it another way, the minimum lag with which it is reasonably safe to use endogenous variables as instruments is at least one less for use in higher-numbered than for use in the same or lower-numbered sectors.

As a matter of fact, our result is a bit stronger than this. It is apparent from (9.31) that the use of endogenous variables lagged θ^* periods as instruments in higher-numbered sectors involves covariances of the order of $\Lambda^{\theta^*+1}H$. Even the use of endogenous variables lagged $\theta^* + 1$ periods as instruments in the same or lower-numbered sectors, however, involves covariances of the order of only Λ^{θ^*+1}. No positive power of H is involved in the first term of the expansion for the latter covariances. Since H is diagonal with diagonal elements less than unity in absolute value, the difference between the minimum lag with which it is safe to use endogenous variables as instruments in the same or lower-numbered sectors and the corresponding lag for use in higher-numbered ones may be even greater than one. This point will be stronger the more stable one believes the dynamic system to be. It arises because the effects of serial correlation in sector and equation implicit disturbances are direct in the case of lagged endogenous variables used in the same or lower-numbered sectors and are passed through a damped dynamic system in the case of lagged endogenous variables used in higher-numbered sectors.[21]

To sum up: So far as inconsistency is concerned, it is likely to be safer to use endogenous variables with a given lag as instruments in higher-numbered sectors than to use them in the same or lower-numbered sectors. For

---

[21] All this is subject to the minor reservation discussed above concerning sums each term of which approaches zero monotonically. In practice, one tends to ignore such reservations in the absence of specific information as to which way they point.

the latter use, the endogenous variables should be lagged by at least one more period to achieve the same level of consistency.[22]

Now, it may be thought that this result is a rather poor return for all the effort we have put into securing it. While one can certainly conceive of stronger results, the usefulness of the present one should not be underestimated. We remarked at the beginning of this section that one important *desideratum* of an instrumental variable was a close causal connection with the variables appearing in the equation to be estimated. In general, economy-wide (and most other) econometric models have the property that variables with low lags are often (but not always) more closely related to variables to be explained than are variables with high ones. There may be, therefore, a considerable gain in efficiency in the use of recent rather than relatively remote endogenous variables as instruments, and it is important to know that in certain reasonable contexts this may be done without increasing the likely level of resulting inconsistency. We now turn to the discussion of the causal criterion for instrumental variables.

CAUSALITY AND RULES FOR THE USE OF ELIGIBLE INSTRUMENTAL VARIABLES

The Causal Criterion for Instrumental Variables

We stated above that a good instrumental variable should directly or indirectly causally influence the variables in the equation to be estimated in a way independent of the other instrumental variables, and that the more direct such influence is, the better. This statement requires some discussion. As far as the limiting example of an instrument completely unrelated to the variables of the model is concerned, the lesson to be drawn might equally well be that instrumental variables must be correlated in the probability limit with at least one of the included variables. While it is easy to see that *some* causal connection must therefore exist, the question naturally arises why it must be one in which the instrumental variables cause the included ones. If correlation is all that matters, surely the causal link might be reversed or both variables influenced by a common third one.

This is not the case. Consider first the situation in which the proposed instrumental variable is caused in part by variables included in the model. To the extent that this is the case, no advantage is obtained by using the proposed instrumental variable over using the included variables them-

---

[22] The reader should be aware of the parallel between this result and the similar result for the use of *current* endogenous variables that emerges when (BR. 1)–(BR. 3) are assumed. Essentially, we have replaced (BR. 2) with (9.24) and have dropped (BR. 3).

selves. Obviously, the included variables are more highly correlated with themselves than with the proposed instrument. Further, correlation with the disturbance will be maintained if the proposed instrument is used. To the extent that the proposed instrument is caused by variables unrelated to the included variables, correlation with the disturbances will go down, but so also will correlation with the included variables.

The situation is similar if the proposed instrumental variable and one or more of the included ones are caused in part by a third variable. In this case, it is obviously more efficient to use that third variable itself as an instrument, and, if this is done, no further advantage attaches to the use of the proposed instrumental variable in addition. (The only exception to this occurs if data on the jointly causing variable are not available. In such a case, the proposed instrument could be used to advantage.)

In general, then, an instrumental variable should be known to cause the included variables in the equation, at least indirectly. The closer such a causal connection is the better. As can easily be seen from our discussion of block-recursive systems, however, the closer is that connection in many cases, the greater the danger of inconsistency through high correlations with the relevant disturbances. In such systems, for example, current endogenous variables in low-numbered sectors directly cause current endogenous variables in high-numbered sectors, while the same endogenous variables lagged are likely to be safer in terms of inconsistency but are also likely to be more remote causes.[23] The value of the result derived at the end of the last section is that it provides a case in which one set of instrumental variables is likely to dominate another set on both criteria.

Available Instruments and Multicollinearity

There is obviously one set of variables that has optimal properties on several counts. These are the exogenous variables explicitly included in the model. Such variables are (by assumption) uncorrelated in the probability limit with the disturbances and are also in close causal connection to the current variables in any equation; indeed, they *are* some of those variables in some cases.[24] In the happy event that such exogenous variables are adequate in number, that their variance – covariance matrix is nonsingular, and that no lagged endogenous variables appear, there is no need to seek further for instrumental variables to use.

Unfortunately, this is unlikely to be the case in an economy-wide econometric model. Such models tend to be almost self-contained with relatively

---

[23] On causation in general and in decomposable systems (or our block-recursive systems) in particular, see Simon [13].

[24] They may not cause all such variables even indirectly if the dynamic system is decomposable. Such cases are automatically treated in the rules given below.

few truly exogenous variables entering at relatively few places. This is especially the case if government policies obey regular rules, follow signals from the economy and are therefore partly endogenous for purposes of estimation.[25] In estimating any equation, all variables not used as instruments (except the variable explained by the equation) must be replaced by a linear combination of instruments and the dependent variable regressed on such linear combinations. If the second stage of this procedure is not to involve inversion of a singular matrix, then (counting instrumental variables appearing in the equation) there must be at least as many instruments used as there are parameters to be estimated. Further, the linear combinations employed must not be perfectly correlated. Current exogenous variables are simply not generally sufficient to meet this requirement in economy-wide models. Moreover, they do not cause lagged endogenous variables that are likely to be present in a dynamic system.

Clearly, however, if the system is dynamic, it will be possible to use *lagged* exogenous variables as well as current ones. Such use may be especially helpful if lagged endogenous variables are to be treated as endogenous and replaced by linear combinations of instruments that can be taken as causing them in part. Indeed, if lagged endogenous variables *are* to be taken as endogenous, then exclusive use of current exogenous variables as instruments will not satisfy the causal criterion for instrumental variables already discussed. Since we have already seen that lagged endogenous variables should be used as instruments only with caution, it follows that lagged exogenous variables may well provide a welcome addition to the collection of available instruments.

Unfortunately, this also is unlikely to suffice. While it is true that one can always secure a sufficient number of instruments by using exogenous variables with larger and larger lags, such a procedure runs into several difficulties. In the first place, since rather long lags may be required, there may be a serious curtailment of available observations at the beginning of the time period to be used. Second, exogenous variables in the relatively distant past will be relatively indirect causes of even the lagged endogenous variables appearing in the equation to be estimated; it follows that their use will fail the causal criteria given, and that it may be better to accept some inconsistency by using endogenous variables with lower lags. Finally, after going only a few periods back, the chances are high in practice that adding an exogenous variable with a still higher lag adds a variable that is very highly correlated with the instruments already included and therefore adds little independent causal information.[26] While the use of lagged exogenous vari-

---

[25] This is to be sharply distinguished from the question of whether governmentally controlled variables can be used as *policy* as opposed to estimation instruments.

[26] This is especially likely if the exogenous variables are ones such as population which are mainly trends.

ables is therefore highly desirable, it may not be of sufficient practical help to allow the search for instrumental variables to end.

Whatever collection of current exogenous, lagged exogenous, and (none, some, or all) lagged endogenous variables are used, however, the multicollinearity difficulty just encountered tends to arise. Some method must be found for dealing with it.

One set of interesting suggestions in this area has been provided by Kloek and Mennes [11]. Essentially, they propose using principal component analysis in various ways on the set of eligible instruments in order to secure orthogonal linear combinations. The endogenous variables are then replaced by their regressions on these linear combinations (possibly together with the eligible instruments actually appearing in the equation to be estimated), and the dependent variable regressed on these surrogates and the instruments appearing in the equation. Variants of this proposal are also examined.

This suggestion has the clear merit of avoiding multicollinearity, as it is designed to do. However, it may eliminate such multicollinearity in an undesirable way. If multicollinearity is present in a regression equation, at least one of the variables therein is adding little causal information to that already contained in the other variables. In replacing a given endogenous variable with its regression on a set of instruments, therefore, the prime reason for avoiding multicollinearity is that the addition of an instrument that is collinear with the included ones adds little causal information while using up a degree of freedom. The elimination of such multicollinearity should thus proceed in such a way as to conserve causal information. The Kloek–Mennes proposals may result in orthogonal combinations of instruments that are not particularly closely causally related to the included endogenous variables. Thus, such proposals may well be inferior to a procedure that eliminates multicollinearity by eliminating instruments that contribute relatively little to the causal explanation of the endogenous variable to be replaced.[27] Clearly, this may involve using different sets of instruments in the replacement of different endogenous variables. Proposals along these lines are given below.

Rules for the Use of Eligible Instrumental Variables

We have pointed out several times that the causal criterion and that of no correlation with the given disturbance may be inconsistent, and that one may only be able to satisfy one more closely by sacrificing the other to a greater extent. In principle, a fully satisfactory treatment of the use of instrumental variables in economy-wide models would involve a full-scale

---

[27] This seems to have been one of the outcomes of experimentation with different forms of principal component analysis in practice. See Taylor [16].

Bayesian analysis of the losses and gains from any particular action. Such an analysis is clearly beyond the scope of the present chapter, although any recommended procedure clearly has some judgment of probable losses behind it, however vague such judgment may be.

We shall proceed by assuming that the no-correlation criterion has been used to secure a set of eligible instrumental variables whose use is judged to involve only tolerable inconsistencies in the estimation of a given equation. Note that the set may be different for different equations. Within that set are current and lagged exogenous variables and lagged endogenous variables sufficiently far in the past that the effects of serial correlation are judged to be negligible over the time period involved. As shown in the preceding section, the time period will generally be shorter for endogenous variables in sectors lower-numbered than that in which the equation to be estimated appears than for endogenous variables in the same or higher-numbered sectors.[28] Clearly, other things being equal, the use of current and lagged exogenous variables is preferable to the use of lagged and endogenous variables and the use of lagged endogenous variables from lower-numbered sectors is preferable to the use of endogenous variables with the same (or possibly even a slightly greater) lag from the same or higher-numbered sectors than that in which the equation to be estimated occurs. We shall suggest ways of modifying the use of the causal criterion to take account of this. For convenience, we shall refer to all the eligible instrumental variables as "predetermined" and to all other variables as "endogenous."

We shall assume that each endogenous variable is associated with a particular structural equation (either in current or lagged form) in which it appears on the left-hand side with a coefficient of unity. This is not an unreasonable assumption, as such normalization rules are generally present in model building, each variable of the model being naturally associated with that particular endogenous variable that is determined by the decision makers whose behavior is represented by the equation. The normalization rules are in a real sense part of the specification of the model, and the model is not completely specified unless every endogenous variable appears (at least implicitly) in exactly one equation in normalized form. For example, it is not enough to have price equating supply and demand, equations should also be present that explain price quotations by sellers and buyers and that describe the equilibrating process. (For most purposes, of course, such additional equations can remain in the back of the model builders' mind; however, the rules for choosing instrumental variables about to be discussed may require that they be made explicit.) For another example, it is always clear which equation is the "consumption function," even though consumption is one of a set of jointly determined variables whose values are

---

[28] It will not have escaped the reader's notice that very little guidance has been given as to the determination of the absolute magnitude of that time period.

determined by the system as a whole.[29] The fact that one may not be certain as to the proper normalization rules to use in a given case should be taken as a statement of uncertainty as to proper specification and not as a statement that no proper normalization rules exist.

We are now ready to discuss the procedure recommended for selecting instrumental variables from the eligible list. Consider any particular endogenous variable in the equation to be estimated, other than the one explained by that equation. That right-hand endogenous variable will be termed of *zero causal order*. Consider the *structural* equation (either in its original form or with all variables lagged) that explains that variable. The variables other than the explained one appearing therein will be called of *first causal order*. Next, consider the structural equations explaining the *first causal order* endogenous variables.[30] All variables appearing in those equations will be called of *second causal order* with the exception of the *zero causal order* variable and those endogenous variables of first causal order the equations for which have already been considered. Note that a given predetermined variable may be of more than one causal order. Take now those structural equations explaining endogenous variables of second causal order. All variables appearing in such equations will be called of *third causal order* except for the *endogenous* ones of lower causal order, and so forth. (Any predetermined variables never reached in this procedure are dropped from the eligible set while dealing with the given zero causal order variable.)

The result of this procedure is to use the *a priori* structural information available to subdivide the set of predetermined variables according to closeness of causal relation to a given endogenous variable in the equation to be estimated. Thus, predetermined variables of first causal order are known to cause that endogenous variable directly; predetermined variables of second causal order are known directly to cause other variables that directly cause the given endogenous variable, and so forth. Note again that a given predetermined variable can be of more than one causal order, so that the subdivision need not result in disjunct sets of predetermined variables.

We now provide a complete ordering of the predetermined variables relative to the given endogenous variable of zero causal order.[31] Let p be the

---

[29] This argument is, of course, closely related to that of Strotz and Wold [14] in which simultaneity is the approximate or equilibrium version of a system with very small time lags. For an illuminating discussion that bears directly on our discussion, see Basmann [5] and [6] and cf. Strotz and Wold [15]. The issue of whether normalization rules are in fact given in practice is of some consequence in the choice of two-stage least squares or limited-information, maximum-likelihood. See Chow [7].

[30] Observe that endogenous variables appearing in the equation to be estimated other than the particular one with which we begin may be of positive causal order. This includes the endogenous variable to be explained by the equation to be estimated.

[31] I am indebted to J. C. G. Boot for aid in the construction of the following formal description.

largest number of different causal orders to which any predetermined variable belongs. To each predetermined variable, we assign a p-component vector. The first component of that vector is the lowest numbered causal order to which the given predetermined variable belongs; the second component is the next lowest causal order to which it belongs, and so forth. Vectors corresponding to variables belonging to less than p different causal orders have infinity in the unused places. Thus, for example, if $p = 5$, a predetermined variable of first, second, and eighth causal order will be assigned the vector: $(1, 2, 8, \infty, \infty)$. The vectors are now ordered lexicographically. That is, any vector, say f, is assigned a number, $\beta(f)$, so that for any two vectors, say f and h

$$\beta(f) > \beta(h)$$

if and only if either $f_1 > h_1$ or for some   (9.32)

$$j(1 < j \leqq p)f_i = h_i \ (i = 1, \ldots, j-1) \text{ and } f_j > h_j$$

The predetermined variables are then ordered in ascending order of their corresponding β-numbers. This will be called the β-ordering.

Thus, predetermined variables of first causal order are assigned lower numbers than predetermined variables of only higher causal orders; predetermined variables of first and second causal order are assigned lower numbers than predetermined variables of first and only causal orders higher than second (or of no higher causal order), and so forth.[32]

The procedure just described gives an *a priori* preference order on the set of instrumental variables relative to a given zero causal order endogenous variable. This order is in terms of closeness to causal relation. Alternatively, one may wish to modify that order to take further account of the danger of inconsistency. This may be done by deciding that current and lagged exogenous variables of a given causal order are always to be preferred to lagged endogenous variables of no lower causal order, and that lagged endogenous variables from sectors with lower numbers than that of the equation to be estimated are always to be preferred to endogenous variables with the same lag and causal order from the same or higher-numbered sectors. One might even go further and decide that *all* current and lagged exogenous variables of finite causal order are to be preferred to *any* lagged endogenous variable.

---

[32] This is only one way of constructing such an order. If there is specific *a priori* reason to believe that a given instrument is important in influencing the variable to be replaced (for example, if it is known to enter in several different ways with big coefficients) then it should be given a low number. In the absence of such specific information, the order given in the text seems a natural way of organizing the structural information.

However the preference ordering is decided upon, its existence allows us to use *a posteriori* information to choose a set of instruments for the zero causal order endogenous variable in the way about to be described. Once that set has been chosen, that endogenous variable is replaced by its regression on the instruments in the set and the equation in question estimated by least-squares regression of the left-hand endogenous variable on the resulting right-hand variables.[33]

We use *a posteriori* information in combination with the *a priori* preference ordering in the following manner. Suppose that there are T observations in the sample. Regress the zero causal order endogenous variable on the first $T - 2$ instruments in the preference ordering (a regression with one degree of freedom). Now drop the least preferred of these instruments from the regression. Observe whether the multiple correlation of the regression drops significantly as a result. (The standard here may be the significance level of R^2 or simply its value corrected for degrees of freedom.) If correlation does drop significantly, then the $T - 2$nd instrument contributes significantly to the causation of the zero-order endogenous variable even in the presence of all instruments that are *a priori* more closely related to that variable than it is. It should therefore be retained. If correlation does not drop significantly, then the variable in question adds nothing and should be omitted.

Now proceed to the $T - 3$rd instrument. If the $T - 2$nd instrument was retained at the previous step, reintroduce it; if not, leave it out. Observe whether omitting the $T - 3$rd instrument reduces the multiple correlation significantly. If so, retain it, if not, omit it and proceed to the next lower-numbered instrument.

Continue in this way. At every step, a given instrument is tested to see whether it contributes significantly to multiple correlation in the presence of all instruments that are *a priori* preferred to it and all other instruments that have already passed the test. When all instruments have been so tested, the ones remaining are the ones to be used.

Discussion of the Rules

The point of this procedure (or the variants described below) is to replace the right-hand endogenous variables in the equation to be estimated by their regression-calculated values, using instruments that satisfy the causal criterion as well as possible while keeping inconsistency at a tolerable level. Certain features require discussion.

In the first place, multicollinearity at this stage of the proceedings is automatically taken care of in a way consistent with the causal criterion. If some set of instruments is highly collinear, then that member of the set that

---

[33] An important modification of this procedure is described below.

is least preferred on *a priori* grounds will fail to reduce correlation significantly when it is tested as just described. It will then be omitted, and the procedure guarantees that it will be the *least* preferred member of the set that is so treated. If the β-ordering is used, this will be the one most distantly structurally related to the endogenous variable to be replaced. Multicollinearity will be tolerated where it should be, namely, where despite its presence each instrument in the collinear set adds significant causal information.

Second, it is evident that the procedure described has the property that no variable will be omitted simply because it is highly correlated with other variables already dropped. If two variables add significantly to correlation when both are present but fail to add anything when introduced separately, then the first one to be tested will not be dropped from the regression, as omitting it in the presence of the other instrument will significantly reduce correlation.[34] While it is true that variables may be dropped because of correlation with variables less preferred than the $T - 2$nd, which are never tested, the exclusion of the latter variables seems to be a relatively weak reliance on *a priori* information.

This brings us to the next point. Clearly, it is possible in principle that instruments less preferred than the $T - 2$nd would in fact pass the correlation test described if that test were performed after some lower-numbered instruments were tested and dropped. Similarly, an instrument dropped at an early stage might pass the test in the absence of variables *later* dropped because of the increased number of degrees of freedom. One could, of course, repeat the entire procedure in order to test every previously dropped variable after each decision to omit; it seems preferable, however, to rely on the *a priori* preference ordering in practice and to insist that instruments that come late in the β-ordering pass a more stringent empirical test than those that come early. The rationale behind the β-ordering is the belief that it is the earlier instruments in that order that contribute most of the causal information, so that it seems quite appropriate to calculate the degrees of freedom for testing a given instrument by subtracting the number of its place in the order from the total number of observations (and allowing for the constant term).[35]

Turning to another issue, it may be objected that there is no guarantee that the suggested procedures will result in a nonsingular moment matrix to be inverted at the last stage. That is, there may be some set of endogenous

[34] This property was missing in the procedure suggested in an earlier draft of this chapter in which variables were added in ascending order of preference and retained if they added significantly to correlation. I am indebted to Albert Ando for helpful discussions on this point.

[35] Admittedly, this argument loses some of its force when applied to the modifications of the β-order given above.

variables to be replaced whose regressions together involve less than r predetermined variables. Alternatively, counting the instruments included in the equation to be estimated, there may not be so many instruments used in the final stage as there are parameters to be estimated. This can happen, of course, although it is perhaps relatively unlikely. If it does occur, then it is a sign that the equation in question is unidentifiable from the sample available, and that the causal information contained in the sample is insufficient to allow estimation of the equation without relaxing the inconsistency requirements. To put it another way, it can be argued that to rectify this situation by the introduction in the first-stage regressions of variables failing the causal test as described is an *ad hoc* device which adds no causal information. While such variables may in fact appear in such regressions with non-zero coefficients in the probability limit, their use in the sample adds nothing to the quality of the estimates save the ability to secure numbers and disguise the problem.

Of course, such an argument is a bit too strong. Whether a variable adds significantly to correlation is a question of what one means by significance. The problem is thus a continuous rather than a discrete one and should be treated as such. For the criterion of significance used, in some sense, the equation in question cannot be estimated from the sample in the circumstance described; it may be estimable with a less stringent significance criterion. In practice, if the significance requirements are relaxed, the moment matrix to be inverted will pass from singularity to near-singularity, and estimated asymptotic standard errors will be large rather than infinite. The general point is that if multicollinearity cannot be sufficiently eliminated using causal information, little is to be gained by eliminating it by introducing more or less irrelevant variables.

A somewhat related point is that the use of different variables as instruments in the regressions for different endogenous variables in the same equation may result in a situation in which the longest lag involved in one such regression is greater than that involved in others. If data are available only from an initial date, this means that using the regressions, as estimated, involves eliminating some observations at the beginning of the period that would be retained if the longest-lagged instrument were dropped. In this case, some balance must be struck between the gain in efficiency from extra observations and the loss from disregarding causal information if the lagged instrument in question is dropped. It is hard to give a precise guide as to how this should be done. (My personal preference would be for retaining the instrument in most cases.) Such circumstances will fortunately be relatively infrequent as the period of data collection generally begin further than those of estimation, at least in models of developed economies. Further, the reduction in available observations attendant on the use of an instrument with a large lag renders it unlikely that the introduction of that instrument adds significantly to correlation.

Finally, the use of different instruments in the regressions replacing different endogenous variables in the equation to be estimated reintroduces the problem of inconsistency. When the equation to be estimated is rewritten with calculated values replacing some or all of the variables, the residual term includes not only the original structural disturbance but also a linear combination of the residuals from the regression equations used in such replacement. When the equation is then estimated by regressing the left-hand variable on the calculated right-hand ones and the instruments explicitly appearing, consistency requires not only zero correlation in the probability limit between the original disturbance and all the variables used in the final regression but also zero correlation in the probability limit between the residuals from the earlier-stage regression equations and all such variables. If the same set of instruments is used when replacing every right-hand endogenous variable, and if that set includes the instruments explicitly in the equation, the latter requirement presents no problem since the normal equations or ordinary least squares imply that such correlations are zero even in the sample.[36] When different instruments are used in the replacement of different variables, however, or when the instruments so used do not include those explicitly in the equation, the danger of inconsistency from this source does arise.

There are several ways of handling this without sacrificing the major benefits of our procedures. One way is simply to argue that these procedures are designed to include in the regression for any right-hand endogenous variable any instrument that is correlated with the residuals from that regression, computed without that instrument. The excluded instruments are either those that are known *a priori* not to be direct or indirect causes of the variable to be replaced, or those that fail to add significantly to the correlation of the regression in question. The former instruments are known *a priori* not to appear in equations explaining the variable to be replaced and cannot be correlated in the probability limit with the residual from the regression unless both they and the replaced variable are affected by some third variable not included in that regression.[37] Such a third variable cannot be endogenous, however, since in that case the excluded instruments in question would also be endogenous; moreover, our procedure is designed to include explicitly any instrument significantly affecting the variable to be replaced. Any such third variable must therefore be omitted from the model, and it *may* not be stretching things too far to disregard correlations between residuals and excluded instruments stemming from such a source.

As for instruments that are indirectly causally related to the endoge-

---

[36] This is the case when the reduced form equations are used, for example, as in the classic version of two-stage least squares.

[37] If they were nonnegligibly caused by the replaced variable itself, they would be endogenous, contrary to assumption.

nous variable involved but that fail to add significantly to the correlation of the regression in question, these cannot be significantly correlated with the *sample* residual from that regression. One can therefore argue that the evidence is against their being significantly correlated with that residual in the probability limit.

Such an argument can clearly be pushed too far, however. If there are strong *a priori* reasons to believe that the excluded instruments should be included in view of the causal structure of the model, one may not want to reject correlation in the probability limit because multicollinearity (for the long continuance of which there may be no structural reason) leads to insignificant correlation in the sample. A modified course of action then is to include in the regression for any replaced variable any instrument that one believes *a priori* to be important in that regression *and* that appears either in the equation to be estimated or in the regression for any other replaced variable, as computed by the procedures described above.[38] Clearly, not much is lost by doing this since the added variables will not contribute much to the equation in the sample.

Alternatively, one may go the whole way toward guarding against inconsistency from the source under discussion and include in the regression for any replaced variable all instruments that appear in the equation to be estimated or in the regression for any other replaced variable, as computed by the described procedures, whether or not such instrument is thought *a priori* to be important in explaining the replaced variable. This alternative clearly eliminates the danger under discussion. It may, however, reintroduce multicollinearity and may involve a serious departure from the causal criterion if *a priori* noncausal instruments are thus included. Nevertheless, it does retain the merit that every instrumental variable used is either explicitly included in the equation or contributes significantly to the causal explanation of at least one variable so included. In practice, there may not be a great deal of difference between these alternatives, and the last one described may then be optimal (unless it is unavailable because of the degrees of freedom required).

Whatever variant of our procedures is thought best in practice, they all have the merit of using information on the dynamic and causal structure of the model in securing estimates. The use of such information in some way is vital in the estimation of economy-wide econometric models where the ideal conditions for which most estimators are designed are unlikely to be encountered in practice.[39]

---

[38] Omitting instruments that do *not* so appear does not cause inconsistency.

[39] The use of the causal structure of the model itself to choose instrumental variables as described in the text is closely akin to the methods used by Barger and Klein [4] to estimate a system with triangular matrix of coefficients of current endogenous variables.

ACKNOWLEDGMENTS

This chapter, which forms part of a longer contribution to *The Brookings-SSRC Quarterly Econometric Model of the United States*, J. Duesenberry, G. Fromm, E. Kuh, and L. R. Klein, eds., published by North-Holland Publishing Co., 1965, a volume describing the Social Science Research Council model of the United States economy, was largely written during my tenure of a National Science Foundation Postdoctoral Fellowship at the Econometric Institute of the Netherlands School of Economics. I am indebted to T. J. Rothenberg for helpful conversations and to L. R. Klein and E. Kuh for criticism of earlier drafts, but I retain responsibility for any errors.

REFERENCES

[1] Anderson, T. W. "On Asymptotic Distributions of Estimates of Parameters of Stochastic Difference Equations." *Annals of Mathematical Statistics* 30 (1959): 676–687.

[2] Ando, A., and Fisher, F. M. "Near-Decomposability, Partition and Aggregation, and the Relevance of Stability Discussions." *International Economic Review* 4 (1963): 53–67. Reprinted as Chapter 3 of [3].

[3] Ando, A., Fisher, F. M., and Simon, H. A. *Essays on the Structure of Social Science Models.* Cambridge, Massachusetts: M.I.T. Press, 1963.

[4] Barger, H., and Klein, L. R. "A Quarterly Model for the United States Economy." *Journal of the American Statistical Association* 49 (1954): 413–437.

[5] Basmann, R. L. "The Causal Interpretation of Non-Triangular Systems of Economic Relations." *Econometrica* 31 (1963): 439–448.

[6] Basmann, R. L. "On the Causal Interpretation of Non-Triangular Systems of Economic Relations: A Rejoinder." *Econometrica* 31 (1963): 451–453.

[7] Chow, G. C. "A Comparison of Alternative Estimators for Simultaneous Equations." IBM Research Report, RC-781, 1962.

[8] Fisher, F. M. "On the Cost of Approximate Specification in Simultaneous Equation Estimation." *Econometrica* 29 (1961): 139–170. Reprinted as Chapter 2 of [3].

[9] Harrod, R. F. *Towards a Dynamic Economics.* London: Macmillan, 1948.

[10] Hicks, J. R. *A Contribution to the Theory of the Trade Cycle.* Oxford: Clarendon, 1950.

[11] Kloek, T., and Mennes, L. B. M. "Simultaneous Equation Estimation Based on Principal Components of Predetermined Variables." *Econometrica* 28 (1960): 45–61.

[12] Mann, H. B., and Wald, A. "On the Statistical Treatment of Linear Stochastic Difference Equations." *Econometrica* 11 (1943): 173–220.

[13] Simon, H. A. "Causal Ordering and Identifiability." In *Studies in Econometric Method,* edited by Wm. C. Hood and T. C. Koopmans. Cowles Commission Monograph 14. New York: John Wiley, 1953. Reprinted as Chapter 1 of H. A. Simon, *Models of Man,* New York: John Wiley, 1957, and as Chapter 1 of [3].

[14] Strotz, R. H., and Wold, H. "Recursive vs. Nonrecursive Systems: An Attempt at Synthesis." *Econometrica* 28 (1960): 417–427.

[15] Strotz, R. H., and Wold, H. "The Causal Interpretability of Structural Parameters: A Reply." *Econometrica* 31 (1963): 449–450.

[16] Taylor, L. D. "The Principal-Component-Instrumental-Variable Approach to

the Estimation of Systems of Simultaneous Equations." Mimeographed and unpublished paper, 1963, and Ph.D. thesis by same title, Harvard University, 1962.

[17] Wold, H. "Ends and Means in Econometric Model Building." In *Probability and Statistics,* edited by U. Grenander, The Harald Cramér Volume. New York: John Wiley, 1960, 354–434.

[18] Wold, H., and Faxér, P. "On the Specification Error in Regression Analysis." *Annals of Mathematical Statistics* 28 (1957): 265–267.

[19] Wold, H., and Juréen, L. *Demand Analysis.* New York: John Wiley, 1953.

Two Theorems on *Ceteris Paribus* in the Analysis of Dynamic Systems*

Franklin M. Fisher

Albert Ando

10

Analysis of the dynamic properties of two or more interrelated systems is a recurrent problem in social science theory. A particular problem frequently arises (although it often goes unrecognized) in assessing the validity of an analysis of a system, some variables of which are causally related to other variables, which latter, in turn, are either not explicitly taken into account or are assumed constant. Examples are easy to find: Economists may study the behavior of a single country's economy with only secondary regard for the rest of the world; studies of group behavior may pay only secondary attention to the other roles played by the group members in other contexts; two more examples are worked out below, and others may be found in the works about to be cited. Indeed, in a larger sense, the division of social science itself (or of natural science, for that matter) into separate disciplines is an example, for the variables taken as given by one discipline are the very subject matter of another and *vice versa*. In all these examples, the very real problem is present that if variables taken as given are causally affected by the variables of the system being analyzed, or if variables assumed not to affect that

* Reprinted by permission of the authors and publisher from the *American Political Science Review* 108–113. Copyright 1962, The American Political Science Association.

system actually do affect it, the results of the analysis may have little relevance for the study of real problems.

The principal purpose of this chapter is to call the attention of social scientists outside the field of mathematical economics to two related theorems in this area that have recently been proved.[1] We shall not attempt a technical discussion of the theorems here but instead give a general description of them in the following text. We then present two illustrative, albeit somewhat simplified, examples of the sort of results that these theorems can be used to obtain in political science.

Suppose a number of variables such that at any time, t, the value of each of them is a function of the values of some or all of the same set of variables at some past time $t - 1$.[2] Suppose further that it is possible to collect the variables into subsets such that the variables within each subset are functions only of the past values of variables in the *same* subset but not of the past values of any variables in any *different* subset.[3] Then the system is said to be *completely decomposable,* and it is obvious that it really consists of several independent systems each one of which can be analyzed separately without reference to any of the others. In the examples mentioned above, this would be the case if every country's economy were really closed so that there were no intercountry effects and if every member of a group had no other roles to play outside the group or if those other roles had no effect whatsoever on actions within the group.

While the assumption of complete decomposability (or *ceteris paribus*) is often convenient to make, it is seldom likely to be fully satisfied in practice. Thus, there *are* some exports and imports affecting every country; there *are* outside roles for group members, and so forth. The question thus naturally arises: What remains of the results of an analysis carried out under an assumption of complete decomposability if the system being studied is in fact embedded in a larger system that is not truly completely decomposable,

---

[1] Simon and Ando [8] and Ando and Fisher [2] in detail present and prove the theorems. The relation of the sort of system involved to ordinary notions of causation is discussed in Simon [7], while the bearing of this sort of problem on the estimation of parameters in one of a set of interrelated systems is covered in Fisher [3].

[2] If values in the further past are relevant, a formal redefinition can always be made to eliminate time before $t - 1$ without changing anything. Simultaneous dependencies can also be treated, though, for reasons of clarity and intuitive appeal, we restrict ourselves to cases where there is some time lag somewhere in the system. Further, we could consider time as continuous without essential changes. Strictly speaking, however, the theorems under discussion are only known to hold for cases in which the relations involved are linear. Since this was written, the principal results of the theorems have been shown to hold for at least one important class of nonlinear systems in Fisher [5].

[3] Hereafter, to simplify terminology, by the term *set* instead of *subset*, we mean a subset within the set of all variables in the system under consideration.

so that variables assumed to be causally irrelevant do in fact matter? Of course, if the assumption of irrelevance — of complete decomposability — is a very bad one, it is unreasonable to expect much to be left of results obtained under it; but what if variables assumed irrelevant do not in fact matter very much? Is there any sense in which complete decomposability and the great simplification that it permits can be used as a good approximation, so that results obtained under it can be expected to be approximately valid when it is relaxed? This is where the first of our two theorems comes in.

Suppose that our system, instead of being completely decomposable, is in fact only *nearly* completely decomposable. In other words, suppose that the variables within each set do depend on the past values of variables outside that set but that this dependence is small relative to within-set dependencies. In such a case, the Simon–Ando theorem [8] asserts the following: Carry out the analysis of the system on the assumption that it is really completely decomposable (i.e., ignore interset dependencies altogether). Provided that interset dependencies are sufficiently weak relative to intraset ones, *in the short run* that analysis will remain approximately valid in all respects — that is, the system will behave almost as if it *were* completely decomposable.[4]

Now, if this were all, it would be useful but not very remarkable, for it would merely mean that if neglected influences are sufficiently weak, they take a long time to matter much; however, the theorem does not stop here but asserts a far stronger result.

Consider the long-run behavior that the system would exhibit if it were truly completely decomposable. In particular, consider the *relative* behavior of the variables within each set (i.e., consider the long-run behavior of their ratios). Provided again that interset dependencies are sufficiently weak relative to intraset ones, that same relative behavior will show up approximately in the long-run behavior of the true only *nearly* completely decomposable system. It is important to fully grasp this result. It asserts that even when influences that have been neglected have had time to make themselves fully felt, the *relative* behavior of the variables within any set will be approximately the same as would have been the case had those influences never existed, and this despite the fact that the *absolute* behavior of the variables — their levels and rates of change — may be very different indeed. To recapitulate: If a nearly completely decomposable system is analyzed as

---

[4] Here and later what is meant by "sufficiently weak" depends on what standard of approximation one wishes to impose on the results. The more closely one insists that the behavior of the system must approximate that of the corresponding completely decomposable one, the weaker must "sufficiently weak" be. What the theorem guarantees is that *whatever* standard of approximation is required (so long as it is an approximate and not an exact standard), a non-zero degree of weakness always exists that is sufficient to produce results satisfying that standard.

though it were really completely decomposable, the results obtained will remain approximately valid even in the long run as regards the *relative* behavior of the variables within any one set.[5]

The detailed examples of the next section will make all this clearer; however, let us pause briefly over the examples already cited. In one variant of the international trade case, the Simon–Ando theorem asserts that if a country trades very little with the rest of the world, its short-run production of every commodity will be approximately the same as if it did not trade at all. Further, in the long run, the *ratio* in which it produces any two commodities will also be approximately the same, although the *levels* and *rates of growth* of production may be very different in the two cases. Similarly, in the group-behavior case, the Simon–Ando theorem asserts that, if outside roles for group members are relatively unimportant, the short-run behavior of the group will be approximately the same as if those roles did not exist, while in the long run, the *relative* behavior of the group members *vis-à-vis* each other will also be approximately the same, even though the behavior of the group *as a whole*[6] *vis-à-vis* the rest of the world may be quite different.

Unfortunately, however, useful as completely decomposable or nearly completely decomposable systems are, the assumptions inherent in near complete decomposability are grossly unrealistic for many systems that are of great interest to social scientists. We often study systems that we do not believe, even as an approximation, to be causally uninfluenced by outside forces. In many cases, we deal with systems whose variables are believed to be influenced by outside forces but not to influence these outside variables, so that the latter may be taken as givens of the problem. This is the case of causal influences running only one way among sets of variables (although there may be "whorls" of causation within each set). Thus, the economist frequently takes tastes or technology as causally influencing the variables he studies but as being causally uninfluenced by them. So, too, a sociologist may take the existing means of production as given independently of the variables he studies but as influencing the latter in a significant way.

More formally, this kind of assumption is equivalent to saying that our variables can be collected into sets *numbered* (from 1 to N, say) so that the

---

[5] Even before the full system settles down to its ultimate behavior, variables within any one set will move proportionally, so that interset influences can be analyzed as influences among indices, each index representing all variables in a particular subset of the system, rather than as among the individual variables themselves. This point, while a little aside from the main drift of our discussion in the text, is useful and important and will show up in our examples below. It is true of both theorems.

[6] This phrase is not used accidentally. The influence of the outside roles will eventually be on group behavior as a whole rather than on individual behavior within the group (remember that such influences are small relative to within-group forces). See footnote 5.

variables in any given set are functions of their own past values *and* of the past values of the variables in any lower-numbered set but not of the past values of the variables in any higher-numbered one. This sort of system is called *decomposable*. (The reader should go back to the definition of *complete* decomposability to be sure he understands the difference.) Clearly, the dynamic behavior of any set of variables in a decomposable system can be studied taking the behavior of variables in lower-numbered sets as given and without any regard for higher-numbered sets. (Note, however, that the influence of lower-numbered sets *must* be explicitly recognized.)

Again, as in the case of complete decomposability, however, the assumption of decomposability is frequently justified only as a working approximation. Thus, tastes and technology are not *really* independent of the workings of the economic system; the means of production are not *really* given independently of sociological factors.[7] The question then arises as before of the validity of results obtained under such assumptions. Indeed, this question is perhaps even more important here than in the completely decomposable case, for we have already pointed out (and the examples just cited illustrate) that the very separation of social science into the social sciences is a use of the assumption of decomposability.

Fortunately, the same things are essentially true here as in the nearly completely decomposable case. For nearly decomposable systems that are not nearly completely decomposable — in other words, for systems in which *one-way* interset influences are too large to ignore — the Ando–Fisher theorem [2] yields substantially the same results as the Simon–Ando theorem does for the nearly completely decomposable case. Thus, the economist who takes tastes and technology as influencing but uninfluenced by economic variables will find — provided that such an assumption is *nearly* correct — that his results will be approximately valid in all respects in the short run and that even in the long run, when the full effects of feedbacks in the causal structure are felt, the *internal, relative* behavior of the variables he studies will be approximately the same. The importance of this result for the usefulness of intradisciplinary studies in an interrelated world needs no emphasis.[8]

We come now to the application of these theorems to two examples in political science that we shall work out in moderate detail, although without any attempt to argue that these examples are more than illustrations.

First, let us consider an oversimplified model of armament races. Suppose that at any time, the stocks of armaments of any country (as measured

[7] Indeed, it would be odd (although not impossible) for both statements to hold exactly.

[8] This is not to deny, of course, the usefulness of interdisciplinary work. The point is that the results of intradisciplinary studies need not be vitiated because the real world is not so neatly divided as the academic one.

by the number of units of each type of weapon in its arsenal) are a function of the stocks of armaments of all countries (including itself) at some past date.[9] The influences of such past stocks may be of different kinds. Thus, armaments may be higher the higher a prospective enemy's armament stocks; lower the higher the stocks of prospective allies; effects on different weapons may be different; stocks of one type of weapon may influence stocks of others positively or negatively; and so forth.

Now suppose that the world is divided into several arms races in such a way that the armaments of a country in any one arms race depend principally on those of the other countries in the same arms race and relatively little on those of countries in other arms races. (This may not be a bad approximation; one might think of the Arabs and Israelis, on the one hand, and the East–West arms race, on the other.) The system is then nearly completely decomposable and the Simon–Ando theorem predicts the following.

From the initiation of the various arms races (since they are at least slightly interdependent they will all begin at least nominally at the same time) until some time thereafter, each arms race will proceed approximately as it would in the absence of the other races. Indeed, this will continue until after the transient effects of different starting stocks of armaments have effectively disappeared. Once this has happened, there will come a time after which the stocks of various weapons for all the countries in each arms race will grow (or shrink or fluctuate) approximately proportionally, so that the influence of any one arms race on any other may be considered as that of an aggregate "country" with one aggregate stock of weapons on another, despite the fact that the influences among the countries and weapons making up the aggregates may be quite varied. Finally, in the long run, the size and rate of growth of every country's armament stocks at any period will be determined by the rate of growth of that arms race that would grow fastest in the absence of the other arms races, but the *relative* sizes of the weapon stocks within any one arms race will remain approximately the same as if no other arms races existed.

Now consider a somewhat different situation in which the world is again divided into arms races, and the arms races are numbered such that armament stocks in any given country are functions of past stocks of other

---

[9] Again, the use of past time is not a restriction; the past date can be last year or yesterday or an hour ago. Similarly, influences in the further past can be formally subsumed under this model by redefinition. Finally, the formulation includes the case in which it is the *rate of change* rather than the *level* of the armament stocks that are dependent on past armaments. This sort of treatment of arms races originates with Richardson. See [6]. Of course, the stocks of armaments depend on other things such as economic resources; we are trying to keep the example as simple as possible. The functions in question are supposed to represent the strategic choices made by governments.

countries in the same arms race and of countries in lower-numbered arms races but only slightly of stocks of countries in higher-numbered arms races.[10] (This, too, may be somewhat realistic. On the one hand, the level of Pakistani armaments is affected by United States military aid; on the other hand, the level of United States armaments is hardly affected by Pakistani arms levels.) Then the system is nearly decomposable and the Ando–Fisher theorem yields results similar to those just presented in the nearly completely decomposable case. Perhaps it is worth remarking here that the theorem thus shows that a very rapid long-run rate of arms accumulation in a high-numbered arms race (say India–Pakistan or North and South Vietnam) can force the long-run rate of arms accumulation in a lower-numbered race (say U.S.–U.S.S.R.) to the same level but cannot more than negligibly affect the *relative* positions in the latter race.

For the second example, we turn to the voting strength of political parties. Suppose that there are several political parties and that the number of votes for any one party at a given time is a function of the number of votes for each party at some past time. For convenience, we shall assume a constant total population and that every individual votes for one and only one party. We shall further assume that the effect of past votes on present party strength can be represented for each party by a sum of terms, each term representing the probability of an individual who votes initially for a particular party changing his vote to the party in question, times the number of people voting for the initial party. Of course, individuals who stay in the same party are counted as going from that party to itself. It follows then that if we add the coefficients that give the probabilities of going from a particular party to all parties in existence, the sum must be unity, as every individual must end up somewhere.[11]

Thus, in equation form, let X_{1t} be the number of votes for the first party at time, t, X_{2t} be the number of votes for the second party at the same time, and so forth. We have (where there are n parties):

---

[10] They need not be functions of armament stocks in *all* lower-numbered races; one will do. Similarly, not every country in the given arms race need have such links; all that is required is that at least one does, so that the system, while nearly decomposable, is not nearly *completely* decomposable. It is also possible to analyze a case in which a part of a nearly decomposable system happens to be nearly completely decomposable.

[11] A discussion of this general sort of model which aggregates from the individual level is given in Anderson [1]. It would be possible to abandon the probability interpretation and to let higher strength in one party lead to lower strength in another, but this would lead to an example without some of the nice features of the present one. It would also be possible to build in other influences on party voting strength. Again, we are striving for an illuminating example rather than for a full-blown "realistic" theory. The assumption that every individual votes for some party is innocuous, since we can count all those not voting for any formal party as voting for a party of their own, the "Non-such" Party.

$$X_{1t} = A_{11}X_{1t-1} + A_{12}X_{2t-1} + \ldots + A_{1n}X_{nt-1}$$

$$X_{2t} = A_{21}X_{1t-1} + A_{22}X_{2t-1} + \ldots + A_{2n}X_{nt-1}$$

$$\cdot$$
$$\cdot \hspace{10cm} (10.1)$$
$$\cdot$$

$$X_{nt} = A_{n1}X_{1t-1} + A_{n2}X_{2t-1} + \ldots + A_{nn}X_{nt-1}$$

Here, A_{11} is the probability of an individual voting for party 1 continuing to do so; A_{12} is the probability of an individual voting for party 2 changing to party 1; A_{21} is the probability of an individual voting for party 1 changing to party 2; and so forth. Clearly, all the A's must lie between zero and one, and the sum of the A's in any column must be unity.

Now consider any set of parties (a single party can count as a set). Ignore the fact that such a set may gain votes from outside itself and consider only its own internal workings. There will be in the long run a single probability that a randomly chosen individual voting for some party in the set will continue to vote for some party in the set at the next time period (he may move between parties in the set). Equivalently, without ignoring the fact that the set may gain votes from outside, we may consider the long-run probability that one of the people *originally* voting for some party in the set who has remained so voting through some particular time will continue so to vote for yet one more period; this comes to the same thing. (In either definition, we assume that every party in the set begins with a very large number of votes so that a zero vote does not become a problem save for parties that lose all their votes every time period.) We call that long-run probability the *cohesiveness* of the set. Clearly, it must lie between zero and one; the cohesiveness of any single party is the A coefficient in its own row and column (thus, for example, A_{11} is the cohesiveness of the first party taken alone); and the cohesiveness of all parties taken together is unity (since nobody can leave the system as a whole). In general, the cohesiveness of a set lies between the smallest and the largest probability of an individual voting for any given party in the set staying put or changing to another party in the same set (this latter probability is given by the sum of the A's in the column corresponding to the party in question and in the rows corresponding to all parties in the set including the initial one).[12]

Now suppose that the parties can be divided into sets such that the probability of an individual moving between sets is small relative to the probability of his remaining in the same set. Clearly, each set then has

---

[12] *Technical footnote.* Precisely, cohesiveness is given by the largest characteristic root of the submatrix of A's corresponding to the parties in the set. See Fisher [4]. The fact that the root lies between the largest and smallest column sums is well known — see Solow [9].

cohesiveness close to unity. The system is then nearly completely decomposable, and the Simon–Ando theorem yields the following results:

Begin with any distribution of voting chosen at random. In the short run, the voting strength of each party will behave approximately as it would if there were no movements between sets of parties whatsoever. Moreover, there will come a time after which the ratio of the votes for any party to the votes for any other party in the same set will remain approximately constant. Thereafter, each set of parties can be treated as a single party for purposes of analyzing interset movements, with the cohesiveness of the set acting as the probability of remaining therein. Finally, even in the long run, the equilibrium distribution of relative voting strength *within* each set of parties will be approximately the same as if there were no interset movements, although the absolute number of votes involved may be quite different.

In this example, however, the more interesting case is that of near decomposability. Let us first consider what a truly decomposable system would be. This would occur if the parties could be divided into sets where the sets are numbered (from 1 to N, say) in such a way that nobody voting for a party in any set ever changes to a party in a lower-numbered set but such that there are nonnegligible movements into higher numbered sets. Thus, anybody in set N stays there; anybody in set $N - 1$ either stays there or moves to set N; anybody in set $N - 2$ either stays there, moves to set $N - 1$, or moves to set N; and so forth. (Note that set N has cohesiveness 1; no other single set has cohesiveness 1, but set N and set $N - 1$ together, set N, set $N - 1$, and set $N - 2$ together, and so forth all have cohesiveness 1.) The sets are thus "nested," in the sense that all interset movements go toward the center which consists of set N.

Now suppose that all sets begin with a very large number of votes (so that the fact that the lower-numbered sets will eventually lose all their votes is not a difficulty). Consider any set. A time will come such that thereafter the *relative* strength of the parties in that set is determined primarily by their tendency to attract voters from parties in the set with the highest cohesiveness among those sets that are not higher numbered than the set in question. (In the case of set N or of set 1, this is the set in question itself; in any other case, it need not be.) This is a reasonable result, because as time goes on the parties in the sets with the greatest cohesiveness will acquire more and more votes, so that, given the probabilities of any *single* voter moving among the sets, what happens to these sets becomes of increasing importance to all those sets of parties that acquire voters from them.

We now alter the assumptions and suppose that although the dominant interset movements are to higher-numbered sets, there are small movements to lower-numbered ones. (This may not be too unrealistic. We might think of set N as the major parties; there may be a far higher tendency to move into or stay in them than there is to move out.) The system is now only nearly decomposable, and the Ando–Fisher theorem yields the following results:

Begin with any distribution of voting chosen at random. In the short run, the actual number of votes for each party will behave approximately as it would if the only interset movements were to higher-numbered sets. Moreover, there will come a time after which the *relative* distribution of voting among the parties within any set will remain approximately the same as would have been the case in the truly decomposable case earlier described. After this time, interset movements can be analyzed without regard for the voting distribution within the various sets. Finally these *relative* intraset distributions will be approximately maintained, even in the long run, although the absolute number of votes eventually cast for the parties in each set will depend almost wholly on the tendency of that set as a whole to gain voters from set N (which by this time is far larger than any other set).

Note that this result implies that an analysis of the *relative* strength of "minor" parties can proceed without explicit account of the tendency of such parties to gain voters from or even to lose voters to major ones, despite the fact that the latter tendency is large. So far as *relative* voting strength is concerned, all that matters are the various tendencies of the minor parties to lose voters to and gain voters from each other (and even some of these may be negligible). Even wide differences in tendencies to lose voters to (or gain them from) major parties need not be explicitly taken into account, provided only that the former tendencies are very strong relative to the latter, a rather surprising result.[13]

REFERENCES

[1] Anderson, T. W. "Probability Models for Analyzing Time Changes in Attitudes." In *Mathematical Thinking in the Social Sciences*, edited by P. F. Lazarsfeld. Glencoe, Ill: The Free Press, 1954.

[2] Ando, A., and Fisher, F. M. "Near-Decomposability, Partition and Aggregation, and the Relevance of Stability Discussions." *International Economic Review* 4(1963):53–67.

[3] Fisher, F. M. "On the Cost of Approximate Specification in Simultaneous Equation Estimation." *Econometrica* 29 (1961): 139–170.

[4] Fisher, F. M. "An Alternate Proof and Extension of Solow's Theorem on Nonnegative Square Matrices." *Econometrica* 30(1962).

[13] We are perhaps making this result sound a bit more paradoxical than it is. Different tendencies to lose voters to major parties do in fact matter and get taken into account in an implicit fashion. Since every party's voters end up *somewhere* (the sum of the A's in any column is 1), a minor party with a high propensity to lose voters to major ones will have — *other things being equal* — a lower propensity to retain voters than a minor party with a lower propensity to lose voters to major ones. This will show up in considering their relative strength in isolation; however, it can be nullified by other tendencies if other things are *not* equal.

[5] Fisher, F. M. "Decomposability, Near Decomposability and Balanced Price Change under Constant Returns to Scale." *Econometrica* 31(1963):67–89.

[6] Richardson, L. F. *Arms and Insecurity.* Chicago: Quadrangle Books, 1960.

[7] Simon, H. A. "Causal Ordering and Identifiability. "In Studies in Econometric Method, edited by W. C. Hood and T. C. Koopmans. New York: John Wiley & Sons, 1953. Cowles Commission Monograph No. 14. Reprinted in H. A. Simon, *Models of Man,* New York: John Wiley & Sons, 1957.

[8] Simon, H. A., and Ando, A. "Aggregation of Variables in Dynamic Systems." *Econometrica* 29 (1961):111–138.

[9] Solow, R. M. "On the Structure of Linear Models." *Econometrica* 21 (1952): 29–46.

Sensitivity Analysis of Arbitrarily Identified Simultaneous-Equation Models*

Kenneth C. Land

Marcus Felson

11

Through the work of Blalock [1] and Duncan [7], sociologists were introduced to the practice of building recursive models of interdependencies among several variables and to the use of path analysis to provide a flow graph representation of such models and to interpret their estimation. This type of formalization proved to be particularly useful in the representation of the socioeconomic life cycle (see, e.g., Blau and Duncan [2]; Duncan *et al.* [10]) as well as in many other areas of sociology. Not only is this class of models widely used in current sociological research, but most of its known statistical problems have been resolved.[1]

* Reprinted from Kenneth C. Land and Marcus Felson, "Sensitivity Analysis of Arbitrarily Identified Simultaneous-Equation Models," *Sociological Methods and Research* 6, No. 3 (February 1978), pp. 283–307. Copyright 1978 by Sage Publications, Inc. Reprinted by permission of Sage Publications, Inc. and the authors.

[1] Although there was some initial confusion over the statistically efficient way in which to estimate recursive models (Boudon [3]; Goldberger [15]), Land [30] has recently given a systematic treatment of the statistical theory for the identification, parameter estimation, and hypothesis testing of this class of models. In addition, McPherson and Huang [31] have given a small sample version of Land's likelihood ratio test for the simultaneous evaluation of all of the overidentifying restrictions on a recursive model, Specht [36] has interpreted the likelihood ratio test in terms of Wilk's generalized variance and the generalized multiple correlation coefficient, and Mayer and Younger [33] have presented unbiased estimators of the standardized regression coefficients of such models.

More recently, however, sociologists (see, e.g., Duncan *et al.* [11] 1968; Land [29]; Hauser [17,18]; Kohn and Schooler [27]; Waite and Stolzenberg [39]) have begun to use nonrecursive simultaneous-equation systems to represent social processes, and it is reasonable to anticipate the appearance of more examples of nonrecursive models as sociologists learn to formulate their theories in these terms. Statistically, nonrecursive systems have been the mainstay of econometrics for more than thirty years, due to the intrinsically nonrecursive formulation of many economic models.[2] Economic theory, particularly macroeconomic theory, often leads to nonrecursive models with a rather heavy degree of overidentification of component equations (cf. Theil [37, p. 450]), but this is not a necessary characteristic of the mathematics of such systems. In fact, a review of extant nonrecursive sociological models leads to the conclusion that the just-identification, not to say the overidentification, of the equations is achieved by a rather arbitrary assignment of zeros to certain coefficients. In brief, while nonrecursive economic models typically have a large degree of overidentification, sociologists often must cope with nonrecursive models containing underidentified equations.

The standard recipe for dealing with an underidentified model is to extend one's set of data to include more exogenous variables in such a way as to produce a just-identified or overidentified system. However, this prescription often is not an optimal solution to the problem of bringing a given set of data to bear upon a given model. For one thing, relevant exogenous variables for a given model may not be easy to find or simply may not exist. Even if they do exist, the researcher may be forced to go back into the field to collect a new set of data in order to incorporate them into his model. But this ignores the possibility of distilling some information out of already available data. Therefore, the usage of nonrecursive models in sociology leads often to a concern for the evaluation of underidentified equations.[3]

In this chapter, we apply the term *sensitivity analysis* to a class of procedures for exploring the extent to which numerical estimates of parameters from an arbitrarily identified model depend upon the identifying assumptions. Such procedures have previously been used on an *ad hoc* basis by

[2] An exposition of the standard results on the identification, parameter estimation, and hypothesis testing of such models can be found in any of the several extant econometrics textbooks (e.g., Klein [25]; Goldberger [14]; Christ [4]; Kmenta [26]; Malinvaud [32]; Theil [37]; Johnston [22]; Wonnacott and Wonnacott [42]).

[3] While we emphasize the problem of underidentification which occurs in nonrecursive simultaneous-equation models, it should be noted that recursive models in observed variables often become underidentified if one drops the assumption of stochastically independent disturbance or error terms (cf. Land [30, pp. 47–48]). Moreover, if unmeasured variables are introduced into a recursive model, then more stringent conditions must be met to achieve identification (cf. Wiley [41, pp. 77–78]).

sociologists to explore the consequences of various theoretical assumptions prior to final parameter estimation under some specific set of identifying restrictions. To give these and other related procedures an explicit recognition as sensitivity analysis techniques, we first review standard notation and identification concepts for simultaneous-equation models in the chapter's second and third sections, respectively. In the fourth section, we relate our concept to previous work and review extant methods for sensitivity analysis of underidentified models. We also suggest mathematical programming, which employs functional maximization under inequality constraints, as an operational procedure for such an examination. We next illustrate this procedure with an application to a specific underidentified sociological model in our fifth section. The chapter concludes with some suggestions on how the sensitivity analysis of underidentified models might be made more statistically rigorous in the future.

NOTATION FOR SIMULTANEOUS-EQUATION MODELS

To review identification concepts within the context of simultaneous-equation models, it is necessary first to establish some notation. Suppose that the general notational system allows for some definite number G of equations determining $G \geq 2$ endogenous variables, denoted $y_1, \ldots, y_G$, by $K \geq 1$ exogenous variables, denoted $z_1, \ldots, z_K$. We employ β's as coefficients measuring the effects of the y's on each other, the γ's as coefficients measuring the effects of the exogenous z's on the endogenous y's, and the u's are stochastic disturbances representing the effects of all other sources of variation in the determination of the y's. A constant term can, of course, be allowed by setting one of the z's identically equal to one. In any specific model, a number of the β's and γ's in each equation will, of course, be set equal to zero on *a priori* theoretical grounds, so that the corresponding variables will not appear in the determination of particular endogenous variables. Such equations can be conveniently put into the standard matrix format of a "system of simultaneous linear equations," with all of the y's and z's on the left side of the equalities as follows:

$$y_1 + \beta_{12}y_2 + \ldots + \beta_{1G}y_G + \gamma_{11}z_1 \ldots + \gamma_{1K}z_K = u_1$$

$$\beta_{21}y_1 + y_2 + \ldots + \beta_{2G}y_G + \gamma_{21}z_1 \ldots + \gamma_{2K}z_K = u_2 \qquad (11.1)$$

$$\beta_{G1}y_1 + \beta_{G2}y_2 + \ldots + y_G + \gamma_{G1}z_1 \ldots + \gamma_{GK}z_K = u_G$$

where we employ the unit diagonal normalization rule in equations 11.1. This assures that the gth equation still measures the determination of the gth variable in terms of unit changes in the measurement metric of that variable. If we now define the following vectors and matrices,

$$
B = \begin{bmatrix} 1 & \beta_{12} \cdots \beta_{1G} \\ \beta_{21} & 1 \cdots \beta_{2G} \\ & \cdot \quad \cdot \quad \cdot \\ & \cdot \quad \cdot \quad \cdot \\ & \cdot \quad \cdot \quad \cdot \\ \beta_{G1} & \beta_{G2} \cdots 1 \end{bmatrix} \qquad \Gamma = \begin{bmatrix} \gamma_{11} & \gamma_{12} \cdots \gamma_{1K} \\ \gamma_{21} & \gamma_{22} \cdots \gamma_{2K} \\ & \cdot \quad \cdot \quad \cdot \\ & \cdot \quad \cdot \quad \cdot \\ & \cdot \quad \cdot \quad \cdot \\ \gamma_{G1} & \gamma_{G2} \cdots \gamma_{GK} \end{bmatrix} \qquad (11.2)
$$

$$
y = \begin{bmatrix} y_1 \\ y_2 \\ \cdot \\ \cdot \\ \cdot \\ y_G \end{bmatrix} \qquad z = \begin{bmatrix} z_1 \\ z_2 \\ \cdot \\ \cdot \\ \cdot \\ z_K \end{bmatrix} \qquad u = \begin{bmatrix} u_1 \\ u_2 \\ \cdot \\ \cdot \\ \cdot \\ u_G \end{bmatrix}
$$

then equations 11.1 can be rewritten as

$$
By + \Gamma z = u \qquad (11.3)
$$

Subtracting Γz from both sides, we get

$$
By = -\Gamma z + u \qquad (11.4)
$$

If B is nonsingular, B^{-1} exists and we can premultiply Eq. 11.4 to obtain the solution

$$
y = -B^{-1}\Gamma z + B^{-1}u \qquad (11.5)
$$

In brief, this solves for the values of the endogenous y's solely as a function of the exogenous z's and the stochastic disturbances.

STANDARD CONCEPTS OF IDENTIFIABILITY

For nonrecursive systems of equations such as 11.1 and 11.3, the identification problem essentially revolves around whether the list of variables appearing in each equation is sufficiently unique so that no linear combination of the remaining equations of the system can produce an equation with exactly the same list of variables. More precisely, in a complete linear system of G equations, the parameters of the gth equation are *not estimable* when there exists a linear combination of the other G-1 equations that contains only the variables that occur in the gth equation; in this case, the gth equation is usually said to be *not identifiable* or *underidentified* (cf. Theil [37, pp. 448–449]).[4]

Econometricians have explored various ways in which to identify linear equation systems. The basic procedures involve (1) constraining various coefficients to be zero so that certain variables do not appear in certain equations (exclusion restrictions) and (2) constraining the variances and covariances of the disturbances of the equations. For example, the recursive models so widely used in sociology are usually identified (cf. Land [30, pp. 27–32]) by constraining all of the coefficients above the main diagonal of the matrix B in equation 11.3 to be zero (the recursive condition) and by constraining the variance–covariance matrix of the disturbances to be diagonal (the independent disturbances condition).

Using these procedures, various conditions for the identification of equations have been formulated in terms of the order and rank properties of matrices formed from the variables excluded from each equation of a model.[5] The *order condition* is usually stated as follows: A necessary, but not sufficient, condition for the identifiability of the gth equation of a G-equation linear model under exclusion restrictions is that there be at least G-1 variables (endogenous or exogenous) excluded *a priori* from that equation (Fisher [13, p. 40]). As an application of this condition, consider the following two-equation system determining two endogenous variables in terms of three exogenous variables:

$$y_1 + \beta_{12}y_2 + \gamma_{11}z_1 + \gamma_{12}z_2 + \gamma_{13}z_3 = u_1 \qquad (11.6)$$

$$\beta_{21}y_1 + y_2 + \gamma_{21}z_1 + \gamma_{22}z_2 + \gamma_{23}z_3 = u_2. \qquad (11.7)$$

According to the order condition, it is necessary to exclude at least one variable from each of these equations in order for it to be identified. Clearly, either $\beta_{12} = 0$ or $\beta_{21} = 0$ would eliminate the nonrecursivity of the model, so that one must specify zero coefficients for some of the z's if this is to be a nonrecursive system. Thus, for example, an a priori theoretical specification of $\gamma_{12} = 0$ and $\gamma_{21} = 0$ would identify each equation. Moreover, if the theoretical specification for the first equation was $\gamma_{11} = \gamma_{12} = 0$, then the choice

---

[4] More general concepts of identification were provided by Wald [40] and Hurwicz [20].

[5] The original derivation of the rank and order conditions was given by Koopmans *et al.* [28]. More general statements of these conditions, together with many other results, were given in Fisher's [13] excellent treatise.

of either $\gamma_{11} = 0$ or $\gamma_{12} = 0$ would suffice to identify it, whereas both of these conditions are more than sufficient. This possibility leads to the distinction between a just-identified and an overidentified equation (cf. Fisher [13, pp. 28–29]). In brief, an equation is said to be *just-identified* under a set of *a priori* restrictions if and only if that equation is identified under that set of restrictions but is not identified under any proper subset of the restrictions. On the other hand, an equation is said to be *overidentified* if and *only if* there exist two different, but not necessarily disjoint, linearly independent sets of *a priori* restrictions, each of which suffices to just-identify the equation.

ARBITRARY IDENTIFYING ASSUMPTIONS AND METHODS OF SENSITIVITY ANALYSIS

The essence of the identification concept is that it is a theoretical issue that must be addressed prior to statistical estimation, because numerical values of parameter estimates from a particular data set depend on the parameter space of the model to be estimated and, therefore, on the identifying restrictions utilized. Consequently, the fact that social theory often is not strong enough to provide a set of identifying restrictions, without the use of arbitrary, *ad hoc* assumptions, should be of concern to sociologists.

To address this concern, we suggest that researchers should engage in a sensitivity analysis of the extent to which numerical estimates of the parameters of an arbitrarily identified model depend upon the particular identifying assumptions used in obtaining the estimates. That is, in the absence of theory sufficiently strong to identify a model, we propose that researchers should explore systematically the extent to which the estimated values of parameters obtained from an arbitrarily identified model depend upon the identifying assumptions. Suppose, for instance, that parameter α of a given model can be identified and estimated only by arbitrarily assuming that some other parameter, β, is equal to 0. Suppose, however, that sociological theory is only strong enough to support the assumption that $\beta \geq 0$. The latter assumption, while theoretically more defensible than the former, leaves α underidentified and, therefore, not estimable in the traditional sense. On the other hand, it may be possible to show that, as β takes on various nonnegative values, the corresponding estimates of α are not greatly affected, at least within the context of the particular sample on which the estimates are based. Such a sensitivity analysis may lend some credibility to a final estimation of the model under the arbitrary identifying assumption that $\beta = 0$.

Although not always explicitly recognized as such, several methods of sensitivity analysis exist in the sociological literature. In this section, we summarize three of these methods, and we present a fourth that generalizes one of the three. None of these methods is based on a rigorous statistical decision theory. Rather, they are best viewed as empirical procedures for exploring the implications of alternative identifying assumptions for parameter estimates. At the end of the chapter, we indicate some possible avenues of research that could be utilized to make these procedures statistically rigorous.

Method 1—Estimation under Alternative Specifications

One of the most common strategies for dealing with an underidentified model is to try several alternative specifications to find out how well each one works. A good illustration of this strategy is found in Hauser's study [17] of socioeconomic background and educational performance. In particular, one of Hauser's models deals with how socioeconomic background influences mathematical and reading abilities as measured by standardized tests. Hauser's model provides a good example of a nonrecursive simultaneous-equation model which is identified by arbitrary assumptions. The path diagram of this model is given in Figure 11.1.

The representation in Figure 11.1 follows the usual conventions of path analysis for standardized variables: Population values of path coefficients along the arrows are denoted by small p's with first subscript corresponding to the dependent variable and second to the independent variable, and unanalyzed population correlations along the curved lines denoted by ρ's

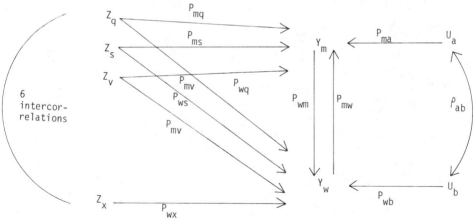

Figure 11.1. Representation of Hauser's simultaneous-equation model of reading and mathematics achievement, white public secondary school students: Nashville SMSA, 1957. Z_x = father's occupation; Z_v = father's education; Z_s = number of siblings; Y_m = Stanford mathematics grade equivalent; Y_w = Stanford reading grade equivalent; Z_q = intelligence; U_a = unmeasured determinants of Y_m; U_b = unmeasured determinants of Y_w. SOURCE: Based on Figure 5.5 in Hauser [17, p. 78].

with identifying subscripts. Observe that the diagram indicates the *a priori* constraint that father's occupation does not affect mathematical performance. Under the additional assumption that the disturbances are uncorrelated, Hauser was able to obtain estimates of the parameters of this model. That is, under the *a priori* assumptions:

$$p_{mx} = 0 \qquad (11.8)$$

$$\rho_{ab} = 0 \qquad (11.9)$$

he obtained the following estimates of the equations of this model:

$$y_m = 0.463 Y_w + 0.290 Z_q - 0.001 Z_s + 0.007 Z_v + 0.732 U_a \quad (11.10)$$

$$y_w = 0.018 Y_m + 0.528 Z_q - 0.073 Z_s + 0.064 Z_v \\ + 0.041 Z_x + 0.806 U_b \qquad (11.11)$$

Substantively, these estimates indicate that (1) most of the effects of family background variables on reading and mathematics skills are transmitted via intelligence, (2) that higher reading scores contribute to better mathematics scores, and (3) that higher mathematics scores have little feedback effect on reading scores.

In order to examine the sensitivity of these estimates to the specification in Eqs. 11.8–11.9, Hauser performed several analyses under other identifying assumptions. Specifically, to determine the impact of assumption 11.8 on the results, Hauser computed the estimates by successively setting p_{mv}, p_{ms}, and p_{mq} individually equal to zero, from which he concluded that "the numerical solution [of 11.10] is not very sensitive to [the decision about] which of the paths from the background variables to achievement in mathematics is dropped" (Hauser [17, p. 78]). Next, to ascertain whether or not assumption 11.9 is crucial to the results reproduced above, Hauser examined the alternative specification that $p_{wm} = 0$, $\rho_{ab} = 0$, with no constraint on p_{mx}, concluding that "the results [11.10–11.11] are almost identical with those we would have obtained by ordinary regression of Y_w on Z_x, Z_v, Z_x, and Z_q and of Y_m on Z_x, Z_v, Z_s, Z_q, and Y_w" (Hauser [17, p. 79]).

Hauser's analysis of this model can be interpreted as a sensitivity analysis of the identifying assumptions. Both of the *a priori* assumptions, 11.8 and 11.9, under which the model is estimated, are somewhat arbitrary. For instance, it is reasonable to argue that such unmeasured variables as hard work, test-taking ability, maturity, and interest in school would help students do well on both mathematics and reading scores. In such a case, ρ_{ab} would be, at most, nonnegative rather than zero. Similarly, sociological theory is hardly strong enough to specify that $p_{mx} = 0$, although a number of theoretical arguments might lead to the specification that $p_{mx} \geq 0$. However, if the equalities in Eqs. 11.8 and 11.9 are relaxed, then the coefficients in equations 11.10 and 11.11 are underidentified. Thus, the best that can be done is to show that the numerical estimates obtained under assumptions Eqs. 11.8 and 11.9 are insensitive to these assumptions. We interpret Hauser's experimentation with estimation under alternative specifications as constituting such a demonstration.

Method 2—Estimation with Arbitrary Values

A second method of sensitivity analysis, for which some precedent exists in the literature, consists of trying out a series of values of certain under-identified parameters and tabulating (or graphing) the corresponding esti-mated values of other underidentified parameters to ascertain how sensitive the latter are to variation in the former. This method was used by Duncan [8] to estimate several correlations and path coefficients involving unmeasured motivational variables in an underidentified recursive model of the process of socioeconomic achievement. The path diagram of the model is given in Fig. 11.2. The model was constructed to determine what inferences can be

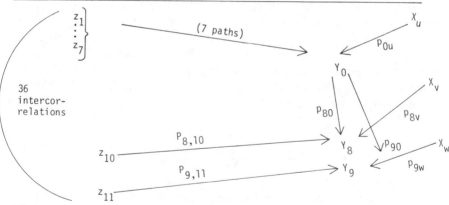

Figure 11.2. A representation of Duncan's model for inferring correlations involving unmeasured variables, "ambition" and "work orien-tation." Z_1 = father's occupation; Z_2 = number of siblings; Z_3 = education; Z_4 = occupation at marriage; Z_5 = income at marriage; Z_6 = current occupation; Z_7 = current income; Z_{10} = "ambition" (unmeasured); Z_{11} = "work orientation" (un-measured); Y_0 = subjective achievement; Y_8 = importance of getting ahead; Y_9 = commitment to work; X_u, X_v, X_w include unmeasured determinants of Y_0, Y_8, Y_9, respectively. SOURCE: Modification of Duncan [8, p. 89].

made concerning the role of motivational variables in a model of the achievement process when the indicators of such variables must be presumed to be subject to contamination by the dependent variables of the model (Duncan [8, p. 74]). This problem arises in the interpretation of data from a cross-sectional study in which indicators of motivation necessarily are secured at the same time as are indicators of achievement. Such a possibility is indicated in Fig. 11.2 by the causal arrow from variable Y_0, "Subjective achievement," to the two motivational items Y_8, "Importance of getting ahead," and Y_9, "Commitment to work."

The objective of this model is to make inferences from sample correlations among the observed variables to implied sample correlations between the hypothetical motivational variables, "Ambition" and "Work orientation," and to the estimated paths between these variables and the measured ones. However, by allowing for non-zero paths p_{80} and p_{90}, the model becomes underidentified. To obtain estimates, Duncan tried several arbitrary values of p_{8v} and p_{9w}, the error paths for the measured motivational variables, noting that "because there was no conceptual or empirical basis for estimating these residual paths, several solution sets were obtained" (Duncan [8, p. 93]). Duncan found that all combinations of values for the residuals paths imply a positive correlation between the hypothetical motivation variables, whereas there is "a nil correlation between the two indicators" (Duncan [8, p. 95]). Because this and other analyses conform, for the most part, to *a priori* specifications, Duncan [7, p. 103] argues that further analyses should be based on the hypothetical, rather than the observed, variables. In other words, Duncan "tried out" different values for those residual path coefficients not of direct interest to him in order to assess the sensitivity of other parameter estimates to these values. On the basis of this sensitivity analysis, Duncan made identifying assumptions that led to the analysis of other models.

Method 3—Use of Algebraic Inequalities

As a third method of sensitivity analysis, inequality constraints on selected parameters can be used in conjunction with sample statistics to set informative numerical bounds on sample estimates for other parameters in a model. To illustrate how this method can be applied to the Hauser model of Fig. 11.1, suppose we constrain ρ_{ab} only to be nonnegative, that is,

$$\rho_{ab} \geq 0 \tag{11.12}$$

and use this in conjunction with constraint 11.8 to determine the upper bound on p_{wm} that is theory-admissible under these constraints and data-admissible with respect to Hauser's data. Under assumptions in Eqs. 11.8

and 11.12, equation 11.10 is just-identified and its coefficients can be estimated by indirect least squares. For the determination of Y_m, these indirect least-squares estimates are the same as those in equation 11.10, except for p_{ma} which can be estimated as follows. First, expanding population correlations in terms of the postulated model (see, e.g., Duncan [9]), we have

$$(\rho_{ma}) = (p_{ma}) + (\rho_{wa})p_{mw} \tag{11.13}$$

$$p_{mw} = \rho_{wq}p_{mq} + \rho_{ws}p_{ms} + \rho_{wv}p_{mv} + p_{mw} + (\rho_{wa}p_{ma}) \tag{11.14}$$

$$1.0 = \rho_{mm} = \rho_{mq}p_{mq} + \rho_{ms}p_{ms} + \rho_{mv}p_{mv} + \rho_{mw}p_{mw} + (\rho_{ma}p_{ma}) \tag{11.15}$$

Next, multiplying Eq. 11.13 by p_{ma} gives

$$(\rho_{ma}p_{ma}) = (p_{ma}^2) + (\rho_{ma}p_{wa})p_{mw} \tag{11.16}$$

In these four equations, we have placed the unknown coefficients in parentheses; all of other terms are estimated either by sample correlations or coefficients estimated from the identified part of the model. In particular, substituting estimated values, rearranging Eqs. 11.13–11.15, and using r's and $\hat{p}$'s to denote sample estimates of corresponding population correlation and path coefficients, respectively, yields

$$(r_{ma}) = (\hat{p}_{ma}) + (r_{wa})(.463) \tag{11.17}$$

$$(r_{wa}\hat{p}_{ma}) = .640 - [(.290)(.570) + (-.001)(-.169) \\ + (.007)(.209) + .463] = .010 \tag{11.18}$$

$$(r_{ma}\hat{p}_{ma}) = 1.0 - [(.209)(.556) + (-.001)(-.123) \\ + (.007)(.164) + (.463)(.640)] = .586 \tag{11.19}$$

where the computations have been carried out to more decimal places than are reported. Using the results of 11.17–11.19 in 11.16, we find that

$$.586 = \hat{p}_{ma}^2 + (.010)(.463) \tag{11.20}$$

so that

$$\hat{p}_{ma} = \sqrt{.582} = .763 \tag{11.21}$$

Further substitution shows that

$$r_{ma} = .769 \tag{11.22}$$

and

$$r_{wa} = .013 \tag{11.23}$$

This last estimate can now be used to narrow down the range of possible values of the path coefficient p_{wm}. To see this, note that expansion of ρ_{wa} yields

$$\rho_{wa} = \rho_{ma} p_{wm} + \rho_{ab} p_{wb} \tag{11.24}$$

Since ρ_{ab} and p_{wb} are *a priori* considered to be nonnegative, it follows that $\rho_{wa} \geqslant \rho_{ma} p_{wm}$ and, substituting sample values,

$$\hat{p}_{wm} \leqslant r_{wa}/r_{ma} = .013/.769 \tag{11.25}$$

We therefore obtain the following sample estimates of the upper bound on p_{wm} under this specification:

$$\hat{p}_{wm} \leqslant .017 \tag{11.26}$$

The significance of this computation is that, under very weak assumptions, we have narrowed down the possible range of p_{wm} allowed by Hauser's data. Even without assuming that $\rho_{ab} = 0$, we are able to assert that p_{wm} is probably considerably smaller in value than p_{mw}, which Hauser found to be 0.463.[6] Thus, on the basis of a weaker specification, we have corroborated Hauser's finding that reading skills probably have more effect on mathematics skills than vice versa, at least so far as these standardized tests indicate. In other words, we have shown that this finding is insensitive to the specification on ρ_{ab}.

This use of assumptions about the relative sizes of coefficients in sensitivity analysis has several antecedents in the sociological and econometric

---

[6] The bound on p_{wm} given by inequality 11.25 is itself a sample statistic, the standard error of which can be estimated approximately by applying the standard Taylor series expansion (Kendall and Stuart [24, pp. 231–236]). Of course, for a sample as large as Hauser's, the standard error will be quite small.

literature. On the sociological side, Hodge and Siegel [19] utilized assumptions about the relative magnitudes of path coefficients in their model of socioeconomic measurement error. In econometrics, Zellner [43] has not only derived bounds on the values of parameters of simultaneous-equation models from the usual specifying assumptions, but he has also shown that such bounds can be estimated consistently and are formally analogous to those encountered in the classical errors-in-the-variables regression model.[7]

Method 4—Mathematical Programming

All of the foregoing methods have been used previously in the literature to perform sensitivity analyses on arbitrary identifying assumptions. But the use of algebraic inequalities, as in the method just discussed, if suitably generalized, suggests that a systematic algorithm for incorporating inequalities, as found in mathematical programming, might also be a useful method. Although mathematical programming has long been used in statistics and econometrics to *estimate* the coefficients of a model (see, e.g., Judge and Takayama [23]; Rothenberg [34, pp. 43–58]), it has not generally been used to perform sensitivity analyses in the sense in which that term is used here. There is, to our knowledge, only one previous use of mathematical programming related to that proposed here. That is, Felson [12] maximized and minimized the values of coefficients and sums of coefficients in an underidentified model—similar to that in Fig. 11.3—by use of linear programming.

Mathematical programming is a branch of mathematics that has been developed in the past three decades due primarily to the stimulus of applications in economics and industrial engineering (see, e.g., Dantzig [6]; Vajda [38]). It is employed to solve problems of the following generic type: maximize or minimize a function of n variables, $f(x_1, \ldots, x_n)$, subject to certain inequalities or equations that constrain the arguments of the function to satisfy various constraints. Symbolically, the general form of the problem is:

$$\text{maximize (minimize) } z^* = f(x_1, \ldots, x_n) \tag{11.27}$$

$$\text{subject to: } g_i(x_1, \ldots, x_n)\{\leqslant = \geqslant\}b_i, \; i = 1, \ldots, m \tag{11.28}$$

---

[7] On first glance, it appears that Zellner's bounds, which are based on asymptotic variances and covariances, are applicable to sensitivity analysis for underidentified parameters. In fact, however, they can be applied only to find bounds on just-identified or overidentified parameters. This means that the Zellner bounds could be used to place intervals on the reduced-form equations (corresponding to an underidentified model) in which the coefficients consisted of combinations of the underidentified parameters which combinations are themselves identified. But this would not provide bounding information on the underidentified structural parameters taken by themselves.

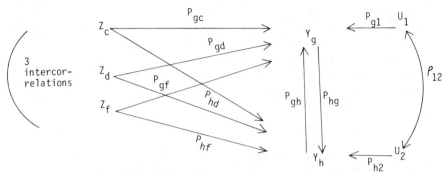

Figure 11.3. Representation of a simultaneous-equation model of age of first job and educational attainment. Y_g = son's age at first job; Y_h = son's education; Z_c = father's education; Z_d = father's occupation; Z_f = number of siblings; U_1 = unmeasured determinants of Y_g; U_2 = unmeasured determinants of Y_h.

where the values of $x_1, \ldots, x_n$ are to be determined that maximize (minimize) the function $f(x_1, \ldots, x_n)$ subject to the m inequalities or equalities indicated in expression 11.28. These restrictions, 11.28, are usually called *constraints*, and the function to be maximized (minimized) is called the *objective function*. If the objective function and constraints can be written as linear functions:

maximize (minimize):

$$z^* = \sum_{j=1}^{n} k_i x_j \tag{11.29}$$

subject to:

$$\sum_{j=1}^{n} a_{ij} x_j \; \{\leqslant = \geqslant\} \; b_i, \, i = 1, \ldots, m \tag{11.30}$$

where the a_{ij} and k_j are known constants, then the programming problem is called a *linear programming* problem (see, e.g., Hadley [16]). Usually, in the formulation of a linear programming problem, it is also specified that each variable must be nonnegative:

$$x_j \geqslant 0, j = 1, \ldots, n \tag{11.31}$$

since this facilitates numerical computations.[8] Any programming problem that involves either a nonlinear objective function or nonlinear constraints is called a *nonlinear programming* problem.

We suggest that mathematical programming may be usefully employed to perform sensitivity analyses of models that have been arbitrarily identified. We present this method by illustrating its application to a specific substantive model in the next section.

AN APPLICATION OF LINEAR PROGRAMMING TO SENSITIVITY ANALYSIS OF AN UNDERIDENTIFIED EDUCATIONAL ATTAINMENT MODEL

Among the best-established positive correlations in sociological literature is that between the educational attainments of father and son. Although some excellent attempts have been made to partial out the mechanisms whereby this correlation is generated (see, e.g., Blau and Duncan [2]; Duncan *et al.* [11]; Duncan *et al.* [10]; Jencks *et al.* [21]; Sewell and Hauser [35]), many important issues remain unresolved. For instance, Duncan *et al.* [10] report a sample correlation of .449 between son's educational attainment and age at first job. This is not surprising, since it takes time to obtain more education and this in turn leads to later labor force entry. However, working at a younger age may also feed back upon educational attainment by keeping people away from educational institutions during the years in which it is most convenient to go to school. Indeed, dropping out of school and going to work at a young age would seem to reduce ultimate educational attainment and make it somewhat harder to go back to school. Furthermore, persons of lower status origins as well as those from families with many siblings may find greater economic, social, or cultural support for leaving school to go to work at a younger age. We will not take space for an elaborate discussion of what determines age at first job, but we note that several alternative processes ought to be incorporated into any model specification.

[8] If one wishes to allow some variable x_j to be of either sign, it is sufficient to define two nonnegative variables x_j' and x_j'' such that

$$x_j = x_j' - x_j''$$

If $x_j'' \geq x_j'$, then $x_j \leq 0$. Any coefficient multiplied by x_j can be expanded thus

$$kx_j = kx_j' - kx_j''$$

Thus, the restriction that $x_j \geq 0$ poses no problem for use of variables taking on negative values.

The path diagram in Fig. 11.3 represents one such model, where father's occupation (Z_d), father's education (Z_c), and number of siblings (Z_f) directly affect son's age at first job (Y_g) and son's education (Y_h). The latter two variables are allowed "direct" effects upon one another, which means that all three background variables affect the endogenous variables both directly and indirectly. In addition, the stochastic errors or disturbances are represented by U_1 and U_2.

The model represented in Fig. 11.3 can be expressed by the following pair of simultaneous equations:

$$Y_g = p_{gh}Y_h + p_{gc}Z_c + p_{gd}Z_d + p_{gf}Z_f + p_{g1}U_1 \qquad (11.32)$$

$$Y_h = p_{hg}Y_g + p_{hc}Z_c + p_{hd}Z_d + p_{hf}Z_f + p_{h2}U_2 \qquad (11.33)$$

By the order condition reviewed above ("Standard concepts of identifiability"), a necessary exclusion restriction for identifying each of these equations is that one exogenous or endogenous variable be excluded from each. Since neither equation satisfies this condition, both equations are underidentified. Moreover, the stipulation that the correlations between endogenous variables and the disturbances, ρ_{h1} and ρ_{g2}, are zero would contradict the specified reciprocal relationship. It might be reasonable to assume that the following condition is approximately true:

$$\rho_{12} = 0 \qquad (11.34)$$

but this assumption would be no help in identifying the model unless one of the causal paths were specified to be zero. Furthermore, one suspects that whatever unmeasured economic, social, or cultural traits lead to lower educational attainment may also produce early labor force participation, so that positive values of ρ_{12} are likely. Nevertheless, it may be possible to utilize constraints on model parameters and sample statistics to study the plausibility of various combinations of arbitrary identifying assumptions.

The data for this analysis are taken from the 1962 OCG study (Occupational Change in a Generation), as reported in Duncan et al. [10; Table A.1]. We utilize correlations among the relevant variables for a subsample of 14,357 native non-black males aged 25 to 64 in the experienced civilian labor force in March 1962. This study is well documented (see Blau and Duncan [2]; Duncan et al. [10]), and probably one of the best available data sets for testing basic models of the socioeconomic achievement process.

Consistent with the discussion at the end of the preceding section, one can formulate the following programming problem for bounding the model parameters:

Table 11.1. Linear programming estimates of sample min – max intervals for 12 coefficients[a]

Objective function	Minimum	Maximum	Interval
$z_1^* = \hat{p}_{gh}$	0	.3437	.3437
$z_2^* = \hat{p}_{gc}$	.0756	.1537	.0781
$z_3^* = \hat{p}_{gd}$	.0622	.1518	.0896
$z_4^* = \hat{p}_{gf}$	0	.0729	.0729
$z_5^* = \hat{p}_{hg}$	0	.3218	.3218
$z_6^* = \hat{p}_{hc}$	.1778	.2772	.0994
$z_7^* = \hat{p}_{hd}$	.2119	.2608	.0489
$z_8^* = \hat{p}_{hf}$	.1887	.2122	.0235
$z_9^* = (r_{h1}\hat{p}_{g1})$	.0468	.2936	.2468
$z_{10}^* = (r_{g1}\hat{p}_{g1})$	.8114	.9123	.1009
$z_{11}^* = (r_{g2}\hat{p}_{h2})$	0	.2936	.2936
$z_{12}^* = (r_{h2}\hat{p}_{h2})$	.6234	.7179	.0945

[a] Data relate son's education and age of first job for native non-black males aged 25 – 64 in the experienced civilian labor force, 1962.

$$\text{maximize (or minimize) } z_i^*$$
$$\text{(as specified on the left hand side of Table 11.1)} \quad (11.35)$$

subject to these constraints:

$$1.0 = p_{gg} = p_{gh}(p_{gh}) + p_{gc}(p_{gc}) + p_{gd}(p_{gd}) + p_{gf}(p_{gf}) + (p_{g1}p_{g1}) \quad (11.36)$$

$$p_{gh} = (p_{gh}) + p_{hc}(p_{gc}) + p_{hd}(p_{gd}) + p_{hf}(p_{gf}) + (p_{h1}p_{g1}) \quad (11.37)$$

$$p_{gc} = p_{hc}(p_{gh}) + (p_{gc}) + p_{cd}(p_{gd}) + p_{cf}(p_{gf}) \quad (11.38)$$

$$p_{gd} = p_{hd}(p_{gh}) + p_{cd}(p_{gc}) + (p_{gd}) + p_{df}(p_{gf}) \quad (11.39)$$

$$p_{gf} = p_{hf}(p_{gh}) + p_{cf}(p_{gc}) + p_{df}(p_{gd}) + (p_{gf}) \quad (11.40)$$

and

$$1.0 = p_{hh} = p_{hg}(p_{hg}) + p_{hc}(p_{hc}) + p_{hd}(p_{hd}) + p_{hf}(p_{hf}) + (p_{h2}p_{h2}) \quad (11.41)$$

$$p_{hg} = (p_{hg}) + p_{gc}(p_{hc}) + p_{gd}(p_{hd}) + p_{gf}(p_{hf}) + (p_{g2}p_{h2}) \quad (11.42)$$

$$p_{hc} = p_{gc}(p_{hg}) + (p_{hc}) + p_{cd}(p_{hd}) + p_{cf}(p_{hf}) \quad (11.43)$$

$$p_{hd} = p_{gd}(p_{hg}) + p_{cd}(p_{hc}) + (p_{hd}) + p_{df}(p_{hf}) \qquad (11.44)$$

$$p_{hf} = p_{gf}(p_{hg}) + p_{cf}(p_{hc}) + p_{df}(p_{hd}) + (p_{hf}) \qquad (11.45)$$

Again, parentheses set off coefficients for which there is no direct sample estimate, and the constraint equations are derived by expanding population correlations in terms of the postulated model. If we allow residual correlations $(p_{g1}\, p_{g1})$, $(p_{h1}\, p_{g1})$, $(p_{h2}p_{h2})$, and $(p_{g2}\, p_{h2})$ to remain as products, we have 12 unknowns and only 10 equations with which to estimate them. Applied to these constraints, linear programming is not sufficient for placing meaningful bounds on the parameters. Fortunately, social theory helps us to specify additional constraints: that the impact of father's education and occupation on both endogenous variables will be positive and that the impact of number of siblings on both endogenous variables will be negative. Furthermore, the impact of the two endogenous variables on one another should be positive. If we multiply X_f by (-1), so that each sibling is worth a negative unit, then all coefficients will be expected to be nonnegative:

$$p_{gh},\ p_{gc},\ p_{gd},\ p_{gf},\ p_{hg},\ p_{hc},\ p_{hd},\ p_{hf}$$
$$(p_{g1}\, p_{g1}),\ (p_{h1}\, p_{g1}),\ (p_{h2}\, p_{h2}),\ \text{and}\ (p_{g2}\, p_{h2}) \geqslant 0 \qquad (11.46)$$

Given constraints 11.36–11.46 and the sample correlation matrix, we can obtain sample min–max intervals for each unknown coefficient by minimizing and maximizing each objective function shown on the left-hand side of Table 11.1. Observe that all 12 of these objective functions assign a weight of 1 or 0 to the unknown coefficients. For example, the objective function z^*, can be more fully described as:

$$\begin{aligned}
z_1^* = &(1.0)\hat{p}_{gh} + (0)\hat{p}_{gc} + (0)\hat{p}_{gd} + (0)\hat{p}_{gf} \\
&+ (0)\hat{p}_{hg} + (0)\hat{p}_{hc} + (0)\hat{p}_{hd} + (0)\hat{p}_{hf} \\
&+ (0)(r_{h1}\,\hat{p}_{g1}) + (0)(r_{g1}\,\hat{p}_{g1}) \\
&+ (0)(r_{g2}\,\hat{p}_{h2}) + (0)(r_{h2}\,\hat{p}_{h2})
\end{aligned} \qquad (11.47)$$

Those unknowns that are products of pairs of coefficients are each taken as single unknowns in equations 11.36, 11.37, 11.41, and 11.42, so that all of these constraints are linear. If we had sought to bound the individual coefficients rather than their products, nonlinear programming would have been necessary. In the present example, linear programming was sufficient for reducing the feasible set of solutions for 11.35 subject to constraints 11.36–11.46, within a reasonable range, but future researchers may wish to employ nonlinear programming to further narrow down the solutions.

As the right-hand side of Table 11.1 indicates, 8 of 12 unknown coefficients can be narrowed down to a sample min–max interval of .1 or less, with $\hat{p}_{gh}$ and $\hat{p}_{hg}$ having the largest intervals.[9] However, the inability of the constraints to place more narrow intervals about $\hat{p}_{gh}$ and $\hat{p}_{hg}$ does not prevent us from narrowing down the values of the other theoretically important coefficients and determining, for instance, that family background factors are likely to affect both son's education and son's age at first job. Nevertheless, the background factors do not uniformly have strictly positive min–max intervals in Table 11.1. In particular, the sample min–max interval for $\hat{p}_{gf}$ includes zero, which suggests that one plausible restriction is that $p_{gf} = 0$. Moreover, the product $\hat{r}_{g2}\hat{p}_{h2}$ has a sample lower bound of zero, and the product $\hat{r}_{h1}\hat{p}_{g1}$ is not much greater than zero. Since the residual path coefficients of such a model are likely to be rather large, these estimates suggest that the correlation between the disturbances may be rather small. Therefore, one pair of identifying assumptions under which the model can be estimated by standard procedures is $p_{gf} = 0$ and $\rho_{12} = 0$. Moreover, since both of the lower bounds of the sample min–max intervals for p_{gh} and p_{hg} are zero, another plausible pair of identifying assumptions would constrain either p_{gh} or p_{hg} to be equal to zero and $\rho_{12} = 0$.

In general, mathematical programming offers a pragmatic means of isolating plausible constraints that enable the estimation of theoretically interesting coefficients in otherwise underidentified equations.[10]

CONCLUSION

This chapter has described four methods for analyzing the sensitivity of parameter estimates in arbitrarily identified simultaneous-equation models to the identifying assumptions of those models. None of these methods is based on a rigorous statistical theory. Although we have suggested (in footnote 6) ways to approximate the standard errors of the statistical bounds derived in the preceding sections, we have not reported such approximations because of the essentially *ad hoc* nature of the bounds. We recommend that these bounding methods should be applied only to data from large samples if standard errors are unknown.

In spite of the *ad hoc* nature of sensitivity analysis techniques, it is likely that these techniques will continue to be used in the absence of stronger social theory. Therefore, it seems prudent to refine the statistical theoretic basis of sensitivity analysis. Because this calls for a rational basis for making

[9] Again, the min–max intervals that we give in Table 11.1 are subject to sampling variation, although the standard errors are probably not large because of the large size of the sample on which this example is based. In any case, the sampling errors could be approximated as mentioned in footnote 6.

[10] Computer programs for mathematical programming are widely available.

decisions about the plausibility of alternative identifying restrictions in light of their empirical implications, the proper statistical basis may be in Bayesian or statistical decision theory. Describing such a basis is beyond the scope of this chapter, which has only identified currently available methods.

Though this chapter is mainly methodological, it has some implications for social theorists as well. For example, if theories can specify that certain pairs of coefficients are of the same or opposite sign, that one coefficient must be larger or equal to another, and so forth, such information may prove very useful in specifying inequality constraints. We believe that researchers often fail to employ all available *a priori* information. The techniques of sensitivity analysis can render use of such information well worth the effort.

ACKNOWLEDGMENTS

An earlier version of this chapter was presented at the Seventy-First Annual Meeting of the American Sociological Association, New York, August 30–September 3, 1976. The research reported here was supported, in part by a grant from the Russell Sage Foundation. We are indebted to Robert M. Hauser and anonymous reviewers for comments on an earlier version of the chapter. Any remaining errors are, of course, the responsibility of the authors.

REFERENCES

[1] Blalock, H. M., Jr. *Causal Inferences in Non-experimental Research.* Chapel Hill: Univ. of North Carolina Press, 1964.
[2] Blau, P., and Duncan, O. D. *The American Occupational Structure.* New York: John Wiley, 1967.
[3] Boudon, R. "A New Look at Correlational Analysis." In *Methodology in Social Research,* edited by H. M. Blalock, Jr. and A. B. Blalock, pp. 199–235. New York: McGraw-Hill, 1968.
[4] Christ, C. F. *Econometric Models and Methods.* New York: John Wiley, 1966.
[5] Coleman, J. S., Campbell, E. Q., Hobson, C. J., McPartland, J., Mood, A. M., Weinfeld, F. D., and York, R. L. *Equality of Educational Opportunity.* Washington, D.C.: Government Printing Office, 1966.
[6] Dantzig, G. B. *Linear Programming and Extensions.* Princeton: Princeton University Press, 1963.
[7] Duncan, O. D. "Path Analysis: Sociological Examples." *American Journal of Sociology* 72 (1966): 3–16.
[8] Duncan, O. D. "Contingencies in Constructing Causal Models." In *Sociological Methodology 1969,* edited by E. F. Borgatta, pp. 79–112. San Francisco: Jossey-Bass, 1969.
[9] Duncan, O. D. *Introduction to Structural Equation Models.* New York: Academic Press, 1975.
[10] Duncan, O. D., Featherman, D. L., and Duncan, B. *Socioeconomic Background and Achievement.* New York: Seminar Press, 1972.

[11] Duncan, O. D., Haller, A. O., and Portes, A. "Peer Influences on Aspirations: A Reinterpretation." *American Journal of Sociology* 74 (1968): 119–137.

[12] Felson, M. "Conspicuous Consumption and the Swelling of the Middle Class in America." Ph.D. dissertation, University of Michigan, 1973.

[13] Fisher, F. M. *The Identification Problem in Econometrics.* New York: McGraw-Hill, 1966.

[14] Goldberger, A. S. *Econometric Theory.* New York: John Wiley, 1964.

[15] Goldberger, A. S. "On Boudon's Method of Linear Causal Analysis." *American Sociological Review* 35 (1970): 97–101.

[16] Hadley, G. *Linear Programming.* Reading, Mass.: Addison-Wesley, 1962.

[17] Hauser, R. M. *Socioeconomic Background and Educational Performance.* Washington, D.C.: American Sociological Association, Rose Monographs Series, 1971.

[18] Hauser, R. M. "Disaggregating a Social–Psychological Model of Educational Attainment." In *Structural Equation Models in the Social Sciences,* edited by A. S. Goldberger and O. D. Duncan, pp. 255–284. New York: Seminar Press, 1973.

[19] Hodge, R. W., and Siegel, P. M. "A Causal Approach to the Study of Measurement Error." In *Methodology in Social Research,* edited by H. M. Blalock, Jr. and A. B. Blalock, pp. 28–59. New York: McGraw-Hill, 1968.

[20] Hurwicz, L. "Generalization of the Concept of Identification." In *Statistical Inference in Dynamic Economic Models,* edited by T. Koopmans, pp. 245–257. New York: John Wiley, 1950.

[21] Jencks, C., Smith, M., Acland, H., Bane, M. J., Cohen, D., Gintis, H., Heyns, B., and Michelson, S. *Inequality: A Reassessment of the Effect of Family and Schooling in America.* New York: Harper & Row, 1972.

[22] Johnston, J. J. *Econometric Methods.* New York: McGraw-Hill, 1972.

[23] Judge, G. G., and Takayama, T. "Inequality Restrictions in Regression Analysis." *Journal of the American Statistical Association* 61 (1966): 166–181.

[24] Kendall, M. G., and Stuart, A. *The Advanced Theory of Statistics.* Vol. 1. Distribution Theory. London: Griffin, 1958.

[25] Klein, L. R. *An Introduction to Econometrics.* Englewood Cliffs, N.J.: Prentice-Hall, 1962.

[26] Kmenta, J. *Elements of Econometrics.* New York: Macmillan, 1971.

[27] Kohn, M., and Schooler, C. "Occupational Experience and Psychological Functioning: An Assessment of Reciprocal Effects." *American Sociological Review* 38 (1973): 97–118.

[28] Koopmans, T. C., Rubin, H., and Leipnik, R. B. "Measuring the Equation Systems of Dynamic Economics." In *Statistical Inference in Dynamic Economic Models,* edited by T. C. Koopmans, pp. 53–237. New York: John Wiley, 1950.

[29] Land, K. C. (1971). "Significant Others, the Self-Reflexive Act and the Attitude Formation Process: A Reinterpretation." *American Sociological Review* 36 (1971): 1085–1098.

[30] Land, K. C. "Identification, Parameter Estimation, and Hypothesis Testing." In *Structural Equation Models in the Social Sciences,* edited by A. S. Goldberger and O. D. Duncan, pp. 19–49. New York: Seminar Press, 1973.

[31] McPherson, J. M., and Huang, C. J. "Hypothesis Testing in Path Models." *Social Science Research* 3 (1974): 127–140.

[32] Malinvaud, E. *Statistical Methods of Econometrics.* Chicago: Rand McNally, 1966.

[33] Mayer, L. S., and Younger, M. S. "Procedures for Estimating Standardized Regression Coefficients from Sample Data." *Sociological Methods and Research* 2 (1974): 431–454.

[34] Rothenberg, T. J. *Efficient Estimation with A Priori Information.* New Haven: Yale University Press, 1973.

[35] Sewell, W. H., and Hauser, R. M. *Education, Occupation and Earnings: Achievement in the Early Career.* New York: Academic Press, 1975.

[36] Specht, D. A. "On the Evaluation of Causal Models." *Social Science Research* 4 (1975): 113–134.

[37] Theil, H. *Principles of Econometrics.* New York: John Wiley, 1971.

[38] Vajda, S. *Mathematical Programming.* Reading, Mass.: Addison-Wesley, 1961.

[39] Waite, L. J., and Stolzenberg, R. M. "Intended Childbearing and Labor Force Participation of Young Women: Insights from Non-Recursive Models." *American Sociological Review* 41 (1976): 235–251.

[40] Wald, A. "Note on Identification of Economic Relations." In *Statistical Inference in Dynamic Economic Models,* edited by T. C. Koopmans, pp. 238–244. New York: John Wiley, 1950.

[41] Wiley, D. E. "The Identification Problem for Structural Equation Models with Unmeasured Variables." *Structural Equation Models in the Social Sciences,* edited by A. S. Goldberger and O. D. Duncan, pp. 69–84. New York: Seminar Press, 1973.

[42] Wonnacott, R. J., and Wonnacott, T. H. *Econometrics.* New York: John Wiley, 1970.

[43] Zellner, A. "Constraints Often Overlooked in Analyses of Simultaneous Equation Models." *Econometrica* 40 (1972): 849–853.

THE CAUSAL APPROACH
TO MEASUREMENT
ERROR AND
AGGREGATION

III

Part III deals with complications produced whenever variables have been imperfectly measured or whenever one finds it necessary to utilize aggregate data to test theories about more micro-level processes. In both instances, we are concerned with how to handle situations involving missing pieces of information that introduce additional unknowns into the system. There is now a substantial and growing body of literature on both of these topics, and therefore the editor has decided to supplement the necessarily small number of chapters in the present volume with a second volume that will focus primarily on measurement-error complications in panel and experimental designs.

Measurement errors generally produce biases in one's estimates that can only be corrected by providing a measurement-error model that contains specific assumptions about the sources of these measurement errors and that will also enable one to estimate measurement-error variances and covariances. In the simplest of cases where the measured variable $X' = X + u$, where the measurement-error term, u, is assumed uncorrelated with the true value X, the resulting biases will be a function of the ratio of s_u^2 to s_x^2. But this ratio will be unknown since both X and u will be unmeasured. We encounter an identification problem produced by these unmeasured variables, even in instances of recursive systems. If there are sources of measurement-error biases, as well as strictly random errors, there will be still more unknowns, and without a theory the situation may become intractable.

Whenever there are multiple measures of each variable, there may be sufficient empirical information to produce overidentified systems, provided relatively simple assumptions can be made about sources of random and nonrandom measurement errors. As a general rule, the more indicators one has and the simpler his assumptions, the greater the number of excess equations that will be available to test the compatibility of the data with the model. There will always be a number of alternative models that are consistent with any given set of data, but one may proceed by rejecting inadequate models. Having tentatively settled on a particular model, one may then estimate path coefficients, though it may not be possible to estimate unstandardized coefficients without imposing additional restrictions.

The simplest models involving multiple indicators obviously stem from the literature on factor analysis and related approaches, but they afford a somewhat different perspective that permits the introduction of various kinds of complications involving different sources of nonrandom errors. The general strategy suggested is to construct "auxiliary theories" that explicitly link each indicator variable with the unmeasured variables of interest.[1] In factor analysis, the unmeasured variables or "factors" are taken as causes of the indicators, and each indicator is taken as an endogenous

[1] See Blalock [1].

variable that is a function of the unmeasured factors alone, plus a unique disturbance term. If the indicators are represented as I_i and the factors as F_j, we in effect have a set of reduced-form equations of the type

$$I_i = b_{i1}F_1 + b_{i2}F_2 + \cdots + b_{ik}F_k + u_i$$

where we assume that none of the indicators appears in any of the equations for the remaining indicators. In other words, we rule out any direct causation among the indicators. In many sociological and political science applications, however, this simple model is obviously inappropriate, and we need a more complex auxiliary theory.

The notion of an auxiliary theory implies that we construct causal models linking indicators and unmeasured variables, just as though there were no fundamental difference between the two. We then examine the situation to see if, using only the measured variables, we have available enough empirical information to estimate all of the path or regression coefficients in the system. Usually the system will be underidentified, and we must then consider the nature of the simplifications that will be necessary to achieve identification and to produce excess equations that can be utilized for testing purposes.

In the first chapter in Part III, Costner begins with a very simple model involving only two indicators of X and two of Y, assuming strictly random measurement error and a recursive context in which X affects Y. With a total of four measures, there are six correlations that can be obtained from the data, and these may be used to estimate the five path coefficients of the model. The excess equation can then provide a test criterion in this overidentified system. The principle can be extended in a number of directions, as Costner and others have shown. First, it can be extended to the general k-equation case, provided there are at least two measures of each variable, and it may also be utilized in connection with certain kinds of simple causal chains when there is a single measure for the intervening variables. If additional indicators of each variable are used, the system becomes highly overidentified so that multiple tests can be made. This also permits one to introduce certain relatively simple kinds of nonrandom measurement errors in some of the indicators, as demonstrated in the chapters by Costner (12), Werts, Linn, and Jöreskog (13), Herting (14), and Herting and Costner (15).

The chapter by Werts, Linn, and Jöreskog, which was specifically written for the earlier edition of this volume, anticipates a line of development that has culminated in a very useful series of LISREL programs developed by Jöreskog and Sörbom and designed to facilitate the testing of measurement-error models that may be superimposed upon structural-equation systems of the type that are discussed in Part II. Herting's chapter provides a useful expository discussion of this rather complicated LISREL procedure. Unfortunately, it will usually be the case that one's initial measurement-error models will not provide very close empirical fittings to one's data because of errors of specification that may be difficult to pin down in the absence of a well-grounded measurement-error theory. Herting and Costner examine a number of different types of possible misspecifications and provide a series of simulation studies designed to suggest possible strategies that may be used to help locate alternative sources of errors by utilizing the tests and measures provided in the LISREL program. In doing so, their aim is to stress the fact that this important new tool can be used in an exploratory fashion to improve upon our causal models of measurement-error processes.

The chapter by Sullivan (16) involves a very different practical strategy for utilizing multiple – partial controlling techniques as an alternative to the multiple-indicator approach. In many instances, one will be dealing with blocks of highly intercorrelated variables, with no clear idea as to the causal connections within these blocks. Some variables within each block may be perfectly measured, whereas others may not. Some indicators may be causes of the variables of interest, whereas others may be effects. Therefore, it may not be possible to specify auxiliary measurement theories or to justify the use of a sophisticated estimating procedure such as LISREL. Nor will it be practical to use each indicator separately because of the multiplicity of tests and estimates. A more *ad hoc* procedure of the type suggested by Sullivan therefore may be advisable.

The final chapter by Hannan (17) represents a summary and synthesis of the very extensive literature on aggregation that has appeared in the econometrics literature and that has had certain parallels within the sociological literature (where the focus has been more on disaggregation, however). Clearly, most of our macrolevel analyses are very much affected by choices of units of analysis (e.g., states or counties versus nations, census tracts versus blocks), and we often lack definitive theories to serve as guidelines in making these choices. If one is to expect "consistency" across several levels of aggregation, it is necessary to assume relatively simple linear models as well as homogeneity of the causal processes (as measured by the slope coefficients) across the categories into which the individual elements have been grouped. In many practical situations involving aggregation, as when persons are grouped according to their geographic proximity to each other, we often lack an adequate understanding of the relationship between the criterion used for grouping (e.g., proximity within a single county) and the other variables in the theoretical system. If we are to crystallize our thinking across levels of analysis, it will obviously become necessary to deal more systematically with the problems discussed in Hannan's chapter.

The chapters in this final section should make it abundantly clear that there is nothing rigid or closed about the causal approach to modeling and formalization and that many diverse methodological problems can be conceptualized in causal terms. But although the approach itself is flexible, all specific applications point to the fact that the construction of really testable theories, as well as the estimation of the parameters in these theories, is an exceedingly difficult task. More often than not, realistic theories will contain too many unknowns for solution, so that compromises involving simplifying assumptions must be built in at numerous points. When one adds to this the fact that the collection of adequate data is both expensive and time-consuming, so that any given analysis will necessarily involve a number of unmeasured variables, it becomes obvious that the road ahead will be difficult indeed. Nevertheless, it is highly desirable to construct tentative theories that can be evaluated with partly inadequate data, so that a cumulative process can be set in motion. If we merely hide our assumptions and fail to face up to the complexities of theory construction and data analysis, this may increase our sense of accomplishment and reduce our personal anxieties, but it will not resolve our methodological problems. In short, there are many more chapters to be written.

REFERENCES

[1] Blalock, H. M. "The Measurement Problem: A Gap between the Languages of Theory and Research. In *Methodology in Social Research,* edited by H. M. Blalock and Ann B. Blalock, Chapter 1. New York: McGraw-Hill, 1968.

Theory, Deduction, and Rules of Correspondence

Herbert L. Costner*

The requirement that scientific theories include both abstract concepts and concrete implications, and that the two be logically connected, has been treated rather casually by sociologists. Traditionally, sociological theorists have focused on abstractions with loose and ill-defined implications about matters of fact. More recently, some sociological formulations have shifted to the opposite extreme, stating only connections between measures without any attempt to make more abstract claims. Either of these modes of theory construction is costly, sacrificing either the clarity of empirical implications or the integrating potential of abstract concepts. Although the literature of the philosophy of science has provided us with terms for referring to the gap between abstract conceptions and concrete events — *rules of correspondence, epistemic correlations, operational definitions,* and *indicators of abstract dimensions* — these terms do little more than remind us that the gap is there. They do not provide clear guidelines for bridging the gap and suggest no criteria for determining the adequacy of the more or less arbitrarily devised connections between abstract and empirical levels. Clearly, the empirical testing of abstract theories must remain somewhat loose until

* Reprinted by permission of the author and publisher from the *American Journal of Sociology* 75: 245–263. Copyright 1969, The University of Chicago Press.

some strategies for dealing with this problem are devised. To the degree that rules of correspondence are weak and subject to distorting errors, deductions about matters of fact must be regarded as uncertain and possibly misleading.

This general problem is explored in the present chapter, not as a problem in semantics — which is the common way of treating it — but as a special problem in theory construction. The general strategy to be employed consists of including the rules of correspondence as an auxiliary part of the theory. The auxiliary theory will thus consist of statements connecting abstract dimensions and their empirical indicators, statements that will be treated like other theoretical propositions. The implications that may then be deduced allow, under certain conditions, two different kinds of decisions to be made empirically. First, one may determine whether particular indicators are inadequate for testing the implications of a specific abstract formulation because of artifactual measurement error. Second, if the indicators are not found inadequate, one may determine whether the abstract formulation itself is tenable. Although the second kind of decision — the tenability of the abstract formulation itself — is the crucial decision in the final analysis, some decision on the adequacy of the indicators is a prerequisite.

This attempt to treat the problem of rules of correspondence in a formal way builds quite explicitly on the work of others and owes much to their lead. It represents an extension of their work rather than a major departure from their approach. Blalock [2], following Northrop, has argued convincingly for the necessity of two languages — a theoretical and an operational language — and has suggested that the connections between the two be expressed in an auxiliary theory. I will follow Blalock in representing the auxiliary theory in the form of an explicit causal model. Siegel and Hodge [7], building on the work of Blalock, Duncan, and Wright, have utilized causal models representing auxiliary theories to investigate the effects of measurement error on the correlation between selected variables. Their detailed work, along with that of Blalock, sets the tone for the present discussion.

We will begin with a discussion of specific desiderata that auxiliary theories should help to achieve, illustrating the problems encountered with simple models incorporating highly simplified auxiliary theories. We will then move to a discussion of auxiliary theories more nearly adequate to the tasks outlined in the earlier discussion.

We consider first a relatively simple model proposing a one-way causal relation between two abstract variables; it may be summarized in the proposition that a change in X leads to a change in Y, but not the reverse, and may be represented graphically by an arrow from X to Y. An auxiliary theory providing one indicator for each of the abstract variables is added. We assume that the indicators are "reflectors" of the abstract variables, that is, that a change in the abstract variable will lead to a change in its own

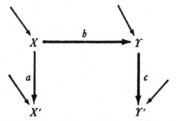

Figure 12.1. Two-variable model with one indicator for each variable.

indicator, and we represent these connections with arrows from abstract variable to indicator.[1] The model representing the basic theoretical proposition and the auxiliary theory is shown in Fig. 12.1. Associated with each of the three arrows in this causal diagram is a coefficient representing the regression of one variable on another, assuming all variables to be in the form of standard measures. These coefficients may take any value from -1 to 1. Arrows from unspecified sources are added in the graphic representation of the model to represent variation not accounted for in the model; specifically, when the coefficients a, b, and c are less than unity in absolute value, these additional arrows may be assigned values such that all of the variance is "accounted for" either by the model or by unspecified sources. In this model, we make the usual assumption for causal models that there are no common sources of error variance, or, referring to the algebraic representation of the model, following Simon [8], that all error terms are uncorrelated.

Ideally, we would be able to estimate the magnitudes of the three unknown coefficients in the model. In this case, however, we have only one observed correlation, $r_{X'Y'}$, and although the model implies that this correlation is a function of the three unknown coefficients ($r_{X'Y'} = abc$, where a, b, and c are path coefficients as shown in Fig. 12.1), this single equation in three unknowns does not provide unique solutions for the coefficients without further assumptions. Assuming knowledge only of the signs of the epistemic coefficients a and c, which is not an unreasonable assumption, it is possible to draw a conclusion about only the sign of the coefficient between abstract variables, b, from knowledge of the sign of the observed correlation, $r_{X'Y'}$. But this is a weak deduction at best and would have limited utility in making further deductions if this miniature model were to be incorporated as a unit into a more complex model involving several abstract vari-

---

[1]In experimental studies, the indicators of independent variables may be treated as "producers" rather than "reflectors" of the abstract variables, and the connection would be represented by arrows from indicator to abstract variable. Somewhat different problems are presented by the experimental case, which is not discussed in this chapter.

ables. Our usual procedure in "testing" causal models with only one indicator for each abstract variable is to take the correlation between indicators as
the correlation between the corresponding abstract variables. In so doing,
we are, in effect, assuming that the epistemic coefficients are either 1 or else
very high and subject only to minor random errors. If both epistemic coefficients are less than unity — and they should usually be assumed to be
so — $r_{X'Y'}$ will be a biased estimate of the coefficient between abstract variables, underestimating that coefficient to the degree that the product of the
epistemic coefficients is less than unity. Such bias of typically unknown
degree in the estimates of the abstract coefficients seriously hampers efforts
to test certain implications of more complex models, although the disad-
vantage of this conservative bias is not made immediately evident by the
very simple model of Fig. 12.1.

There is, however, the possibility of another kind of error more pernicious than random errors of measurement. With an auxiliary theory that
provides only one indicator for each abstract variable, this more pernicious
error remains unrecognizable. If the two indicators in Fig. 12.1 have common sources of error not shown in the model, the observed correlation
between indicators may yield a heavily distorted estimate of the coefficient
between abstract variables even beyond the distortion that is attributable to
random error. This kind of indicator error is represented by the two models
in Fig. 12.2, which do not exhaust the possibilities. Each of the models in Fig.
12.2 represents an alternative to the model of Fig. 12.1, alternatives in which
the error terms for the Fig. 12.1 model would be correlated, contrary to
assumption. The pernicious character of this kind of measurement error is
that if it is present we may be grossly misled, and, with only one indicator for
each abstract variable, its presence cannot be readily recognized. Even if
such error is suspected, its influence cannot be disentangled from the influence of the causal connections explicitly represented in the other parts of the
model.

The traditional language of measurement error includes no term for this
specific kind. The measurement error represented in the models of Fig. 12.2
is not *constant error* as we ordinarily think of it, that is, an identical quantity
added to or subtracted from every measure; such constant error would be

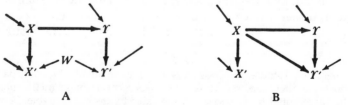

A   B

Figure 12.2. Two-variable model with one indicator for each variable and
        differential bias.

uncorrelated with any other variable and hence of no consequence in a causal model. Neither is it *random error*. And it is not *correlated error* in its usual meaning — that is, error magnitudes correlated with the magnitude of the true value. The measurement error in the models of Fig. 12.2 is a *differential constant error*, that is, an error that would be constant over repeated measurement for a given case but variable over cases so that this constant error is correlated with another indicator in the model. The general notion of such differential constant error, or *differential bias*, as I shall call it, is quite familiar, although a general term for designating it is not. For example, when arrests are used as an indicator of the incidence of crime in different areas of the city, the abstract correlation between social class and crime, thus measured, is presumably exaggerated because of differential bias. The economic characteristics of areas affect not only the incidence of crime but also the degree of error in arrests as an indicator of the incidence of crime. Figure 12.2B represents this general kind of circumstance in abstract form, that is, the independent variable has an effect on the indicator of the dependent variable both through the dependent variable and directly. A different kind of differential bias is represented in Fig. 12.2A. For example, when two abstract variables are measured by responses to verbal statements and both sets of responses are affected by "social desirability" response sets, the correlation between the errors in the two indicators will lead to a distorted estimate of the correlation between the two abstract variables because of this additional common source of variation in the indicators. Differential bias, then, is not a new idea, although this particular way of representing it in the form of unwanted connections between indicators in a causal model may not be familiar. Unlike random error, differential bias does not necessarily lead to an underestimation, on the average, of the correlation between the abstract variables, and its effects cannot be taken into account by utilizing such familiar devices as sampling distributions or a "correction for attenuation," which are based on the assumption of random errors. Ideally, auxiliary theories would be so constructed that differential bias, if present, would be recognizable empirically. This is evidently not the case with such simple auxiliary theories as those represented in Fig. 12.1 and 12.2.

Now we consider a slightly more complex model in which three abstract variables are linked in a causal sequence, with one indicator for each. The model is shown in Fig. 12.3.[2] Reference to this model will allow consideration of still another problem, in testing the implications of causal models, that formal auxiliary theories would, ideally, help to resolve.

With epistemic coefficients a, c, and e of 1, this model would imply that $r_{X'Z'}$, $r_{X'Y'}$, and $r_{Y'Z'}$ are non-zero and that $r_{X'Z'} = r_{X'Y'} r_{Y'Z'}$. The empirical tenability of this implication would provide the clue as to whether addi-

[2] A precisely parallel problem is encountered if the model is changed by reversing the arrow between X and Y, that is, if the relationship between X and Z is spurious.

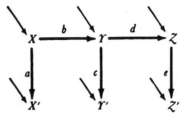

Figure 12.3. Three-variable model with one indicator for each variable.

tional causal connections between abstract variables X and Z should be added to the model. But with epistemic coefficients less than unity in absolute value, this implication will not, in general, be true. In fact, with epistemic coefficients less than unity in absolute value, and with only one indicator for each abstract variable, it is not at all clear how a decision on this implication of the model can be made empirically except by some rather casual rule of thumb. Blalock[1, pp. 148–155] has shown that random error in the measurement of the intervening variable is especially troublesome in this regard, whereas random error in the measurement of the other two variables is much less critical. But an auxiliary theory, ideally, would go further and allow a test of the implications of the abstract model on the abstract plane, uncomplicated by measurement error; the extension to still more complex models would then be relatively straightforward. This is not possible with the auxiliary theory represented in Fig. 12.3, even in the absence of any differential bias; the presence of differential bias would complicate the matter still further.

We have now enumerated and illustrated very briefly three desiderata that should ideally be accomplishable by utilizing an auxiliary theory formally representing the connections between abstract variables and their indicators. First, it should be possible to arrive at an estimate for each of the unknown coefficients, including the epistemic coefficients. Second, it should be possible to recognize differential bias, if present, and thereby recognize the inappropriateness of particular indicators in the test of a specific formulation. And third, it should be possible to test the implications of the causal connections incorporated on the abstract plane of the model, without resorting to the grossly oversimplified assumption that the epistemic coefficients are so close to unity as to be of no concern.

We can achieve these desiderata quite adequately if our auxiliary theory provides at least three indicators for each abstract variable in the model; we achieve them somewhat less completely if our auxiliary theory provides only two indicators for each abstract variable, provided the model on the abstract plane is not unduly complicated. For simplicity, we concentrate here on models that propose only one path between each pair of abstract variables. The general line of reasoning should apply to more complex models, provided they are identifiable (on the identification problem, see

Fisher [4]). We consider first auxiliary models that provide only two indicators for each abstract variable and then proceed to consider models providing three indicators.

TWO-INDICATOR MODELS

Figure 12.4 represents the model identical, on the abstract plane, with the model of Fig. 12.1, but now with an auxiliary theory that provides two indicators for each abstract variable. We assume that the correlation between all pairs of indicators is known; hence, instead of one observed correlation as in Fig. 12.1, we now have six observed correlations. With the assumption of uncorrelated error terms, each of these six observed correlations may be expressed as a function of the five unknown coefficients as follows:

$$r_{X_1 X_2} = ab \tag{12.1}$$

$$r_{Y_1 Y_2} = de \tag{12.2}$$

$$r_{X_1 Y_1} = acd \tag{12.3}$$

$$r_{X_1 Y_2} = ace \tag{12.4}$$

$$r_{X_2 Y_1} = bcd \tag{12.5}$$

$$r_{X_2 Y_2} = bce \tag{12.6}$$

These six equations allow an empirically testable deduction that serves as a clue to the presence of certain kinds of differential bias. It is evident that the model implies non-zero r's and

$$(r_{X_1 Y_1})(r_{X_2 Y_2}) = (r_{X_1 Y_2})(r_{X_2 Y_1}) \tag{12.7}$$

This may be shown by substituting the equivalents for these correlations in

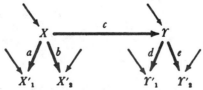

Figure 12.4. Two-variable model with two indicators for each variable.

terms of unknown coefficients, which yields

$$(acd)(bce) = (ace)(bcd) \tag{12.8}$$

$$abc^2de = abc^2de \tag{12.9}$$

If differential bias is present, as illustrated in Fig. 12.5A, equation (12.7) will not hold. In Fig. 12.5A, we have

$$r_{X_1'Y_1'} = acd \tag{12.3'}$$

$$r_{X_1'Y_2'} = ace \tag{12.4'}$$

$$r_{X_2'Y_1'} = bcd + fg \tag{12.5'}$$

$$r_{X_2'Y_2'} = bce \tag{12.6'}$$

and

$$r_{X_1'Y_1'}r_{X_2'Y_2'} \neq r_{X_1'Y_2'}r_{X_2'Y_1'} \tag{12.10}$$

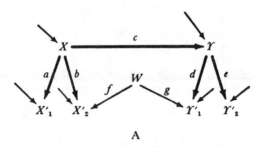

A

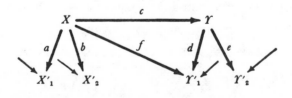

B

Figure 12.5. Two-variable models with two indicators for each variable and differential bias.

since

$$(acd)(bce) \neq (ace)(bcd + fg) \qquad (12.11)$$

In general, when differential bias provides an additional source of common variance between two and only two indicators , or when different amounts of additional common variance between different pairs of indicators are supplied by differential bias, equation (12.7) will not hold. However, equation (12.7) *will* hold for the kind of differential bias illustrated in Fig. 12.5B. In Fig. 12.5B, we have

$$r_{X_1' Y_1'} = acd + af \qquad (12.3'')$$

$$r_{X_1' Y_2'} = ace \qquad (12.4'')$$

$$r_{X_2' Y_1'} = bcd + bf \qquad (12.5'')$$

$$r_{X_2' Y_2'} = bce \qquad (12.6'')$$

and equation 12.7 holds, that is,

$$(r_{X_1' Y_1'})(r_{X_2' Y_2'}) = (r_{X_1' Y_2'})(r_{X_2' Y_1'})$$

since

$$(acd + af)(bce) = (ace)(bcd + bf)$$

$$abc^2 de + abcef = abc^2 de + abcef$$

For two-indicator models, equation (12.7), which we will call the "consistency criterion" for two-indicator models, is thus a necessary, but not a sufficient, condition for the absence of differential bias. If this equation holds exactly, the two estimates for a given path coefficient will be identical; otherwise the two estimates for a given coefficient will be unequal. Failure of the data to satisfy this equation, at least approximately, indicates that, in some respect, the indicators provided in the auxiliary theory are not appropriate for testing the abstract model. With only two indicators for each abstract variable, no test that is sufficient for ruling out all kinds of differential bias has been devised. But if equation (12.7) holds, unique estimates for each coefficient in the model may be computed, and the other desiderata

previously outlined may be fulfilled in a relatively straightforward fashion.
We return to Fig. 12.4 for a discussion of these other desiderata.

The six equations generated by the model of Fig. 12.4 (i.e., equations
[12.1], [12.2], [12.3], [12.4], [12.5], and [12.6]) yield two estimates for each
unknown coefficient; these two estimates will be identical if equation (12.7)
is exactly satisfied. Alternatively, the satisfaction of the consistency criterion
uses one of the six equations, leaving five equations in five unknowns that
may be solved to yield unique solutions for the unknown coefficients. The
solutions[3] are:

$$c^2 = \frac{(r_{X_1Y_2})(r_{X_2Y_1})}{(r_{X_1X_2})(r_{Y_1Y_2})} = \frac{(r_{X_1Y_1})(r_{X_2Y_2})}{(r_{X_1X_2})(r_{Y_1Y_2})} \tag{12.12}$$

$$a^2 = (r_{X_1X_2})\frac{(r_{X_1Y_2})}{(r_{X_2Y_2})} = (r_{X_1X_2})\frac{(r_{X_1Y_1})}{(r_{X_2Y_1})} \tag{12.13}$$

$$b^2 = (r_{X_1X_2})\frac{(r_{X_2Y_2})}{(r_{X_1Y_2})} = (r_{X_1X_2})\frac{(r_{X_2Y_1})}{(r_{X_1Y_1})} \tag{12.14}$$

$$d^2 = (r_{Y_1Y_2})\frac{(r_{X_2Y_1})}{(r_{X_2Y_2})} = (r_{Y_1Y_2})\frac{(r_{X_1Y_1})}{(r_{X_1Y_2})} \tag{12.15}$$

[3] The solutions, stated in terms of squares, leave the signs of the corresponding
path coefficients formally ambiguous. First, it should be noted that it is empirically
possible for the squares representing these solutions to be negative — an outcome
clearly inconsistent with the implications of the model and therefore requiring a
modification in it. Assuming that all squares representing solutions are positive, the
determination of the signs of the coefficients is still ambiguous, since either the
positive or the negative root could be taken. Furthermore, different patterns of signs
in the path coefficients may yield an identical pattern of signs in the observed
correlations. For example, if all indicators are inverse indicators of their respective
abstract variables, the pattern of signs among the observed correlations would be
identical to that obtaining if all indicators are direct indicators of their respective
abstract variables. This ambiguity must be resolved by making *a priori* assumptions
about the signs of the epistemic coefficients, that is, assumptions as to whether each
indicator is a direct or an inverse indicator of its abstract variable. With these assumptions, the sign of the path coefficients connecting abstract variables is no longer
ambiguous; that is, the sign of that path is the same as the sign of the observed
correlation between two direct indicators or two inverse indicators and opposite to
the sign of the observed correlation between a direct and an inverse indicator.
Assumptions about the signs of the epistemic coefficients must, of course, be consistent with the observed correlations between indicators of the same abstract variable; for example, if two indicators of the same abstract variable are negatively
correlated with each other, both cannot be assumed to be direct indicators of that
variable and both cannot be assumed to be inverse indicators of that variable.

$$e^2 = (r_{Y_1'Y_2'})\frac{(r_{X_2'Y_2'})}{(r_{X_2'Y_1'})} = (r_{Y_1'Y_2'})\frac{(r_{X_1'Y_2'})}{(r_{X_1'Y_1'})} \tag{12.16}$$

The potentiality for obtaining empirical estimates for each of the unknown coefficients of the model fulfills another of the desiderata for auxiliary theories outlined above. The possibility of deriving estimates for the epistemic coefficients bears comment, especially since they may be given more importance than they properly deserve. It is not the case that such empirically estimated epistemic coefficients provide a solution to the "semantic problem" frequently alluded to in discussing the problem of devising appropriate indicators. The empirically estimated epistemic coefficients estimate the correlation between each specific indicator and the abstract variable that is assigned a particular role in the model on the abstract plane; they have no bearing whatsoever on the appropriateness, in terms of conventional meanings, of the terms that are attached to the abstract variables. The epistemic coefficients thus do not provide a solution to the semantic aspects of the problem of indicator validity. They provide only an estimate of the degree to which extraneous factors and random error influence an indicator's service as indicator of an abstract variable that takes its meaning both from the role it is assigned in the abstract model and from the total set of indicators that are provided for it in the auxiliary theory.

We now move to the third of the desiderata for auxiliary theories outlined above. Ideally, we have noted, an auxiliary theory would allow a test of the implications of causal models on the abstract plane, uncomplicated by problems of measurement error. Figure 12.6 represents an abstract model in which such an implication on the abstract plane emerges, and reference to that figure will facilitate discussion. For easy reference, Fig. 12.7 presents the model represented in Fig. 12.6 with the intervening variable omitted. It should be clear from the preceding discussion that the consistency criterion must, in this instance, be satisfied in three different ways, that is, with respect to that segment of the model involving X and Y and their indicators, Y and Z and their indicators, and X and Z and their indicators. If the consistency criterion is not satisfied in any one of these three tests, some of the indicators are subject to differential bias and hence not appropriate to test

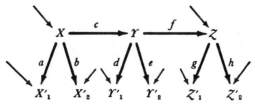

Figure 12.6. Three-variable model with two indicators for each variable.

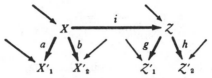

Figure 12.7. The model of Figure 12.6 with the intervening variable omitted.

the abstract model as a whole. If the consistency criterion is satisfied in all three of these tests, it should be evident that we could proceed to derive a solution for each of the three coefficients between abstract variables, c, f, and i. Such solutions are based on equations analogous to equation (12.12) for c^2. With solutions for the coefficients between abstract variables, it is no longer necessary to work with the directly observed correlations in testing the implications of the model on the abstract plane. The solutions for the coefficients on the abstract plane may be used instead. Thus, the implication of the model in Fig. 12.6 is that

$$cf = i \tag{12.17}$$

This, rather than any equation involving the directly observed correlations, provides a test for the abstract model. A high degree of random measurement error in the indicators for the intervening variable no longer has such serious implications as it does when one must work with the directly observed correlations, although such random measurement error will increase the random variation in the solutions for unknown coefficients involving that intervening variable, that is, c and f.

THREE-INDICATOR MODELS

Figure 12.8 represents the model identical, on the abstract plane, with the model of Fig. 12.1, but now with an auxiliary theory that provides three indicators for each abstract variable. Again, we assume that the correlation between all pairs of indicators is known; hence we have fifteen observed correlations, three between indicators of X, three between indicators of Y,

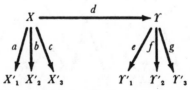

Figure 12.8. Two-variable model with three indicators for each variable.

and nine between indicators of X and Y. There are two general ways to proceed to obtain estimates of the unknown coefficients in this model, and each procedure provides different clues to the presence of differential bias. On the one hand, we may take the indicators for each abstract variable in pairs and proceed exactly as in the two-indicator case with the nine two-indicator models thus formed. On the other hand, we may take advantage of the fact that with three indicators for each abstract variable, the epistemic coefficients may be estimated utilizing only that fragment of the model that constitutes the auxiliary theory for that abstract variable. Using the solutions thus obtained, we may then proceed to obtain additional estimates of the coefficient linking the abstract variables. If the model is uncomplicated by any kind of differential bias, the several estimates derived by the two procedures for each coefficient should all be identical except for random error; inconsistencies will serve as clues to differential bias, as elaborated below.

We first consider the procedure of forming nine two-indicator models from the single three-indicator model. One such two-indicator model may be formed from each set of pairs of indicators for each abstract variable, that is, X'_1 and X'_2 combined with Y'_1 and Y'_2, X'_1 and X'_3 combined with Y'_1 and Y'_2, etc. In a manner analogous to that described above, we obtain a consistency-criterion equation for each of the nine two-indicator models, each analogous to equation (12.7) above. These nine consistency-criterion equations will have the form

$$(r_{X'_h Y'_i})(r_{X'_j Y'_k}) = (r_{X'_h Y'_k})(r_{X'_j Y'_i}) \tag{12.18}$$

where h, i, j, and k each assume the values 1, 2, and 3 subject to the restriction that $h \neq j$ and $i \neq k$.[4]

As before, the failure of any one of these equations to be satisfied indicates the presence of differential bias of the type illustrated in Fig. 12.9A, and the specific consistency-criterion equations that fail to hold will locate the indicators having a common source of variance not represented in the original model. For Fig. 12.9A, for example, showing a common source of variance between X'_3 and Y'_1, all consistency-criterion equations involving both of these indicators would fail to hold. For the model of Fig. 12.9A, all the remaining consistency-criterion equations should be true except for random measurement error.

Each of the nine two-indicator models that can be formed from a single

---

[4] Although thirty-six such equations may be formed, only nine are distinct, since the order of the r's on a given side of the equality sign is irrelevant. Thus, for example, $(r_{X'_1 Y'_1})(r_{X'_2 Y'_3}) = (r_{X'_1 Y'_3})(r_{X'_2 Y'_1})$ is the same equation as $(r_{X'_2 Y'_3})(r_{X'_1 Y'_1}) = (r_{X'_2 Y'_1})(r_{X'_1 Y'_3})$.

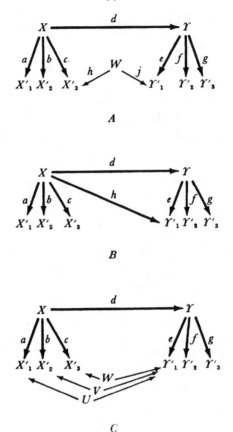

Figure 12.9. Two-variable model with three indicators for each variable and differential bias.

three-indicator model will yield two estimates of the abstract coefficient, d, for a total of eighteen such estimates. For example, in a manner analogous to that for obtaining equation (12.12) above, we obtain

$$d^2 = \frac{(r_{X_1'Y_2'})(r_{X_2'Y_1'})}{(r_{X_1'X_2'})(r_{Y_1'Y_2'})} = \frac{(r_{X_1'Y_1'})(r_{X_2'Y_2'})}{(r_{X_1'X_2'})(r_{Y_1'Y_2'})} \qquad (12.19)$$

from the two-indicator model involving X_1', X_2', Y_1', and Y_2'. These two estimates will be identical if the corresponding consistency-criterion equation holds, that is, if $(r_{X_1'Y_2'})(r_{X_2'Y_1'}) = (r_{X_1'Y_1'})(r_{X_2'Y_2'})$. Eight additional pairs of estimates for d^2 may be analogously obtained. All have the form

$$d^2 = \frac{(r_{X_h Y_i})(r_{X_j Y_k})}{(r_{X_h X_j})(r_{Y_i Y_k})} = \frac{(r_{X_h Y_k})(r_{X_j Y_i})}{(r_{X_h X_j})(r_{Y_i Y_k})} \tag{12.20}$$

where h, i, j, and k each assume the values 1, 2, and 3 subject to the restriction that $h \neq j$ and $i \neq k$.

Although the satisfaction of the nine consistency-criterion equations implies that each pair of such estimates should be equal, the satisfaction of these consistency-criterion equations is not, in itself, sufficient to imply that equality obtains between pairs. As previously noted in the discussion of two-indicator models, the satisfaction of such consistency-criterion equations is not sufficient to imply the absence of differential bias of the kind illustrated in Fig. 12.9B, even though differential bias of the kind illustrated in Fig. 12.9A will lead to the failure of some of these equations to hold. An additional criterion can be specified for three-indicator models that, if satisfied, will be sufficient to imply the absence of the kind of differential bias illustrated in Fig. 12.9B. This requires that we obtain additional estimates of the abstract coefficient, d, by a different route.

Referring once again to Fig. 12.8 and assuming no common sources of variance between the indicators X_1', X_2', and X_3' except their common dependence on the abstract variable, X, we may express the correlations between these three indicators in terms of epistemic coefficients as follows:

$$r_{X_1' X_2'} = ab \tag{12.21}$$

$$r_{X_1' X_3'} = ac \tag{12.22}$$

$$r_{X_2' X_3'} = bc \tag{12.23}$$

By simple algebra we obtain the following estimates for the squares of each of these epistemic coefficients:

$$a^2 = \frac{(r_{X_1' X_2'})(r_{X_1' X_3'})}{r_{X_2' X_3'}} \tag{12.24}$$

$$b^2 = \frac{(r_{X_1' X_2'})(r_{X_2' X_3'})}{r_{X_1' X_3'}} \tag{12.25}$$

$$c^2 = \frac{(r_{X_1' X_3'})(r_{X_2' X_3'})}{r_{X_1' X_2'}} \tag{12.26}$$

In an analogous manner, we obtain the following equations for the correlations between Y indicators:

$$r_{Y_1 Y_2} = ef \tag{12.27}$$

$$r_{Y_1 Y_3} = eg \tag{12.28}$$

$$r_{Y_2 Y_3} = fg \tag{12.29}$$

And the following estimates for the squares of the epistemic coefficients for Y may be obtained as simple algebraic solutions of the three equations above:

$$e^2 = \frac{(r_{Y_1 Y_2})(r_{Y_1 Y_3})}{r_{Y_2 Y_3}} \tag{12.30}$$

$$f^2 = \frac{(r_{Y_2 Y_3})(r_{Y_1 Y_2})}{r_{Y_1 Y_3}} \tag{12.31}$$

$$g^2 = \frac{(r_{Y_1 Y_3})(r_{Y_2 Y_3})}{r_{Y_1 Y_2}} \tag{12.32}$$

We note further that, with the same assumption of uncorrelated error terms, the model of Fig. 12.8 also implies

$$r_{X_1 Y_1} = ade \tag{12.33}$$

$$r_{X_1 Y_2} = adf \tag{12.34}$$

$$r_{X_1 Y_3} = adg \tag{12.35}$$

$$r_{X_2 Y_1} = bde \tag{12.36}$$

$$r_{X_2 Y_2} = bdf \tag{12.37}$$

$$r_{X_2 Y_3} = bdg \tag{12.38}$$

$$r_{X_3 Y_1} = cde \tag{12.39}$$

$$r_{X_3 Y_2} = cdf \tag{12.40}$$

$$r_{X_3 Y_3} = cdg \tag{12.41}$$

Squaring both sides of these equations and substituting in them the solutions given above (equations [12.24], [12.25], [12.26], [12.30], [12.31], and [12.32]) for the epistemic coefficients, we obtain nine additional estimates for d^2. All will have the form

$$d^2 = \frac{(r^2_{X_h Y_i})(r_{X_j X_k})(r_{Y_m Y_n})}{(r_{X_h X_j})(r_{X_h X_k})(r_{Y_i Y_m})(r_{Y_i Y_n})} \tag{12.42}$$

where h, i, j, k, m, and n each assume the values 1, 2, and 3, subject to the restriction that $h \neq j \neq k$ and $i \neq m \neq n$.

In the absence of differential bias (i.e., as in the model of Fig. 12.8) all nine estimates of d^2 should be identical, except for random measurement error. The model of Fig. 12.9A, on the other hand, involves differential bias and will not yield identical estimates. That model yields a set of nine equations identical with equations (12.33)–(12.41), except that equation (12.39) will be

$$r_{X_3 Y_1} = cde + hj \tag{12.39'}$$

and the one estimate of d^2 based on $r_{X_3 Y_1}$ will be an overestimate if hj is positive, that is,

$$d^2 < \frac{(r^2_{X_3 Y_1})(r_{X_1 X_2})(r_{Y_2 Y_3})}{(r_{X_1 X_3})(r_{X_2 X_3})(r_{Y_1 Y_3})(r_{Y_1 Y_2})} \tag{12.43}$$

As previously noted, the presence of differential bias of the type represented in Fig. 12.9A would also have been indicated by the fact that certain of the consistency-criterion equations (having the form of equation [12.18]) would fail to hold. However, the consistency of the nine estimates for d^2 given by equations having the form of equation (12.42) will also be sensitive to the presence of differential bias of the type represented in Fig. 12.9B. In the model of Fig. 12.9B

$$r_{X_1 Y_1} = ade + ah \tag{12.33'}$$

$$r_{X_2 Y_1} = bde + bh \tag{12.36'}$$

and

$$r_{X_3'Y_1'} = cde + ch \qquad (12.39'')$$

Estimates of d^2 having the form of equation (12.42) and based on $r_{X_1'Y_1'}$, $r_{X_2'Y_1'}$, or $r_{X_3'Y_1'}$ will be overestimates if the paths a, b, c, and h are all positive. Thus, these three estimates would diverge from the other six in the same direction but not necessarily to the same degree.

We have now defined two types of consistency criteria for three-indicator models. The first consists of a set of nine equations of the form of equation (12.18) which are analogous to equation (12.7) for the two-indicator model. Failure of any one of these equations to hold indicates the presence of differential bias of the type illustrated in Fig. 12.9A. The second type of consistency criteria for three-indicator models consists of a set of nine estimates of d^2 having the form of equation (12.42). The failure of all of these estimates to be equal to each other indicates the presence of differential bias, either of the type represented in Fig. 12.9A or in Fig. 12.9B. More specifically, the divergence of a single estimate in this set of nine from the remaining estimates indicates the presence of differential bias of the type represented in Fig. 12.9A, that is, extraneous common variance between one indicator of X and one indicator of Y. The divergence (not necessarily equal divergence) of three of these estimates of d^2, each of which is based on a single Y indicator, would indicate the presence of differential bias of the type represented in Fig. 12.9B if the consistency-criterion equations of the form of equation (12.18) were all satisfied but would indicate the type of differential bias suggested by Fig. 12.9C if the consistency-criterion equations involving Y_1' were not satisfied. The three-indicator model thus allows a test for a type of differential bias not possible with the two-indicator model by introducing an additional consistency criterion, namely, that all nine estimates of d^2 given by equations having the form of equation (12.42) be identical.

We may go one step further with the three-indicator model. We have been concerned above with differential bias, that is, a source of common variance extraneous to the model between at least one indicator of X and at least one indicator of Y. There is another kind of nonrandom measurement error which can also distort the estimates of the abstract coefficient (d in the model of Fig. 12.8). This is a source of common variance between the indicators of the same abstract variable other than their common dependence on that abstract variable. This is illustrated in Fig. 12.10. In the model of Figure 12.10, X_1' and X_2' have common variance both because of their common dependence on X and because of their common dependence on W. The correlation $r_{X_1'X_2'}$ will therefore be larger than would be the case without this extraneous common variance, assuming, for simplicity, that all coefficients are positive. As a consequence, all estimates of d^2 having the term $r_{X_1'X_2'}$ in the numerator would be overestimates of d^2, while all estimates

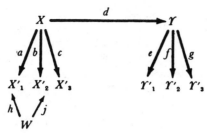

Figure 12.10. Two-variable model with three indicators for each variable and with an extraneous source of common variance between two indicators of the same variable.

having that term in the denominator would be underestimates. The three-indicator model yields twenty-seven estimates of d^2 (i.e., eighteen estimates having the form of equation [12.20] and an additional nine estimates having the form of equation [12.42]). Twelve of these estimates have $r_{X'_1 X'_2}$ in the denominators and will be underestimates (assuming a, b, h, and j all positive), three have that term in the numerators and will be overestimates, while the remaining twelve do not include that term and will not be affected. Clearly, different types of nonrandom measurement error will result in different patterns of divergence among the twenty-seven estimates of d^2 given by the three-indicator model and thus allow the pattern of inconsistency among estimates to be used as a clue, not only to the presence of differential bias, but also to the nature of whatever nonrandom measurement error may be operating. Certain complex combinations of different types of nonrandom error, however, would probably defy an attempt to discern the pattern.

If all the above-named consistency conditions for the three-indicator model are satisfied, it will be possible to derive estimates for all the coefficients and to test the implications of the model on the abstract plane in a manner analogous to that represented in equation (12.7) above.

DISCUSSION

Since the various consistency conditions outlined above are crucial for obtaining unique estimates of the unobservable coefficients in abstract causal models, some brief elaboration of these conditions may be appropriate. The consistency conditions defined here serve to underscore the points made in several insightful discussions of the utility of multiple indicators in testing abstract propositions (Curtis and Jackson [3]; Lazarsfeld [6]; Webb *et al.* [9] Chap. 1). Let it be noted, however, that the consistency criteria here defined do *not* require that all correlations between different pairs of indicators be identical; rather, the criteria require that certain products of correlations shall be identical or that several estimates of a single abstract coefficient shall be consistent. These criteria may be met even though all observed correlations between X and Y indicators are quite different from each other. The expectation that all correlations between different pairs of indicators be identical is unnecessarily restrictive; it assumes, in effect, that all epistemic coefficients between a given abstract variable and its indicators are identical, that is, that all indicators of a given variable are equally good indicators, which is contrary to our common thinking about the uneven quality of indicators.

It may be reasonably asked what is meant when we say that the consistency criterion is satisfied. Do we mean that the two sides of equation (12.7) are exactly identical, that they are approximately identical, or that they should not differ to a degree that is statistically significant at the commonly utilized levels of significance? If all measures are made on the same set of cases, no variation between the two sides of equation (12.7) should be attributable to variation between samples of cases, and since random measurement error has been explicitly taken into account in the model, it may appear that such error would not contribute to variation between the two sides of equation (12.7). While it is true that measurement error has been taken into account in one sense, it has not been taken into account in a way that rules out its effect in producing variation between the two sides of equation (12.7). In effect, equation (12.7) asserts the equality of two estimates of the same abstract coefficient, c. But both estimates are subject to sampling variability to the degree that the epistemic coefficients (a, b, d, and e) are less than unity. What we mean, then, when we say that the consistency criterion is satisfied is that the two sides of equation (12.7) do not differ from each other to a statistically significant degree. This is formally identical to the vanishing of the "tetrad difference" in the classic Spearman factor analysis, and the standard error of the "tetrad difference" is known (Holzinger [5, p. 6]). Satisfying the additional consistency criteria in the three-indicator model presents an additional statistical inference problem, the solution to which does not appear to be found in the factor analysis literature.

If the consistency criterion is satisfied in a particular model or segment of a model, there may be a temptation to interpret it as a validation test for the indicators. The consistency criterion is not a validation of indicators; the absence of differential bias for a given set of indicators in the context of one specific model is no guarantee that differential bias for some of those same indicators will be absent in the context of a different model — or even in another segment of the same complex model. The satisfaction of the consistency criterion is a feature of the model or a segment thereof; it is not a feature of the indicators themselves that can be transferred with them to other models.

The general conclusion to be reached from this discussion is that, although causal models are strictly untestable with a single indicator for each abstract variable unless one assumes very slight measurement error, an auxiliary theory providing multiple indicators for each of the abstract variables will, assuming the consistency criteria are met, allow a test of the implications of the abstract causal model and provide, in addition, estimates of the epistemic coefficients involved in the auxiliary theory itself. A crucial matter in the whole enterprise, however, is the satisfaction of the consistency criteria as a guard against differential bias, and we may find that certain causal models are simply not testable with certain indicators because differential bias is present.

The general strategy of devising and utilizing auxiliary theories that has been outlined in this chapter, with the crucial role it assigns to the intercorrelation between different indicators of a single abstract variable, is probably inappropriate as a guide for dealing with formulations at the highest levels of abstraction. The ties between very highly abstract concepts and the empirical world appear to take a form that is different from that assumed in this discussion. Specifically, highly abstract concepts are frequently designed to encompass a variety of different forms of a given phenomenon that are not necessarily intercorrelated with each other. There is no reason to assume *a priori*, for example, that all of the many forms of deviant behavior are intercorrelated. Similarly, frustrations are many and varied, and the degree to which one suffers frustration in one guise is no clue to the degree of frustration of another kind. The admission of uncorrelated indicators for a given abstract variable renders the strategy discussed in this chapter inapplicable. The problems associated with detailing the connections between such highly abstract concepts, their uncorrelated subforms, and the indicators of each subform — and doing so in a way that allows an unambiguous empirical test of the theory at the highest level of abstraction — will undoubtedly require a more complex and intricate kind of auxiliary theory than the relatively simple type employed in the present discussion. Some progress toward the development of these more complex and intricate auxiliary theories would help provide a needed integration of high levels of abstraction and testable deductions in sociological theory.

ACKNOWLEDGMENTS

Revised version of paper presented at the meeting of the American Sociological Association, Boston, Massachusetts, August 26–29, 1968. I wish to thank Hubert M. Blalock, Jr., Otis Dudley Duncan, Jack P. Gibbs, Arthur S. Goldberger, and Karl F. Schuessler for helpful comments on an earlier version of this chapter.

REFERENCES

[1] Blalock, Hubert M., Jr. *Causal Inferences in Nonexperimental Research.* Chapel Hill: University of North Carolina Press, 1961.
[2] Blalock, Hubert M., Jr. "The Measurement Problem: A Gap between the Lan-

guages of Theory and Research." In *Methodology in Social Research*, edited by H. M. Blalock and A. Blalock. New York: McGraw-Hill, 1968.

[3] Curtis, Richard F., and Jackson, Elton F. "Multiple Indicators in Survey Research." *American Journal of Sociology* 68 (1962): 195–204.

[4] Fisher, Franklin M. *The Identification Problem in Econometrics*. New York: McGraw-Hill, 1966.

[5] Holzinger, Karl John. *Statistical Resumé of the Spearman Two-Factor Theory*. Chicago: University of Chicago Press, 1930.

[6] Lazarsfeld, Paul F. "Problems in Methodology." In *Sociology Today: Problems and Prospects*, edited by R. K. Merton *et al.* New York: Basic Books, 1959.

[7] Siegel, Paul M., and Hodge, Robert W. "A Causal Approach to the Study of Measurement Error." In *Methodology in Social Research*, edited by H. M. Blalock and A. Blalock. New York: McGraw-Hill, 1968.

[8] Simon, Herbert A. "Spurious Correlation: A Causal Interpretation." In *Models of Man*. New York: John Wiley & Sons, 1959.

[9] Webb, Eugene T., Campbell, Donald T., Schwartz, Richard D., and Sechrest, Lee. *Unobtrusive Measures: Nonreactive Research in the Social Sciences*. Chicago: Rand McNally, 1966.

Estimating the Parameters of Path Models Involving Unmeasured Variables

Charles E. Werts

Robert L. Linn

Karl G. Jöreskog

13

Costner [3], Blalock [1], and Heise [6] have demonstrated the application of path analysis to problems involving multiple indicators of underlying constructs. In this type of problem, the researcher is frequently called on to deal with overidentified systems; for example, Costner has demonstrated the use of the "consistency criterion" (Spearman tetrad difference) as a test of the fit of the data to the hypothetical model in simple two construct systems with two independent measures of each construct. The purpose of this note is to discuss the use of a general method for factor analysis (Jöreskog [8]) to obtain a single "best fit" estimate of each parameter and an overall goodness of fit test for the consistency of the data with the model. Factor analysis is a special case of Jöreskog's [10] general method for the analysis of covariance structures that could be used instead of the confirmatory factor analysis procedure discussed in this chapter.

USING CONFIRMATORY FACTOR ANALYSIS TO OBTAIN ESTIMATES FOR PATH MODELS

Jöreskog's confirmatory factor analysis procedures allow the investigator to fix some parameters (called "fixed") of the model and estimate the others (called "free"). This procedure can be illustrated with the problem depicted

251

in Fig. 13.1 in which three measures X_0, X_1 and X_2 of mathematics achievement (T_1) are administered at time 1 and two of these measures X_3 and X_4 are repeated at a later time 2 (T_2). Thus, X_1 and X_3 are obtained from the same test as are X_2 and X_4 and the corresponding errors of measurement are assumed to be correlated for this reason. Suppose the researcher wants to estimate the reliabilities of the tests and the correlation between T_1 and T_2. In order to analyze this model, the path analyst would write the following equations involving the correlation among variables and path coefficients (b^*):

$$\rho_{01} = b^*_{X_0T_1} b^*_{X_1T_1}$$

$$\rho_{02} = b^*_{X_0T_1} b^*_{X_2T_1}$$

$$\rho_{12} = b^*_{X_1T_1} b^*_{X_2T_1}$$

$$\rho_{03} = b^*_{X_0T_1} \rho_{T_1T_2} b^*_{X_3T_2}$$

$$\rho_{04} = b^*_{X_0T_1} \rho_{T_1T_2} b^*_{X_4T_2} \tag{13.1}$$

$$\rho_{13} = b^*_{X_1T_1} \rho_{T_1T_2} b^*_{X_3T_2} + b^*_{X_1e_1} \rho_{e_1e_3} b^*_{X_3e_3}$$

$$\rho_{14} = b^*_{X_1T_1} \rho_{T_1T_2} b^*_{X_4T_2}$$

$$\rho_{23} = b^*_{X_2T_1} \rho_{T_1T_2} b^*_{X_3T_2}$$

$$\rho_{24} = b^*_{X_2T_1} \rho_{T_1T_2} b^*_{X_4T_2} + b^*_{X_2e_2} \rho_{e_2e_4} b^*_{X_4e_4}$$

and

$$\rho_{34} = b^*_{X_3T_2} b^*_{X_4T_2}$$

where

$$b^*_{X_1e_1} = \sqrt{1 - (b^*_{X_1T_1})^2}$$

$$b^*_{X_2e_2} = \sqrt{1 - (b^*_{X_2T_2})^2}$$

$$b^*_{X_3e_3} = \sqrt{1 - (b^*_{X_3T_2})^2}$$

and

$$b^*_{X_4e_4} = \sqrt{1 - (b^*_{X_4T_2})^2}$$

From the perspective of structural analysis, these equations can only be useful for exploring the identification question. If each of the path coefficients can be expressed as a function of the correlations in at least one way,

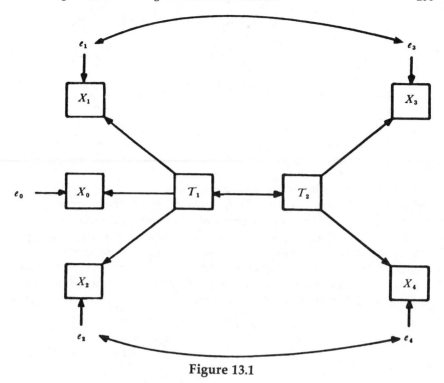

Figure 13.1

then the system is identifiable. The factor analyst who is dealing with this type of problem prefers to explore the identification question in matrix form (Jöreskog [8, p. 186]), checking to see if there is a "unique" solution for each parameter, that is, if all linear transformations of the factors that leave the fixed parameters unchanged also leave the free parameters unchanged. In Jöreskog's approach, finding all possible solutions of these path equations (as in Costner, [3]) has no purpose (except for possibly exploring the identification question). All that is required is that the fundamental structural equations from which the path equations derive be stated in matrix form. To formulate the above model in terms of Jöreskog's general model, one has to write a structural equation for each observed variable. Each equation relates an observed variable to some unobserved variables. In Fig. 13.1, there are even unobserved variables all together, namely, the two true scores T_1 and T_2 and the five error scores e_0, e_1, e_2, e_3, and e_4. For a given observed variable X_i, there are as many terms in the equation for X_i as there are arrows pointing to X_i in the graph. It is convenient to regard each variable as being measured as deviations from its mean so that mean values can be excluded in the equations. The equations corresponding to Fig. 13.1 are

$$X_0 = \beta_0 T_1 + e_0$$

$$X_1 = \beta_1 T_1 + e_1$$

$$X_2 = \beta_2 T_1 + e_2 \tag{13.2}$$

$$X_3 = \beta_3 T_2 + e_3$$

$$X_4 = \beta_4 T_2 + e_4$$

The error terms are assumed to be uncorrelated with T_1 and T_2. The path equations (13.1) may be derived from equations (13.2) using the procedure given by Duncan [4]. If the error terms were all mutually uncorrelated, the equations (13.2) would represent a factor analysis model with two oblique factors and certain factor loadings specified to be zero (Jöreskog [8]) or as a model for two sets of congeneric tests (Jöreskog [9]). However, when some error terms are correlated, it is necessary to formulate the model differently. Equations (13.2) may be written in matrix form as

$$
\begin{pmatrix} X_0 \\ X_1 \\ X_2 \\ X_3 \\ X_4 \end{pmatrix}
=
\begin{bmatrix}
\beta_0 & 0 & 1 & 0 & 0 & 0 & 0 \\
\beta_1 & 0 & 0 & 1 & 0 & 0 & 0 \\
\beta_2 & 0 & 0 & 0 & 1 & 0 & 0 \\
0 & \beta_3 & 0 & 0 & 0 & 1 & 0 \\
0 & \beta_4 & 0 & 0 & 0 & 0 & 1
\end{bmatrix}
\begin{pmatrix} T_1 \\ T_2 \\ e_0 \\ e_1 \\ e_2 \\ e_3 \\ e_4 \end{pmatrix}
\tag{13.3}
$$

It was assumed that e_0, e_1, and e_2 are uncorrelated and so also e_3 and e_4 but that e_1 is correlated with e_3 and e_2 is correlated with e_4. The variance–covariance matrix of $(T_1, T_2, e_0, e_1, e_2, e_3, e_4)$ is therefore of the form

$$
\Phi =
\begin{bmatrix}
1 & & & & & & \\
\rho_{T_1 T_2} & 1 & & & & & \\
0 & 0 & \sigma_{e_0}^2 & \text{symmetric} & & & \\
0 & 0 & 0 & \sigma_{e_1}^2 & & & \\
0 & 0 & 0 & 0 & \sigma_{e_2}^2 & & \\
0 & 0 & 0 & \sigma_{e_1 e_3} & 0 & \sigma_{e_3}^2 & \\
0 & 0 & 0 & 0 & \sigma_{e_2 e_4} & 0 & \sigma_{e_4}^2
\end{bmatrix}
\tag{13.4}
$$

The ones in the diagonal of this matrix indicate that the variances of T_1 and T_2 were for convenience standardized to unit variance, and the error variances in the diagonal indicate that these variances are "free parameters" to be estimated by the program. The off diagonal zeros indicate that the corresponding factors are assumed uncorrelated; the remaining off diagonal covariances ($\rho_{T_1 T_2}$, $\sigma_{e_1 e_3}$, $\sigma_{e_2 e_4}$) between the correlated factors being estimated ("free parameters") by the program. The covariance between T_1 and T_2 is a correlation because of the standardization. The basic input into Jöreskog's computer program is the identification of fixed and free parameters defined by equations (13.3) and (13.4) and the sample variance–covariance matrix or correlation matrix (see Jöreskog, Gruvaeus, and van Thillo [11]).

From (13.3) and (13.4) the population variance–covariance matrix (Σ) of the observed scores (X_0, X_1, . . . X_4) is

$$\Sigma = \Lambda \Phi \Lambda' \tag{13.5}$$

where Λ is the 5×7 matrix of factor loadings from equation (13.3), Λ' is the transpose of this matrix, and Φ is the matrix in (13.4). This is a special case of Jöreskog's [8] confirmatory factor analysis model with five observed scores, seven factors and no residual factors. In this model, there are thirteen parameters to be estimated from the data, namely the five β_i, the five variances $\sigma_{e_i}^2$, the correlation $\rho_{T_1 T_2}$ and the two covariances $\sigma_{e_1 e_3}$ and $\sigma_{e_2 e_4}$. However, there are fifteen (five population variances plus ten population covariances) independent elements in Σ, so that (13.5) represents a system of fifteen equations in thirteen unknowns. In practice, the population matrix Σ is not available and the parameters have to be estimated from a sample variance–covariance matrix. Two different methods of estimation may be used: the least-squares method and the maximum likelihood method (see Jöreskog, Gruvaeus, and van Thillo [11]). The maximum likelihood method is based on the assumption that the observed variables have a multivariate normal distribution, and with this method, a chi-square statistic will be obtained for testing the goodness of fit of the model. Approximate standard errors may also be obtained for each estimated parameter. The least-squares method does not depend on any distributional assumptions, but with this method, no goodness of fit statistic is available.

A question that arises in the analysis of models of this kind is: When is it appropriate to standardize the observed variables and use the sample correlation matrix instead of the sample variance–covariance matrix? Standardization is convenient when the units of measurement are arbitrary or irrelevant but should not be done unless the model is scale-free. A model is said to be scale-free if a change in the unit of measurement in one or more of the observed variables can be appropriately absorbed by a corresponding change in the parameters. The model (13.5) is scale-free. For example, it is seen that if we use aX_1 and bX_3 instead of X_1 and X_3, the new parameters β_1^*, β_3^*, $\sigma_{e_1}^{2*}$, $\sigma_{e_3}^{2*}$, and $\sigma_{e_1 e_3}^{**}$ will be $a\beta_1$, $b\beta_3$, $a^2\sigma_{e_1}^2$, $b^2\sigma_{e_3}^2$ and $ab\sigma_{e_1 e_3}$, so that one can get the estimates from the old with a knowledge about the scale-factors a and b. However, if it is assumed that two or more of the X_i are parallel or tau-equivalent, each of these measures cannot be standardized separately. Standardization can only be done to a common metric for the variables that are assumed to be equivalent.

CAUTIONS CONCERNING CORRELATED ERRORS

For the example in Fig. 13.1, the model may be phrased in terms of the multitrait–multimethod approach (Campbell and Fiske [2]) by postulating the existence of one method (M) factor for each of the repeated tests. The structural equations would then be:

$$X_0 = \beta_0 T_1 + e_0'$$

$$X_1 = \beta_1 T_1 + \alpha_1 M_1 + e_1'$$

$$X_2 = \beta_2 T_1 + \alpha_2 M_2 + e_2' \qquad (13.6)$$

$$X_3 = \beta_3 T_2 + \alpha_3 M_1 + e_3'$$

$$X_4 = \beta_4 T_2 + \alpha_4 M_2 + e_4'$$

where the e_i' are assumed to be mutually uncorrelated and uncorrelated with the T_i or M_i.

It can be shown that $\alpha_1, \alpha_2, \alpha_3$, and α_4 cannot be separately identified but only the products $(\alpha_1)(\alpha_3)$ and $(\alpha_2)(\alpha_4)$ can be identified. This problem might be handled assuming $\alpha_1 = \alpha_3 = \alpha'$ and $\alpha_2 = \alpha_4 = \alpha''$. Given this assumption, (13.6) can be written

$$
\begin{pmatrix} X_0 \\ X_1 \\ X_2 \\ X_3 \\ X_4 \end{pmatrix} =
\begin{bmatrix}
\beta_0 & 0 & 0 & 0 \\
\beta_1 & 0 & \alpha' & 0 \\
\beta_2 & 0 & 0 & \alpha'' \\
0 & \beta_3 & \alpha' & 0 \\
0 & \beta_4 & 0 & \alpha''
\end{bmatrix}
\begin{pmatrix} T_1 \\ T_2 \\ M_1 \\ M_2 \end{pmatrix} +
\begin{bmatrix} e_0' \\ e_1' \\ e_2' \\ e_3' \\ e_4' \end{bmatrix}
\qquad (13.7)
$$

In matrix notation (13.7) is written as

$$X = \Lambda F + e'$$

where F is the vector of factors and e' is the vector of residuals. The variance–covariance matrix of the factors is assumed to be

$$\Phi = \begin{bmatrix} 1 & \rho_{T_1 T_2} & 0 & 0 \\ \rho_{T_1 T_2} & 1 & 0 & 0 \\ 0 & 0 & 1 & 0 \\ 0 & 0 & 0 & 1 \end{bmatrix}$$

The population variance–covariance matrix is then:

$$\Sigma = \Lambda \Phi \Lambda' + \psi^2 \tag{13.8}$$

where ψ^2 is a diagonal matrix of residual variances, that is, indicating residual terms.

Models (13.5) and (13.8) are equivalent. Both have thirteen independent parameters, and there is a one-to-one correspondence between the parameters of each model. The purpose of discussing a multitrait–multimethod formulation of the problem is because the assumption that the methods factors are independent of each other and the trait factors must be made explicitly. Hopefully, the investigator will be reminded that such assumptions must be substantively justified. Unless some theoretical specifications are made about the nature of the methods factors, there is no way of considering the reasonableness of the assumptions. If, for example, the methods factor resulted from memory for the items from the first occasion to the second, then the "method" or memory factor in the first test might well be associated with the "method" or memory factor on the second test, the degree of association depending perhaps on the similarity of item formats. Furthermore, a memory factor might well be associated with the mathematics achievement factor (i.e., "trait" in Campbell and Fiske's language) since memory ability may influence mathematics achievement. On close examination, the investigator may discover that a "methods" factor may be really just another "trait" factor. For these reasons, we strongly recommend against the common practice of drawing correlated error terms on path diagrams without explicit substantive justification. Researchers should try to specify exactly what influences they believe underlie the correlated errors and consider whether such influences may be reasonably assumed to be independent of the true trait factors on which the study has focused.

An important question in multitrait–multimethod models is whether two traits in fact differ. The correlation between T_1 and T_2 is a measure of trait similarity, but there are no rules for knowing how high this correlation should be before the traits are considered to be essentially the same trait for theoretical purposes. If the researcher wishes a significance test of whether $\rho_{T_1 T_2} = 1$, then the procedure given by Jöreskog [9, pp. 14–17] can be used.

UNSTANDARDIZED PATH ANALYSIS

It can be observed that in dealing with the unmeasured variable problem, Costner [3], Blalock [1], and Heise [6] have typically made assumptions involving correlations; for example, Heise assumed that reliability remained constant over time. Path analysis may also be performed using unstandardized regression coefficients, and assumptions may be made at this level. For example, suppose that in the model depicted in Fig. 13.1 that no third measure (X_0) at time 1 was available, which would mean that the model was underidentified (six known versus seven unknown parameters). If the researcher were willing to assume that X_1 and X_2 had the same units of measurement and were measures of the same factor, this would correspond to the assumption that the corresponding unstandardized regression weights were equal, that is, $B_{X_1 T_1} = B_{X_2 T_1}$ or $R_{X_1 T_1} \sigma_{X_1} = R_{X_2 T_1} \sigma_{X_2}$. The assumption means that the covariance of X_1 with T_1 equals the covariance of X_2 with T_1 and that the corresponding path coefficients are:

$$b^*_{X_1 T_1} = \sqrt{R_{12}(\sigma_2 \div \sigma_1)}$$

and
$$b^*_{X_2 T_1} = \sqrt{R_{12}(\sigma_1 \div \sigma_2)}$$

where $\sigma_i^2 =$ variance of X_i.

Given these coefficients, the other unknown coefficients may be computed. In the particular case of repeated tests, it is often more plausible to assume the units of measurement for the same test over time stay constant (e.g., $B_{X_1 T_1} = B_{X_3 T_2}$), although the reliabilities may change (e.g., $R_{X_1 T_1} \neq R_{X_3 T_2}$). This assumption allows one to compute the total regression of T_2 on T_1 as follows:

1. From path analysis of the model in Fig. 13.1 (X_0 excluded):

$$R_{23} = b^*_{X_2 T_1} R_{T_1 T_2} b^*_{X_3 T_2}$$

$$R_{12} = b^*_{X_1 T_1} b^*_{X_2 T_1}$$

2. By substitution:

$$R_{23} = \left(\frac{R_{12}}{b^*_{X_1 T_1}}\right) R_{T_1 T_2} b^*_{X_3 T_2}$$

3. Substituting in the unstandardized weights

$$b^*_{X_1 T_1} = \frac{B_{X_1 T_1} \sigma_{T_1}}{\sigma_1} \text{ and } b^*_{X_3 T_2} = \frac{B_{X_3 T_2} \sigma_{T_2}}{\sigma_3}$$

$$R_{23} = R_{12} \left(\frac{\sigma_1}{B_{X_1 T_1} \sigma_{T_1}} \right) (R_{T_1 T_2}) \left(\frac{B_{X_3 T_2} \sigma_{T_2}}{\sigma_3} \right)$$

4. Since

$$B_{T_2 T_1} = R_{T_1 T_2} (\sigma_{T_2} / \sigma_{T_1})$$

the assumption that

$$B_{X_1 T_1} = B_{X_3 T_2}$$

substituted into step 3 implies that

$$R_{23} = R_{12} \left(\frac{\sigma_1}{\sigma_3} \right) B_{T_2 T_1}$$

or

$$B_{T_2 T_1} = \frac{\sigma_{23}}{\sigma_{12}}$$

5. Path analysis also yields the correlation of T_1 and T_2 as:

$$R_{T_1 T_2} = \sqrt{\frac{R_{14} R_{23}}{R_{12} R_{34}}} = \sqrt{\frac{\sigma_{14} \sigma_{23}}{\sigma_{12} \sigma_{34}}}$$

6. Therefore, the ratio of the true variances is:

$$\frac{\sigma_{T_2}}{\sigma_{T_1}} = \frac{B_{T_2 T_1}}{R_{T_1 T_2}} = \frac{\sigma_{23}}{\sigma_{12}} \sqrt{\frac{\sigma_{12} \sigma_{34}}{\sigma_{14} \sigma_{23}}}$$

or

$$\frac{\sigma_{T_2}}{\sigma_{T_1}} = \sqrt{\frac{\sigma_{23} \sigma_{34}}{\sigma_{12} \sigma_{14}}}$$

When assumptions are to be made at the units of measurement level, it may be simpler for purposes of studying the identification question to write the path equations in terms of covariances and unstandardized regression weights; for example, the equation Fig. 13.1 (X_0 excluded) would be:

$$\sigma_{14} = B_{X_1 T_1} \sigma_{T_1 T_2} B_{X_4 T_2} = B_{X_1 T_1} B_{T_2 T_1} B_{X_4 T_2} \sigma_{T_1}^2$$

$$\sigma_{23} = B_{X_2 T_1} \sigma_{T_1 T_2} B_{X_3 T_2} = B_{X_2 T_1} B_{T_2 T_1} B_{X_4 T_2} \sigma_{T_1}^2$$

$$\sigma_{12} = B_{X_1 T_1} B_{X_2 T_1} \sigma_{T_1}^2$$

and $\qquad\quad \sigma_{34} = B_{X_3 T_2} B_{X_4 T_2} \sigma_{T_2}^2$

The purpose in presenting the algebra of unstandardized systems is that the user of Jöreskog's program needs to understand the consequences of various assumptions, and the question of identifiability needs to be explored prior to actual computations. Thus, the user must know that assuming $B_{X_1 T_1} = B_{X_2 T_1}$ leads to the identification of the corresponding path coefficients in the Fig. 13.1 model (excluding X_0), whereas the assumption that $B_{X_1 T_1} = B_{X_3 T_2}$ does not identify the corresponding path coefficients but fixes the ratio of the variance in T_2 to the variance in T_1. In the former case, it is permissible to standardize both T_1 and T_2, whereas in the latter case, only one of these factors should be standardized (e.g., if T_1 is standardized, then the program estimates the true unstandardized regression weight $B_{T_2 T_1}$).

The assumption that $B_{X_1 T_1} = B_{X_3 T_2}$ is basic to longitudinal growth studies since growth implies a change with time along a particular dimension. If this assumption is inserted into the model in Fig. 13.1 (i.e., in equation [13.2] $\beta_1 = \beta_3$) and only T_1 is standardized, it becomes directly possible to estimate the true correlation of status with gain from estimates generated by the model:

$$\hat{R}_{(T_1)(T_2 - T_1)} = \frac{\hat{\sigma}_{T_1 T_2} - 1}{\sqrt{\hat{\sigma}_{T_2}^2 + 1 - 2\hat{\sigma}_{T_1 T_2}}}$$

where $\hat{\sigma}_{T_1 T_2}$ is the estimated covariance of T_1 and T_2, and $\hat{\sigma}_{T_2}^2$ is the estimated variance of T_2.

If the problem is to estimate the true gain for each person, it is helpful to define a change factor Δ defined as $T_2 = T_1 + \Delta$. Equations (13.2) now become

$$X_0 = \beta_0 T_1 + e_0$$

$$X_1 = \beta_1 T_1 + e_1$$

$$X_2 = \beta_2 T_1 + e_2$$

$$X_3 = \beta_3 T_1 + \beta_3 \Delta + e_3$$

$$X_4 = \beta_4 T_1 + \beta_4 \Delta + e_4$$

In essence, Δ has replaced T_2 and estimating true gain involves the familiar problem of estimating factor scores. Jöreskog [7] gives additional details in applying his general model to growth studies.

REFERENCES

[1] Blalock, H. M., Jr. "Multiple Indicators and the Causal Approach to Measurement Error." *American Journal of Sociology* 75 (1969): 264–272.

[2] Campbell, D. T., and Fiske, D. W. "Convergent and Discriminant Validation by the Multitrait-Multimethod Matrix." *Psychological Bulletin* 56 (1959): 81–105.

[3] Costner, Herbert L., "Theory, Deduction, and Rules of Correspondence." *American Journal of Sociology* 75 (1969): 245–263.

[4] Duncan, O. D. "Path Analysis: Sociological Examples." *American Journal of Sociology* 72 (1966): 1–16.

[5] Guttman, L. A. "A Generalized Simplex for Factor Analysis." *Psychometrika* 20 (1955): 173–192.

[6] Heise, D. R. "Separating Reliability and Stability in Test-Retest Correlation." *American Sociological Review* 34 (1959): 93–101.

[7] Jöreskog, K. G. "Factoring the Multitest-Multioccasion Correlation Matrix." *Research Bulletin* 69–62 (1969). Princeton, N.J.: Educational Testing Service, July.

[8] Jöreskog, K. G. "A General Approach to Confirmatory Maximum Likelihood Factor Analysis." *Psychometrika,* 34 (1969): 183–202.

[9] Jöreskog, K. G. "Statistical Analysis of Sets of Congeneric Tests." *Research Bulletin* 69–97 (1969). Princeton, N.J.: Educational Testing Service, December.

[10] Jöreskog, K. G. "A General Method for Analysis of Covariance Structures." *Biometrika* 57 (1970): 239–251.

[11] Jöreskog, K. G., Gruvaeus, G. T., and van Thillo, M. "ACOVS—A General Computer Program for Analysis of Covariance Structures." *Research Bulletin* 70–15 (1970): Princeton, N.J.: Educational Testing Service, February.

Multiple Indicator Models Using LISREL

Jerald R. Herting

14

The justification for using linear structural equation models with unmeasured variables, or LISREL (Linear Structural Relationships), can be found in the sociological literature addressing measurement error in causal models (Blalock [6,9]; Costner [16]; Siegal and Hodge [42]). Such techniques permit the construction of causal models that incorporate and correct for the effects of measurement error and are, therefore, fundamental in adequately estimating the parameters of causal models and assessing the goodness of fit of the model to the data.

As stressed by Blalock [8,10,11] (also see Costner [16]), models of social processes should include three components: a theoretical language that states causal relations among concepts; an operational language that specifies measured indicators of the concepts; and an auxiliary theory that links the two languages together by expressing the causal relations between indicators and concepts. It is the latter theory that sociologists generally lack; linear structural equation models with unmeasured or latent variables provide a means of including auxiliary theory in model building and testing. Such equations also meet the three desiderata enumerated by Costner [16, p. 246]. What this means to the researcher is that theoretical propositions can be developed and tested unencumbered by the contamination of measurement error that may seriously bias parameter estimates and threaten the validity of conclusions drawn from their values. The model can estimate the parameters among the unobserved concepts of interest, derive the epistemic correlation between measured indicator and concept, and recognize and control for the effects of differential bias between indicators. A simple example will reiterate the value of such models to sociologists interested in testing causal processes.

Let us assume that Fig. 14.1 represents an actual social process that occurs in the "real" world. Concepts A and B cause C which in turn causes D and F. "C" is an intervening variable; controlling for C would reduce to zero the relationship between any exogenous variable (A or B) and any endogenous variable other than C (D or F). Let us further assume that we are led by sociological theories or intuition to postulate the exact relationship as depicted in Fig. 14.1, thus duplicating in our hypothesized model the actual process as it occurs in the "real" world. In order to discover that we have the correct model, we need to devise some form of measurement for the concepts in the model (i.e., we need measured indicators) as represented in Fig. 14.2. We could then apply some statistical technique to test the adequacy of our hypothesized model in portraying the "true" causal processes. Subject to sampling variation but assuming no measurement error in any of our indicators, we would expect to conclude from a simple path analysis that our "hypothesized" model is an adequate representation of the "true" process. However, to the degree that there is random measurement error in any but the final two variables, and particularily in C, we would expect to reject the hypothesized model repeatedly, concluding that it is an inadequate representation of reality even though it is actually the correct formulation. Given the measurement error, our path analysis would be expected to show "direct" effects of A and B on D and F rather than verifying the original (and correct) assumption that the effects of A and B on D and F are mediated entirely by C. Thus, we may be misled testing causal propositions by the intrusion of measurement error—unless we incorporate and correct for

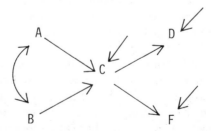

Figure 14.1. An example of a "true" social process.

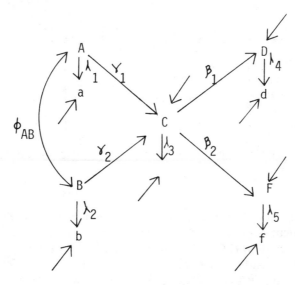

Figure 14.2. Model of Figure 14.1 including measured variables.

such errors. This is precisely what LISREL-type models are designed to do.

Blalock [6] and Costner [16] present detailed discussions demonstrating that we would consistently reject our hypothesized and accurate model in circumstances paralleling the example given above. By focusing on the relationship among A, C, and F and including measurement error (see Fig. 14.3), their conclusions can be briefly stated.

In the figure, "a," "c," and "f" represent the measured variables or indicators of their respective constructs A, C, and F. Basically, we assume that our measures are a function of some "true" value and a random error term; $c = b_{cC} C + e_c$ the classical measurement equation. As the error term e_c approaches zero, the correlation between "c" and C approaches 1.0, that is, a perfect relationship between construct and indicator or no measurement error in the indicator, "c." We may represent the structural equation for "f" and the partial correlation between "a" and "f" controlling for "c" in the conventional way as follows.

$$f = \text{constant} + B_{fa \cdot c}\, a + B_{fc \cdot a}\, c + \text{error}_f \qquad (14.1)$$

$$r_{fa \cdot c} = \frac{r_{fa} - (r_{fc})(r_{ac})}{\sqrt{1.0 - r_{fc}^2}\ \sqrt{1.0 - r_{ac}^2}} = 0 \qquad (14.2)$$

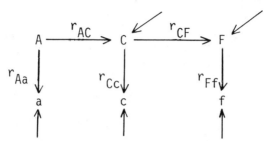

Figure 14.3. The relationships between A, C, and F implied in Figure 14.2.

As shown below, the latter partial correlation is not zero in the presence of measurement error in C, and if the partial is not zero it follows that $B_{fa \cdot c}$ is not zero in equation (14.1). Following general path analytical rules and substituting the path values into equation (14.2), the partial correlation between "a" and "f" may be represented as follows:

$$
\begin{aligned}
r_{fa \cdot c} &= \frac{[r_{Aa} r_{AC} r_{CF} r_{Ff}] - [(r_{Aa} r_{AC} r_{Cc})(r_{Ff} r_{CF} r_{Cc})]}{\sqrt{1.0 - r_{fc}^2} \sqrt{1.0 - r_{ac}^2}} \\
&= \frac{[r_{Aa} r_{AC} r_{CF} r_{Ff}] - r_{Cc}^2 [r_{Aa} r_{AC} r_{CF} r_{Ff}]}{\sqrt{1.0 - r_{fc}^2} \sqrt{1.0 - r_{ac}^2}}
\end{aligned}
\tag{14.3}
$$

If the correlation between C and "c" is equal to one, the two elements in the numerator are equal. Thus, their difference is zero, and the partial is zero. To the degree that the correlation between the concept C and its indicator, "c," is not perfect, the partial will not be zero, and "f" will continue to show some dependence on "a" in equation 14.1.

This example readily extends to more complicated models (see Blalock [6]). Measurement error may not only lead us to include paths that should not be present as illustrated above but may also lead to bias in the estimates of the parameters that are represented by paths included (Blalock [6]; Pindyck and Rubinfeld [39]; Rao and Miller [40].

Thus, the statement above concerning Fig. 14.1 is confirmed; even when it is true, we would generally not accept our hypothesized model in the presence of measurement error in variable C if our analysis is a simple path analysis. When we can reasonably assume no measurement error, a simple path analysis would be adequate, assuming the model is not grossly misspecified. In sociological measurement, however, it is often hazardous to make this assumption.

MULTIPLE INDICATORS

To remedy the problem presented by measurement error, multiple or repeated measures (i.e., multiple indicators) are helpful. Costner [16], supplemented by Blalock's extension [9], presents the case for multiple indicators in structural equation models. Although Costner and Blalock, as well as others (Van Valey [46]; Sullivan [45]; Land [33]) outlined a general strategy for using multiple indicators to correct for measurement error, the methods devised were cumbersome and provided no way to assess the statistical significance of the parameter estimates or of the variation in parameter estimates (i.e. goodness of fit)(Hauser and Goldberger [22]). It is the development of efficient algorithms for maximum likelihood estimation that has fostered the continued and more promising use of multiple indicators by making feasible and practical the analysis of structural equation models with latent variables (see Werts, Linn, and Jöreskog [47]).

Jöreskog [25] can be credited with providing the efficient technique of estimating such models using maximum likelihood techniques for the analysis of covariance structures. The further development and dissemination of the LISREL program has brought this tool to a larger audience. The remaining discussion focuses on Jöreskog's LISREL program (Jöreskog and Sörbom [29]) and its applications in causal modeling.

CAUSAL MODELING WITH MULTIPLE INDICATORS AND LISREL

Long [36] provided an early summary of structural equation models with latent variables and has recently extended the discussion [37]. Bentler [2] has discussed the latent variable causal modeling for psychologists. Most of the presentations fail to link the discussion of the technique to the overall concern for measurement error in the social sciences and the general advantage that the technique brings to the path analytic framework now widely familiar to sociologists. The intent of this chapter is to link the path analytic framework to the techniques of structural modeling with latent variables; to introduce the reader to the LISREL program and to facilitate its use and interpretation; and to provide a general discussion of problems faced by the less technical user. The content of the chapter is aimed at the initiate but will provide a useful review for those somewhat more familiar with the approach and the LISREL program. The overall framework is one of causal modeling, as opposed to factor structures or confirmatory factor analysis. Given this framework, it is assumed that the reader has some familiarity with the basic path analytic framework. Matrix algebra simplifies discussions of this technique. However, the material presented in this matrix form hopefully facilitates the reader's understanding of matrix notation rather than limits the reader's ability to understand the material.

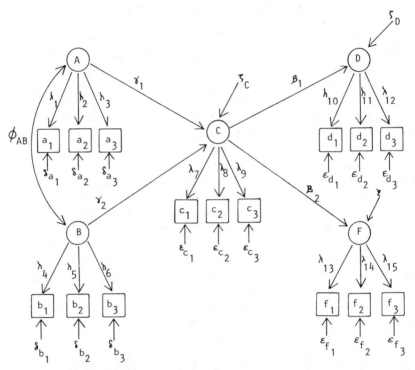

Figure 14.4. A multiple indicator model of the causal process represented in Figure 14.1.

Figure 14.4 is introduced as a multiple indicator model of the causal process depicted in Fig. 14.1. The first part of this section formulates the general framework of LISREL, introducing the terms by drawing similarities and contrasts between the more familiar path model and the less familiar multiple indicator model. At times, the terms and Greek symbols may seem tedious, but they are necessary to understand other work utilizing these models and the LISREL program. Throughout this discussion, all entries in a variance–covariance matrix will be referred to as *covariances*. Our general reasoning also applies to correlations, of course, since correlations can be viewed as standardized covariances. Also for Fig. 14.4, and all figures following, circles represent unmeasured dimensions and squares represent measured variables; upper case letters denote unmeasured dimensions, while lower case letters denote measured variables.

There are three fundamental matrix equations in the LISREL model. The first matrix equation describes the relations at the structural level (i.e., relations among latent unobserved dimensions). This matrix equation—the structural model—is the set of equations that we are ordinarily most interested in estimating since they reflect the causal relationships among the concepts of the basic theory or underlying reasoning. A second matrix equation relates a set of measured variables to the unobserved exogenous dimensions, and a third matrix equation relates measured variables to the unobserved endogenous dimensions. Abstractly, the structural model can be thought of as constituting a path analytical model for the underlying dimensions of interest without measurement error. The other two matrix equations specify which measured indicators are caused by which unobserved dimension(s) (i.e., specify the auxiliary or measurement theory). The latter has some similarity to the familiar exploratory factor analysis, although in this instance, the researcher posits which measured variables have non-zero "loadings" on which unmeasured factors, whereas in exploratory factor analysis there is no such *a priori* specification. The three matrix equations are:

The structural equation:

$$\text{eta} = (\text{beta})(\text{eta}) + (\text{gamma})(\text{ksi}) + \text{zeta} \qquad (14.4)$$
$$\eta = \beta \quad \eta + \quad \Gamma \quad \xi + \zeta$$

The equation relating exogenous dimensions and their indicators:

$$x = (\text{lambda}_x)(\text{ksi}) + \text{delta} \qquad (14.5)$$
$$x = \Lambda_x \quad \xi + \delta$$

The equation relating endogenous dimensions and their indicators:

$$y = (\text{lambda}_y)(\text{eta}) + \text{epsilon} \qquad (14.6)$$
$$y = \Lambda_y \quad \eta + \epsilon$$

If we do not make a distinction between exogenous and endogenous dimensions, either equation (14.5) or (14.6) will suffice to describe the relation between measured variables and unmeasured dimensions.[1]

---

[1] Equation (14.5) or (14.6) is the basic form of a confirmatory factor analysis model. This will be discussed later in the chapter (also see Long [36,37]).

The meaning of the structural equation (14.4) can be clarified by reference to Figure 14.2 or 14.4. The "eta" term refers to the endogenous unmeasured variables or dimensions; these dimensions are at some point in the model dependent dimensions.[2] These are C, D, and F in both figures. The "ksi" term refers to the exogenous (having no prior causes in this model) latent dimensions, A and B in the figures referred to. The "zeta" represents the error term in the structural equation, analogous to the error term in a regression equation. Given that there are three dependent or endogenous dimensions, there are then three zeta error terms. "Beta" refers to the causal parameters among the endogenous dimensions; these are β_1 and β_2 in Fig. 14.2. "Gamma" refers to the causal paths from the exogenous dimensions to the endogenous dimensions. These are represented by γ_1 and γ_2 in the model under discussion.

Two other matrices must be introduced at the structural level. The covariance matrix for exogenous dimensions is represented by "phi"; the "ϕ_{AB}" path in Fig. 14.2 is the covariance between dimensions A and B (i.e., the only off-diagonal non-zero entry in the phi matrix for this model). The diagonal of the phi matrix is constituted by the variances of the unobserved exogenous dimensions; in a standardized matrix (correlation matrix), these would be 1.0. An additional matrix known as "psi" has as its elements covariances of the error terms. The diagonal of the psi matrix is constituted by the error variances of the endogenous dimensions (C, D, and F in this model); off diagonals represent the covariance between errors for each pair of endogenous dimensions. Correlated error is undesirable in any model, but the advantage of LISREL is that at least some such errors can be estimated while all such correlated errors are assumed, without testing, to be zero in the path analytic framework.

In the measurement equations (14.5) and (14.6), the ksi and eta terms have the same meaning as at the structural level. They represent the exogenous (A and B) and endogenous (C, D, and F) dimensions, respectively. The "lambda$_x$" matrix represents the matrix of relationships between the exogenous latent dimensions and the measured variables that they cause, "a" and "b." In the context of Fig. 14.1, these are λ_1 and λ_2, where λ_1 is the relationship between A and "a," and λ_2 is the relationship between B and "b." "Delta" represents the errors in the measures of the indicators of exogenous dimensions, that is, errors in where error refers to variation in the indicator not caused by the underlying dimension. Thus, the variation in each measure is constituted by two components: the part that depends on variation in

---

[2] For those unfamiliar with matrix notation, the symbols merely provide a "short-hand" means of representing a large number of equations; thus, each symbol (e.g., "eta") refers to a set of similar terms rather than just one element. An operation applied to any symbol, say addition, implies that the operation is carried out for all terms in the set referred to by the symbol.

the underlying dimension and the part that is error. This is succinctly stated in the equation: $a = \lambda_1 A + \text{delta}_a$. When measurement error is disregarded in estimating the parameters of a path model this means that λ_1 and λ_2 are assumed to equal one (in standardized form) and delta is assumed to equal zero, implying perfect measurement.

"Lambda$_y$" is the matrix representing the parameters relating the endogenous dimensions to their measured indicators. These are λ_3, λ_4, and λ_5 in the model, representing the relation of C to "c," D to "d," and F to "f," respectively. "Epsilon" is a matrix constituted by the error terms for each indicator of the endogenous dimensions. Two other matrices must be considered at the measurement level. These are the "theta delta" matrix and the "theta epsilon" matrix. Each represents the corresponding variance–covariance matrix for the error terms in the measurement equations. The diagonal of either matrix gives the amount of error variance in its respective indicator. Off diagonals show the degree to which each pair of indicators have common sources of error (i.e., that there is correlated error between two indicators).

The model of Figure 14.1 may be described in matrix form in the following equations. (The precise meaning of the next six equations is clarified below.)

$$
\begin{array}{ccccccc}
\text{eta} & = & \text{beta} & \text{eta} & + & \text{gamma} & \text{ksi} & + & \text{zeta} \\
\end{array}
$$

$$
\begin{bmatrix} C \\ D \\ F \end{bmatrix} = \begin{bmatrix} 0 & 0 & 0 \\ \beta_1 & 0 & 0 \\ \beta_2 & 0 & 0 \end{bmatrix} \begin{bmatrix} C \\ D \\ F \end{bmatrix} + \begin{bmatrix} \gamma_1 & \gamma_2 \\ 0 & 0 \\ 0 & 0 \end{bmatrix} \begin{bmatrix} A \\ B \end{bmatrix} + \begin{bmatrix} \zeta_C \\ \zeta_D \\ \zeta_F \end{bmatrix} \tag{14.7}
$$

$$
\begin{array}{ccccccc}
\eta & = & \beta & \eta & + & \Gamma & \xi & + & \zeta
\end{array}
$$

$$
\begin{array}{ccccc}
y & = & \text{lambda}_y & \text{eta} & + & \text{epsilon}
\end{array}
$$

$$
\begin{bmatrix} c \\ d \\ f \end{bmatrix} = \begin{bmatrix} \lambda_3 & 0 & 0 \\ 0 & \lambda_4 & 0 \\ 0 & 0 & \lambda_5 \end{bmatrix} \begin{bmatrix} C \\ D \\ F \end{bmatrix} + \begin{bmatrix} \epsilon_c \\ \epsilon_d \\ \epsilon_f \end{bmatrix} \tag{14.8}
$$

$$
\begin{array}{ccccc}
y & = & \Lambda_y & \eta & + & \epsilon
\end{array}
$$

$$
\begin{array}{ccccc}
x & = & \text{lambda}_x & \text{ksi} & + & \text{delta}
\end{array}
$$

$$
\begin{bmatrix} a \\ b \end{bmatrix} = \begin{bmatrix} \lambda_1 & 0 \\ 0 & \lambda_2 \end{bmatrix} \begin{bmatrix} A \\ B \end{bmatrix} + \begin{bmatrix} \delta_a \\ \delta_b \end{bmatrix} \tag{14.9}
$$

$$
\begin{array}{ccccc}
x & = & \Lambda_x & \xi & + & \delta
\end{array}
$$

The extension to the multiple indicator model represented by Figure 14.4 is straightforward. The only differences appear in the two measurement equations due to the inclusion of more than one indicator per dimension. The matrix equations for Fig. 14.4 are as follows:

$$
\begin{array}{ccccccc}
\text{eta} & = & \text{beta} & \text{eta} & + & \text{gamma} & \text{ksi} & + & \text{zeta}
\end{array}
$$

$$
\begin{bmatrix} C \\ D \\ F \end{bmatrix} = \begin{bmatrix} 0 & 0 & 0 \\ \beta_1 & 0 & 0 \\ \beta_2 & 0 & 0 \end{bmatrix} \begin{bmatrix} C \\ D \\ F \end{bmatrix} + \begin{bmatrix} \gamma_1 & \gamma_2 \\ 0 & 0 \\ 0 & 0 \end{bmatrix} \begin{bmatrix} A \\ B \end{bmatrix} + \begin{bmatrix} \zeta_C \\ \zeta_D \\ \zeta_F \end{bmatrix} \qquad (14.10)
$$

$$
\begin{array}{ccccccc}
\eta & = & \beta & \eta & + & \Gamma & \xi & + & \zeta
\end{array}
$$

$$
\begin{array}{ccccc}
y & = & \text{lambda}_y & \text{eta} & + & \text{epsilon}
\end{array}
$$

$$
\begin{bmatrix} c_1 \\ c_2 \\ c_3 \\ d_1 \\ d_2 \\ d_3 \\ f_1 \\ f_2 \\ f_3 \end{bmatrix} = \begin{bmatrix} \lambda_7 & 0 & 0 \\ \lambda_8 & 0 & 0 \\ \lambda_9 & 0 & 0 \\ 0 & \lambda_{10} & 0 \\ 0 & \lambda_{11} & 0 \\ 0 & \lambda_{12} & 0 \\ 0 & 0 & \lambda_{13} \\ 0 & 0 & \lambda_{14} \\ 0 & 0 & \lambda_{15} \end{bmatrix} \begin{bmatrix} C \\ D \\ F \end{bmatrix} + \begin{bmatrix} \epsilon_{c_1} \\ \epsilon_{c_2} \\ \epsilon_{c_3} \\ \epsilon_{d_1} \\ \epsilon_{d_2} \\ \epsilon_{d_3} \\ \epsilon_{f_1} \\ \epsilon_{f_2} \\ \epsilon_{f_3} \end{bmatrix} \qquad (14.11)
$$

$$
\begin{array}{ccccc}
y & = & \Lambda_y & \eta & + & \epsilon
\end{array}
$$

$$
\begin{array}{ccccc}
x & = & \text{lambda}_x & \text{ksi} & + & \text{delta}
\end{array}
$$

$$
\begin{bmatrix} a_1 \\ a_2 \\ a_3 \\ b_1 \\ b_2 \\ b_3 \end{bmatrix} = \begin{bmatrix} \lambda_1 & 0 \\ \lambda_2 & 0 \\ \lambda_3 & 0 \\ 0 & \lambda_4 \\ 0 & \lambda_5 \\ 0 & \lambda_6 \end{bmatrix} \begin{bmatrix} A \\ B \end{bmatrix} + \begin{bmatrix} \delta_{a_1} \\ \delta_{a_2} \\ \delta_{a_3} \\ \delta_{b_1} \\ \delta_{b_2} \\ \delta_{b_3} \end{bmatrix} \qquad (14.12)
$$

$$
\begin{array}{ccccc}
x & = & \Lambda_x & \xi & + & \delta
\end{array}
$$

$$
\begin{array}{c}
\text{phi} \\
\begin{array}{cc} A & B \end{array} \\
\begin{array}{c} A \\ B \end{array} \begin{bmatrix} \phi_{AA} & \\ \phi_{AB} & \phi_{BB} \end{bmatrix} \\
\Phi
\end{array} \qquad (14.13)
$$

psi

$$
\begin{array}{c} \\ \zeta_C \\ \zeta_D \\ \zeta_F \end{array}
\begin{array}{ccc} \zeta_C & \zeta_D & \zeta_F \end{array}
\left[\begin{array}{ccc} \psi_{CC} & & \\ 0 & \psi_{DD} & \\ 0 & 0 & \psi_{FF} \end{array} \right]
\qquad (14.14)
$$

Ψ

theta delta

$$
\begin{array}{c} \\ \delta_{a_1} \\ \delta_{a_2} \\ \delta_{a_3} \\ \delta_{b_1} \\ \delta_{b_2} \\ \delta_{b_3} \end{array}
\begin{array}{cccccc} \delta_{a_1} & \delta_{a_2} & \delta_{a_3} & \delta_{b_1} & \delta_{b_2} & \delta_{b_3} \end{array}
\left[\begin{array}{cccccc}
\theta_{\delta_{a_1}} & & & & & \\
0 & \theta_{\delta_{a_2}} & & & & \\
0 & 0 & \theta_{\delta_{a_3}} & & & \\
0 & 0 & 0 & \theta_{\delta_{b_1}} & & \\
0 & 0 & 0 & 0 & \theta_{\delta_{b_2}} & \\
0 & 0 & 0 & 0 & 0 & \theta_{\delta_{b_3}}
\end{array} \right]
\qquad (14.15)
$$

θ_δ

theta epsilon

$$
\begin{array}{c} \\ \epsilon_{c_1} \\ \epsilon_{c_2} \\ \epsilon_{c_3} \\ \epsilon_{d_1} \\ \epsilon_{d_2} \\ \epsilon_{d_3} \\ \epsilon_{f_1} \\ \epsilon_{f_2} \\ \epsilon_{f_3} \end{array}
\begin{array}{ccccccccc} \epsilon_{c_1} & \epsilon_{c_2} & \epsilon_{c_3} & \epsilon_{d_1} & \epsilon_{d_2} & \epsilon_{d_3} & \epsilon_{f_1} & \epsilon_{f_2} & \epsilon_{f_3} \end{array}
\left[\begin{array}{ccccccccc}
\theta_{\epsilon_{c_1}} & & & & & & & & \\
0 & \theta_{\epsilon_{c_2}} & & & & & & & \\
0 & 0 & \theta_{\epsilon_{c_3}} & & & & & & \\
0 & 0 & 0 & \theta_{\epsilon_{d_1}} & & & & & \\
0 & 0 & 0 & 0 & \theta_{\epsilon_{d_2}} & & & & \\
0 & 0 & 0 & 0 & 0 & \theta_{\epsilon_{d_3}} & & & \\
0 & 0 & 0 & 0 & 0 & 0 & \theta_{\epsilon_{f_1}} & & \\
0 & 0 & 0 & 0 & 0 & 0 & 0 & \theta_{\epsilon_{f_2}} & \\
0 & 0 & 0 & 0 & 0 & 0 & 0 & 0 & \theta_{\epsilon_{f_3}}
\end{array} \right]
\qquad (14.16)
$$

θ_ϵ

Of course, equation (14.7) and equation (14.10) are identical for the two models since the structural equation does not change with the addition of more indicators. The "phi" and "psi" matrices for both models are identical as shown in (14.13) and (14.14). The off-diagonal entry in the phi matrix represents the covariance (correlation) between A and B. There are no off-diagonal values in the psi matrix since there are no covariances indicated between the zeta terms in either model; that is, there are no curved, double-headed arrows between the error terms of dimension C, D, or F. The theta delta (14.16) and theta epsilon (14.17) matrices also have only diagonal entries, specifying that there are no correlated errors between any of the measured indicators. Table 14.1 provides a simple summary of the terms used in LISREL.

To facilitate understanding of this matrix representation, it is useful to view the gamma, beta, and lambda matrices as representing the relationships between causes and results, and to view the ksi, delta, epsilon, X and Y vectors — all are column vectors, meaning a one column, many row matrix — as "listing" either *causes* or *results*. Elements in the eta vector may be both causes and results, since there may be a causal structure among the endogenous latent variables (e.g., C is caused by A and B and also C causes D and F in Fig. 14.4). Two steps are required to create the matrix representation of a given causal structure. First, the causes and effects are ordered. In this case, the endogenous concepts (which constitute the eta vector) have been arranged or ordered as C, D, and F. Likewise, the exogenous concepts (which

Table 14.1. List of matrices and terms used in LISREL

S	Observed covariance matrix of measured variables
Σ (sigma)	Model implied covariance matrix

For the structural equations

η (eta)	Endogenous dimensions — the unmeasured dependent variables
ξ (ksi)*	Exogenous dimensions — the unmeasured independent variables
ζ (zeta)	Errors in the structural equations
B, β (beta)**	Causal parameters among the endogenous dimensions
Γ, γ (gamma)	Causal parameters from exogenous to endogenous dimensions
Φ, ϕ (phi)	Variance–covariance matrix for exogenous dimensions
Ψ, ψ (psi)	Variance–covariance matrix of the error terms in the structural equation; diagonal elements are the amount of error variance in the endogenous dimensions

For the measurement equations

(x)	Measured variables/indicators of exogenous dimensions
(y)	Measured variables/indicators of endogenous dimensions
Λ_x, λ_x (lambda$_x$)	Causal parameters between exogenous dimensions and measured variables
Λ_y, λ_y (lambda$_y$)	Causal parameters between endogenous dimensions and measured variables
δ (delta)	Errors in the measures of the indicators of exogenous dimensions
ϵ (epsilon)	Errors in the measures of the indicators of endogenous dimensions
θ_δ (theta delta)	Variance–covariance matrix of the error terms for indicators of exogenous dimensions; diagonal elements are the error variance of the indicators
θ_ϵ (theta epsilon)	Variance–covariance matrix of the error terms for indicators of endogenous dimensions; diagonal elements are the error variance of the indicators

* Ksi is the spelling used in the LISREL program.
** Upper case letters refer to the matrix; lower case refer to the elements in the matrix.

constitute the ksi vector) and the elements in all other matrices, the x vector of measured variables that are reflectors of the two ksi concepts and the y vector of measures that are reflectors of the three eta concepts, are ordered. In this example, we have ordered the x vector as follows: a_1, a_2, a_3, b_1, b_2, and b_3. Although the order is arbitrary, the order must be consistently maintained in the formulation of the model. It is convenient but not necessary to order the endogenous variables (in recursive models) such that a variable is not caused by any variable that appears later (or "to the right") in the ordering. It is also convenient but not necessary to have the order of indicators correspond to the order of the dimension they reflect.

The second step is to formulate the relations between the causes and effects. This step essentially "fills in" the values of the beta, gamma, lambda$_y$, and lambda$_x$ matrices. The number of columns in these matrices corresponds to the number of causes being related to a given vector of results. The number of rows in these matrices corresponds to the number of results to which the causes are related. Columns are associated with causes, rows with results. Following this logic, we formulate equation (14.7), the structural relations between the unmeasured concepts. Results appear to the left of the equal sign. These are concepts C, D, and F. The causes appear to the right-hand side of the equation. The causes include C, D, and F—since there can be a causal structure among the endogenous concepts—, A and B as exogenous factors, and the error vector, zeta.

As stated, the beta matrix relates the endogenous factors to endogenous factors. In this instance, the matrix will have three columns and three rows since there are three endogenous concepts or factors. Filling in the matrix is simply a question of asking: Does the column associated with a given variable cause any one of the other endogenous concepts represented in each row? The first column of beta has a zero in the first row (concepts do not cause themselves), a "β_1" in the second row since concept C affects concept D, and a "β_2" in the third row since C affects concept F. The second column of the beta matrix is 0 in the first row because D does not affect C (no reciprocal relation), zero in the second, and zero in the third because D has no effect on F. The third column has zeroes as well, since F does not affect any other unmeasured concept.

The gamma matrix must also be defined. Since the gamma matrix specifies the relationship between the exogenous/endogenous pairs of concepts and since there are two exogenous concepts, A and B, causing three endogenous concepts (results), C, D, and F, the gamma matrix will have two columns and three rows. The first column corresponds to the effects of A on the three endogenous dimensions in order. The first entry is "γ_1," since A is a direct cause of C. The second and third rows of the first column of gamma are zero indicating that concept A does not directly affect either D or F. The second column will have a "γ_2" in the first row (B has a direct effect on C), and a zero in the second and third rows indicating that B affects only C and not D or F. There are three zeta terms to match the three results; the first is the error in C, the second the error in D and the third the error in F.

The matrix formulations in equations (14.11) and (14.12) can be as easily explained as those above. Note that the symmetry in the lambda matrices are due to our ordering of the indicators. Had we ordered the indicators of the endogenous concepts a_1, a_2, b_3, b_1, b_3, a_3, and b_2, the lambda$_x$ matrix would have appeared as below.

$$x = \text{lambda}_x \ \text{ksi} + \text{delta}$$

$$
\begin{bmatrix} a_1 \\ a_2 \\ b_3 \\ b_1 \\ a_3 \\ b_2 \end{bmatrix}
=
\begin{bmatrix} \lambda_1 & 0 \\ \lambda_2 & 0 \\ 0 & \lambda_6 \\ 0 & \lambda_4 \\ \lambda_3 & 0 \\ 0 & \lambda_5 \end{bmatrix}
\begin{bmatrix} A \\ B \end{bmatrix}
+
\begin{bmatrix} \delta a_1 \\ \delta a_2 \\ \delta b_3 \\ \delta b_1 \\ \delta a_3 \\ \delta b_2 \end{bmatrix}
\qquad (14.17)
$$

$$x = \Lambda_x \quad \xi + \delta$$

How one orders the results or causes — the equations — is arbitrary except that once the order is established it should be consistent throughout the equations (14.10)–(14.16).

The above should clarify how the equations (14.10)–(14.17) were formed from the path diagram in Fig. 14.4. The key is to link the columns of the gamma, beta, and lambda matrices to the ordered causes and to link the rows to the ordered results. Each element of these matrices essentially corresponds to whether a direct path or relationship exists from the cause in the column to the result in the row. For the gamma and beta matrix, causes and results will be concepts, while in the two lambda matrices, the cause will be a concept and the result will be a measured indicator. For certain models, determining each element in each matrix may become tedious, but each such element represents either a causal path (straight, single-headed arrow) or a source of covariation (curved, double-headed arrow) — as in the covariation between a pair of exogenous dimensions or correlated error between two indicators. There is an entry in the matrix for each arrow and a zero otherwise.

As illustrated above, the system of equations implicitly represented by a path diagram with unmeasured variables (i.e., Figs. 14.2 or 14.4) can be described in matrix form by equations (14.3)–(14.5) within the LISREL program. Familiarity with these three matrix equations and the matrix formulation of the covariances of the exogenous dimensions — phi — and the error covariances — psi, theta delta, and theta epsilon — facilitates the translation of a given multiple indicator path diagram into an appropriate specification of the model represented by the diagram. Information in these matrices specifies which parameters are "free" and, thus, to be estimated by the program on the basis of the data, and which are constrained, that is, "fixed" at zero or given some equality constraint.[3] In accord with the definition of parameters as free or constrained (i.e., the model specified by the researcher), the estimation procedure calculates a value for each parameter such that the covariances implied by the model replicates, as closely as possible, the covariances observed in the sample.

The matrix specification, much like path analysis, states which variables enter into the equation for a given measured or unmeasured variable. Unlike path analysis, however, the estimation is done simultaneously using maximum likelihood estimation. However, if we have only one measured variable per concept and, hence, must assume no measurement error (as in Fig. 14.2), the path estimates in the model using either the typical multiple regression (Ordinary Least Squares) technique or the maximum likelihood procedure of LISREL will be the same.

Before providing illustrations of the program, its program content, and results, we will discuss briefly the assumptions underlying both the general causal model and the estimation procedure in LISREL. Although, a formal discussion of the estimation procedure is not presented here, it will be useful to describe the general procedure and the assumptions underlying it so that drawbacks of the method will be highlighted, the circumstances when it is tenuous to apply the technique will be clarified, and necessary cautions and modifications of interpretations will be understood. Also included in the discussion below are two issues that are best introduced prior to the examples but which will become more relevant and clear in the context of the examples.

ESTIMATION OF PARAMETERS

The estimation procedure used in LISREL is maximum likelihood.[4] The basic purpose of the estimation procedure is to find the set of parameter

---

[3] Parameters in the LISREL program are defined by the user as "fixed," "constrained," or "free." Fixed parameters are assigned specific values by the user. Constrained parameters are estimated parameters but are constrained to be equal to one or more other parameters. Free parameters are estimated and not constrained in any manner.

[4] Unweighted least squares (ULS) and generalized least squares (GLS) are alternative methods to the maximum likelihood (ML) procedure. ULS will not always yield efficient estimators, though the estimators will be consistent under much less strenuous assumptions than the other two methods. GLS and ML have been shown to be asymptotically similar. For a discussion of the three techniques see Jöreskog [27].

values that will imply a covariance structure (sigma) that best fits—minimizes the discrepancies from—the observed covariance structure (S), that is, the covariance matrix calculated from the data. The elements in sigma depend on the values Λ_y, Λ_x, β, Γ, Φ, Ψ, θ_δ, and θ_ϵ; hence, the estimation procedure calculates elements in these matrices (except those already fixed) that make sigma (Σ) as nearly identical to S as possible. The procedure is conceptually straightforward. The program begins by setting fixed parameters to their specified values. Thus, the parameters for paths not shown in the path diagram are set to zero. The program also assigns initial values to the "free" parameters; these initial values can be assigned by the program or, if desired, by the researcher. The program then iterates between (1) checking the fit of the covariance structure implied by these parameter estimates (Σ) to the observed covariance matrix, S, and (2) readjusting the values of the parameter estimates in response to the information obtained in checking the fit of model and data.[5] The iteration continues until the fitting function is at a minimum with respect to these parameter values. Being at a "minimum" does not mean that the model necessarily "fits" the data—that S and Σ fully agree—but rather that any other values for the estimates, given the same restrictions specified by the researcher, would lead to larger differences between the covariance structure implied by the model and the actual covariance structure obtained from the sample data.[6] If the model adequately reproduces the sample covariance matrix, as determined by criteria discussed below, we can infer that the specified model is an adequate representation of the "true" process that generated the observed data. For a technical discussion of the procedure, see Jöreskog [25; 27, pp. 73–75]; a less technical but more detailed discussion of the estimation technique can be found in Long [36,37] or in Bentler [2]. For a general discussion of the maximum likelihood technique, see Pindyck and Rubinfeld [39] or Kmenta [32].

The maximum likelihood estimation procedure has convenient large sample properties. First, the estimates are efficient and consistent. Second, the estimates as a set are asymptotically normal. And third, the minimiza-

---

[5] The fitting function is $F = \log |\Sigma| + tr(S\Sigma^{-1}) - \log |S| - (t)$, where t is the total number of indicators. LISREL uses the method of Davidon–Fletcher–Powell as an iterative algorithm in minimizing F, thus estimating the parameters. Adjusting the values of the parameter estimates in this technique relies on evaluating the second derivative of the fitting function with respect to changes in the values of the parameters estimated (for more detail see Gruvaeus and Jöreskog [21]; or Jöreskog [26]).

[6] It is possible for the estimation procedure in LISREL to arrive at a "local minimum" rather than an absolute minimum. What this means is that the program converges and provides estimates, but that these estimates are not optimal. Generally, this is not a problem. The user can test for local minima by providing the program divergent initial "start" values; if the two or more sets of initial values provide the same solution, the user can be confident that the absolute minimum has been reached by the iteration procedure.

tion function has an asymptotic chi-square distribution with degrees of freedom equal to the number of variances and covariances minus the number of independently estimated parameters. The latter is extremely useful in that it provides a statistical test for the fit of model to the data. What this means for the researcher is that for large sample sizes, the parameter estimates are unbiased, that a standard error can be calculated for the estimates,[7] that the standard error of the estimate is as small as other estimation procedures, and that we can form a test that evaluates whether the discrepancies between the covariances implied by the model and the empirical covariances are reasonably attributed to sampling variation or, whether those discrepancies are too large, thus suggesting the specified model fails to represent adequately the causal processes that generated the data. The latter chi-square property, however, is not without problems as discussed below (also see Herting and Costner [23]; Hoelter [24]; Long [36]; Joreskög and Sörbom [29]; Bentler and Bonett [3]).

Two other properties of the maximum likelihood technique are important to note. First, the technique is scale invariant. What this means for the LISREL user is that a standardized covariance structure — a correlation matrix — may be analyzed. More generally, it means that a shift in units of measure, for example from feet to inches, will result in a corresponding shift in the parameter estimates and standard errors by a constant factor determined by the size of the shift in units. The results of the analyses will not be altered by the shift. Since measured variables in the social sciences often have no "natural" unit of measure, being able to work with arbitrarily set units of measure or to shift to standardized units is a useful attribute of the procedure. The estimation procedure is also a full information estimation procedure. It uses all relevant information in the system of equations for the estimation of each single parameter. This implies that misspecifications in one part of the model may affect other parts (see Burt [14]).

The above properties of the maximum likelihood estimation procedure are quite useful and powerful. However, the method, thus the resulting properties, are based on rather demanding assumptions. First, we assume that the relationships among the variables are linear and additive. If the actual causal processes are not linear and additive, then we have misrepresented those processes in a LISREL model and have, therefore, misspecified the model. The resulting parameter estimates are thus inevitably inaccurate to some degree. Second, the program assumes that all variables are interval or ratio level and are measured as deviations from their respective means;

---

[7] The standard errors of the estimates provided by LISREL are approximate. The square root of the diagonals in the "information matrix," calculated during the interaction process, are large-sample estimates of the standard errors (see Jöreskog [27]). The accuracy of this method for computing the standard errors has been questioned (Lee and Jennrich [35]).

measuring the variables as deviations from their mean, as in regression, rids each equation of its constant term but does not alter any of the other estimates of coefficients.[8] Except for "fixed" exogenous variables (as in an experiment), interval measures are necessary to give a clear meaning to the parameters. Assuming an ordinal variable as interval entails violating assumptions of the model and may introduce unknown distortions into parameter estimates. Techniques for incorporating noninterval level variables into structural equation models with latent variables are not well developed (Bentler [2]), although research efforts are being made to incorporate such measures and to examine their impact with the LISREL program (Browne [13]; Clogg [15]; Muthen [38]).

The estimation technique makes a third assumption: that the observed variables and error terms have a multivariate normal distribution. The multivariate assumption is a strong assumption and, perhaps, the most likely to be violated by the nature of social science data. The effect of violating this assumption on the parameter estimates and the chi-square test is not well known (Bielby and Hauser [5]; Bentler [2]). Burt [14] suggests that with large samples the technique is robust with respect to the multivariate assumption. In the literature, the large sample properties of the estimation technique have been given more examination than the small sample properties; hence, the effects of violating the large sample criteria are not well known. A small sample size poses particular problems with respect to the interpretation of the chi-square and the standard errors. In general, violation of multivariate normality suggests that the standard errors about the estimated parameters and the chi square be interpreted with caution.

The general LISREL model further assumes the following relations among the unobserved concepts and error terms in the model:

$$\text{cov}(\zeta, \xi) = 0$$
$$\text{cov}(\epsilon, \eta) = 0$$
$$\text{cov}(\delta, \xi) = 0$$
$$\text{cov}(\zeta, \delta) = 0$$
$$\text{cov}(\zeta, \epsilon) = 0$$
$$\text{cov}(\delta, \epsilon) = 0$$

Respectively, these relations mean that the error terms in the structural equation (zeta) are uncorrelated with the exogenous dimensions; that the

---

[8] The LISREL program does allow the user to examine mean structures, though in the general model the mean structure is ignored (see Jöreskog [27]; Jöreskog and Sörbom [27]). Schoenberg [41] also provides an example of this type of analysis.

errors in indicators of exogenous dimensions are uncorrelated with the exogenous dimensions; that the errors in the indicators of endogenous dimensions are uncorrelated with the endogenous dimensions; and the error terms, zeta, epsilon, and delta, are mutually uncorrelated. These assumptions do not preclude correlated errors within the theta epsilon, theta delta, or psi matrices. At times, it may be possible to incorporate variables in such a manner that correlated error terms previously assumed to be zero may be estimated, but in the general causal model, the above assumptions concerning the error terms and unmeasured variables will be made.

In general, social scientists will find they will often violate some of these assumptions. Particularly problematic are the sample size and multivariate normality assumptions. One should take care in interpreting results in these circumstances and perhaps incorporate other criteria of goodness of fit— besides the chi-square test statistic—to assess the appropriateness of the model and the parameter estimates (see Herting and Costner [23]; Jöreskog [28]). The next two sections will briefly introduce the chi-square goodness of fit measure and hypothesis testing in LISREL and the identification problem in structural equation models.

THE CHI-SQUARE STATISTIC, HYPOTHESIS TESTING, AND GOODNESS OF FIT

Once a causal model is specified and estimated it is important to assess how well it portrays the underlying or "true" causal process. The researcher tests the hypothesis that the specified model (H_0)—the model defined by the free and constrained parameters—fits the data as well as the less restrictive model (H_1) that the matrix is any positive definite matrix, that is, is a unique matrix best described by its unique elements rather than by a more general process. Otherwise stated, we are testing the hypothesis that the observed discrepancies between S and Σ differ from zero only because of sampling variation. Failure to reject this hypothesis supports the porposed model specification, H_0, as an adequate representation of the data. If the discrepancies are larger than would be expected due to sampling variation, the model must be deemed misspecified and, therefore, not an adequate representation of the "true" process.

As noted above, one extremely useful result of the maximum likelihood procedure is that the fitting function has a chi-square distribution with the degrees of freedom determined by the number of variances and covariances present and the number of parameters estimated. Formally, the chi square is equal to N-1 times the minimum value of the fitting function—when the estimation converges at the minimum—and the degrees of freedom are equal to $[1/2\,t\,(t-1)-u]$, where t is the number of measured variables, u is the number of independently estimated parameters, and N is the sample size.[9] The size of the chi-square statistic relative to the degrees of freedom is a means of assessing goodness of fit between model and data. A nonsignifi- cant chi square (e.g., less than twice the degrees of freedom at the .05 level) suggests the model (H_0) is adequately specified. A large chi square relative to the degrees of freedom implies that the model should be rejected, that is, the model is misspecified or the data fail to fulfill other assumptions (e.g. multi- variate normality). The LISREL output provides the chi-square test statistic, the degrees of freedom, and the probability level associated with the chi square. The probability level indicates the likelihood of observing a chi- square value of this magnitude or larger, assuming that the specified model is correct. A probability level greater than .05 suggests that the observed chi square is relatively common and, therefore, the discrepancies observed are due to sampling variation rather than resulting from differences be- tween the specified model and the "true" model.

Another property of the chi-square distribution is that the difference between the chi squares of two "nested" models is also distributed chi square with degrees of freedom equal to the difference in degrees of free- dom between the two alternatives models. If the researcher is interested in (1) testing particular restrictions, that is if a path or a set of paths are zero, rather than a whole model, or (2) if the researcher is in the process of model building rather than model testing, this property is extremely useful.

Nested or hierarchical models are straightforward; for those familiar with the log-linear models (Goodman [20]), the logic is the same. Nesting simply implies that one of the tested models is a subset of the other. A nested model is derived either by adding parameters to a model (making the model less restrictive than the original model) or by removing paths from a model (making the model more restrictive than the original model). In nested models, the chi square of the more restrictive model (χ_m^2) is compared to the chi square of the less restrictive model (χ_l^2); $\chi_m^2 - \chi_l^2$ has a chi-square distri- bution with degrees of freedom equal to $df_m - df_l$. Making a model less restrictive by freeing a parameter or set of parameters provides more infor-

---

[9] The likelihood ratio test criterion is derived from the following: let L_{H_0} be the maximum of L under the hypothesis that the model H_0 represents the "true" process and let L_{H1} be the maximum of L under the hypothesis the observed covariance matrix is any positive definite matrix. The log of each maximum is calculated:

$$\log L_{H_0} = -1/2\, N\, [\log|\Sigma| + tr(S\,\Sigma^{-1})]$$

and

$$\log L_{H_1} = -1/2\, N\, [\log |S| + t]$$

The ratio of the $\log L_{H_0}$ to the $\log L_{H_1}$ equals the likelihood ratio λ. The value $-2 \log \lambda$ is distributed approximately chi square in large samples.

mation for the estimation procedure and, thus, the fit of the model to the data must be as good or better than the more restrictive model — the same or a smaller chi-square value is obtained but at a cost of losing degrees of freedom. The test addresses whether the improvement in fit in the less restrictive model (i.e., the decrease in chi square) is statistically significant.

Figure 14.5A–D provides four examples (the figures will be referred to as A, B, C, and D). The first three figures (A, B, and C) are a sequence of nested models, each becoming less restrictive. The fourth, D, is nested with Figure A but not with B or C. Figure A is a submodel of B if the path from Y to Z is removed in Figure B; B is a submodel of C if the path from W to Z is removed in Figure C. Removing both of these paths from Figure C shows that A is also a submodel of C. Figure D is clearly not nested with B: there are the same number of parameters (five) to be estimated in each figure. Also, there is no path or set of paths that can be removed from Figure C to demonstrate that D is a submodel of C. However, removing the path from Z to Y in Figure D does show that Figure A is a submodel of D.

The researcher may test specific hypotheses concerning portions of a model by establishing a sequence of nested alternative models. For example, in Figure 14.5, we could state that A is the general model (H_0), that B is an alternative (H_1), and that C (H_2) is an alternative to B. We then test the improvement in the chi-square values in sequence from H_0 to H_2. These tests would be:

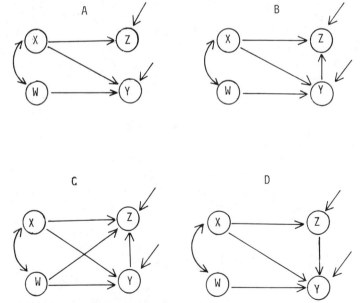

Figure 14.5. An example of "nested" models. (Each dimension has three indicators.)

$$H_0 - H_1: 176.90 - 110.46 = 66.44 \text{ with } 1 \ df \ (50-49)$$
$$H_1 - H_2: 110.46 - \ \ 99.12 = 11.34 \text{ with } 1 \ df \ (49-48)$$

If the first test ($H_0 - H_1$) were not significant, we would not proceed to test the difference in (H_1) and (H_2). Alternatively, we could have specified the original sequence as being only H_0 versus H_2 and simultaneously tested the significance of adding both paths in C — whether adding the paths from Y to Z and W to Z made a significant improvement in fit. These sequential tests do not take into account the dependent nature of the sequence (i.e., each test is dependent upon the rejection of the previous test). This generally is not problematic as the probability of rejection ignoring the dependencies is close to the true probability (Lawley and Maxwell [34]; Bagozzi [1]).

If the researcher is model building rather than testing a specific model, we are likely to have estimated the model implied by A in Figure 14.5 — our best intuitive modeling of the underlying causal process — and found the model not to fit the data; the chi square is large and the probability is less than .05.[10] In this case, we might re-examine the intuitive thinking that led to the development of the first model, or we might examine the empirical evidence in the output that might suggest where the model is misspecified (see Herting and Costner [23]; Sörbom [44]). The new model may be represented by B or C. We estimate the model implied by B (including the path from X to Y) and test the drop in chi-square relative to the loss of one degree of freedom. We still may not fit the data. But if the reduction in the chi-square between A and B were significant, we would include the path from X to Y in the next respecification. We continue testing particular additions of paths until we find a model that fits the data. In this manner, the chi square serves as the basis for judging the goodness of fit rather than serving to test a null hypothesis.[11]

The use of the chi square is not without some hazards so that a few warnings seem appropriate. First, the adequacy of the chi-square test depends upon meeting the assumptions of the estimation procedure. If the N is small or if the data depart from multivariate normality, the chi-square test statistic is suspect. Second, the chi square is sensitive to the sample size and will register significant departures when the sample size is large when in fact the departures are trivial (see Herting and Costner [23]; Hoelter [24]; Bentler and Bonnet [3]). Conversely, if the N is small, a fit may be deemed satisfactory even if some major departures are evident. Finally, as stated

[10] It may also be the case that the model fits the data, and we use the chi-square test to evaluate whether paths initially included are, in fact, necessary, that is, the original model may include some paths that are not significantly different from zero and, thus, are superfluous in achieving a good fit of model and data.

[11] The researcher may employ a split-half design where the model is developed on a portion of the data then tested on the remaining sample.

above, the researcher is often not interested in hypothesis testing but wishes to use the technique to explore or build models. The chi square in this circumstance is not really a test statistic with a given probability distribution but rather a goodness of fit measure so designed that low values relative to the degrees of freedom suggest an adequate fit of model and data and where a large change between nested models implies an improved fit of the model to the data. Jöreskog [28] suggests that this latter use of the chi-square value will often be the most appropriate.

Given these problems with the chi square, the researcher will often find other measures of fit also useful. The LISREL program provides a number of such measures. These are described elsewhere (Herting and Costner [23]) and will only be mentioned here. These measures are a goodness of fit index and an adjusted goodness of fit index; both vary between zero and one, where large values of the index imply a better fit. A mean-square error term that basically assesses the average deviation between the observed (S) and model implied covariance matrix (sigma) is calculated as well. These basically are overall measures of fit. Their major drawback is that they have no statistical distribution or standard errors and, therefore, do not always allow a definitive statement concerning fit of model to data. These measures are, however, useful in comparing models and in helping develop models. In addition to these overall measures of fit, there are also localized assessments of fit. The residual matrix and the normalized residual matrix show specific discrepancies between the observed and model-generated covariances. The first derivatives and modification indices also point to specific fixed parameters that may be impeding the achievement of good fit. These localized measures may be useful in discerning where a given model is misspecified (see Herting and Costner [23]). The reader is also referred to Bentler and Bonnet [3] and Hoelter [24] for measures of fit not produced by the LISREL program. The program output includes measures of the amount of variance explained in the indicators and endogenous dimensions, which enables the user to assess the model in a different way (Jöreskog and Sörbom [29]).

Assessing fit and adjusting the specification to provide a better fit are the most problematic aspects of using LISREL type models. It is sometimes difficult to know just how accurately a model is portraying a causal process and, if not, just where the misspecification in the model is located.

Testing hypotheses, testing nested models, and assessing fit will be demonstrated further in examples below. The next issue is also a fundamental aspect of structural equation models and also has its peculiarities that will be better illustrated in the examples.

IDENTIFICATION

A requirement in estimating a system of equations is that the parameters to be estimated — the unknowns — be identified. What this means is that there must be enough information uniquely to determine values for each unknown parameter. A model can be underidentified, just-identified and overidentified. A user who fails to recognize the necessity to estimate only identified models in LISREL can be led astray by assuming that a parameter value is uniquely determined when, in fact, it is one of an infinite number of solutions.

As stated above, the sigma matrix is determined by the set of values of Λ_y, Λ_x, β, Γ, Φ, Ψ, θ_ϵ, and θ_δ; for convenience, we can call this set of values the "parameter set," represented by Θ_1. Accordingly, once the covariances are calculated, the parameter estimates can be derived from the covariances. The issue of identification is whether the same sigma matrix can be generated by some other "parameter set" (i.e., Θ_2) or whether the sigma matrix is a unique product of Θ_1. The model is said to be identified if for any two possible parameter sets, Θ_1 not equal to Θ_2, the two generated sigmas, Σ_1 and Σ_2 are not equal. If this holds, then the model is identified and the parameters within the parameter set are identified. If this does not hold, the model is not identified; there are multiple solutions for a given parameter or set of parameters.

Even when the model as a whole is not identified, some of the parameters may be "locally" identified. These are the parameters that have the same values in the two or more parameter sets (i.e., in Θ_1 and in Θ_2) that have generated the same sigma. In this case, these locally identified parameter estimates will be consistent even if the model as a whole is not identified.

If the model is not completely identified, restrictions may be made on parameters such that identification is established. For example, one may set the relationship between an indicator and dimension to a specified value or some correlated error between indicators to some reasonable estimate. Of course, these restrictions must apply to the appropriate parameters that were not initially identified or the model will remain underidentified.

Identification, then, depends upon the specification of the model—the free and constrained parameters. For any model, a necessary condition for identification is that the number of equations implicit in the model be as great or greater than the number of unknowns being estimated. This necessary condition can be calculated by the following:

$$u \leqslant (1/2)(p + q)(p + q + 1) \qquad (14.19)$$

where u is the number of parameters to be estimated, p the number of indicators of exogenous dimensions, and q the number of indicators of endogenous dimensions. But this is a necessary condition, not a sufficient condition for identification. The necessary and sufficient conditions for identification can be demonstrated only by showing algebraically that each parameter can be determined. Numerous examples of how to establish identification algebraically exist in the literature (see Jöreskog and Sörbom

[29]; Jöreskog [26]; Duncan [18]; Long [36]). Examples are not shown here due to limitations of space and because for reasonably complex models typical of LISREL analysis, the complexity ordinarily makes algebraic solutions intractable. Other techniques that increase one's confidence that the model is identified are discussed below, although these techniques can not, with certainty, demonstrate the identification of the model.

Overidentification occurs when a given parameter or set of parameters can be determined in more than one way from the covariances in sigma. This leads to overidentifying restrictions that must hold true (i.e., certain relations among the covariances must be consistent) (see Costner [16]). The advantage of a model that is overidentified as compared to a just-identified model is that with an overidentified model, the researcher may free additional parameters, since the model will remain identified, and thus test hypotheses that parameters originally set equal to zero are in fact zero. In a just-identified model, freeing one more parameter implies a loss of identification (i.e., one would be trying to estimate an additional parameter without enough information).

The implications of being underidentified can easily be demonstrated. Refer to Fig. 14.3 and the relationship between the unmeasured variables A and C. If we substitute λ_1, λ_2, and γ_1 for the unobserved correlations r_{Aa}, r_{Cc} and r_{AC}, respectively, the algebra of path analysis shows that the observed correlation r_{ac} is equal to:

$$r_{ac} = \lambda_1 \gamma_1 \lambda_2 \qquad (14.20)$$

One can readily see that the unknown path "λ_1" is equal to r_{ac} divided by the product of "γ_1" and "λ_2." If any two values are chosen for "γ_1" and "λ_2," a value for "λ_1" can be derived. There are an infinite number of choices for "γ_1" and "λ_2," thus, an infinite number of solutions for "λ_1." Of course, applying the necessary condition given above would have established that the model is not identified. The basic problem in equation (14.20) is that there are three unknowns and only one equation giving information about the three unknowns. The equation is underidentified and, therefore, the values of "λ_1," "γ_1," and "λ_2" are not uniquely determined.

In complex structural equation models with unmeasured variables of the type ordinarily analyzed by LISREL, the algebraic formulations demonstrating identification are often tedious and sometimes intractable. Identification poses a special problem since it is not often demonstrable yet must be assumed in order to accept the values the program creates for the estimated paths.

Meeting the necessary condition given above is a first step in assessing the identification of a model. The LISREL program, however, also assesses identification in two additional ways. First, the program checks the "information" matrix (the "information" matrix is discussed in more detail in Jöreskog [27]). If the information matrix is positive definite, then the model is probably identified (Jöreskog [27]; Wiley [48]). If the matrix is not positive definite, the rank of the information matrix provides a clue as to which parameters are not identified. The program output warns the user when this occurs and specifies the parameter most likely not identified.

Jöreskog and Sörbom [29] suggest a second procedure to establish identification. The user can fix all parameters with reasonable values (i.e., the user fixes epistemic correlations at previously determined reliability levels and provides reasonable values for the relations among the unobserved dimensions) and has the program calculate a sigma matrix from these fixed values. In turn, this sigma matrix could be used as the "sample" covariance structure and used as data input to estimate the same model. If the parameter estimates are equal to the parameter estimates used in creating the "sample" covariance structure, the model is most likely identified. If they are not, those parameters that have different values are probably not identified. Neither of these procedures is problem free, nor does either guarantee that the model is identified (see Bentler [2, p. 443]). However, for complex models, these may be the only practical means of establishing some degree of confidence that the model being estimated is indeed identified.

EXAMPLES OF THE LISREL PROGRAM

The following are examples of the LISREL program. These examples are derived from simulated models designed to illustrate a few basic applications of the LISREL program and to demonstrate some of the common issues that may arise when specifying models. The examples also provide an interpretation of results and a clarification of certain issues discussed earlier in this chapter. The discussion of the indices for assessing fit will be short, and the reader is referred to the chapter by Herting and Costner [23] for a more detailed discussion (also see Jöreskog and Sörbom [29]).

Five basic examples are discussed in some detail. The first is a straightforward application of the program to the causal model represented in Fig. 14.4. The next two examples illustrate how "confirmatory factor analysis" can be used to estimate a measurement model,[12] where the emphasis is on specifying the relationships of indicators to dimensions, rather than on the causal structure among the dimensions. A fourth example shows the se-

---

[12] Confirmatory factor analysis is a special model within the general framework of the LISREL program. Programs specific to confirmatory factor analysis are also available (e.g., COFAMM); see Sörbom [43] or Long [36,37] for a more detailed discussion of the specifics of confirmatory factor analysis. The confirmatory factor analyses in the illustrations below demonstrate the basic nature of such models and how the results of such models may be used in specifying the measurement equations in a causal model.

quential use of the chi-square test for hypothesis testing with nested models. Finally, a fifth illustration will show how LISREL can be used to examine and test a causal structure across groups.

Example 1: A Simple Causal Model with Unmeasured Variables

Having discussed the basic model using Fig. 14.4 in some detail, it is convenient to use the same model in the first illustration of the use of the program. The LISREL program is developed in three stages. The outline of the program presented here will follow these stages, emphasizing the second stage. In the first stage, the researcher describes the data—that is, the number of cases, the form of the input data (correlation or covariances or "raw" data), and sets the desired type of analysis—that is, the program analyzes a covariance matrix. In the second stage, the researcher specifies the model as set up in equations (14.4)–(14.6). In the third stage, the researcher indicates the choice of output one wants from the program.

First, for any model, the LISREL program requires that the researcher define the data that will be used by the program to calculate S and to estimate and the parameter estimates. The data input will depend, in part, on the desired form of results that the user wishes the program to provide (e.g., whether standardized or unstandardized values or both are desired). The estimation procedure used by the LISREL routine requires only a covariance structure (i.e., variance–covariance matrices or correlations matrices) to make parameter estimates; whatever form of data one may have initially is changed into the form of a covariance structure by the program before estimation occurs. The user may either enter "raw" data from which LISREL calculates the covariance structures necessary, or one can enter the desired covariance structure produced by some other program (e.g., SPSS, BMDP). If one enters covariances, the program can calculate both unstandardized and standardized parameters; if a correlation matrix is entered, only standardized results may be obtained unless the user also provides the standard deviations for each variable so that the covariances can be calculated from the correlations (i.e., $\text{cov}_{12} = r_{12}\sqrt{sd_1{}^2\, sd_2{}^2}$). It is most convenient to input as data either a covariance matrix or a correlation matrix with a corresponding list of standard deviations for each variable. This allows the user to choose between analyzing standardized values (e.g., the correlation matrix) or unstandardized values. The examples used in this chapter will always input a covariance matrix, though estimating standardized solutions will be illustrated.

The type of data matrix one chooses to analyze depends partly upon the nature of the data and partly on the problem at hand and the intent of the researcher. If the units of measure are arbitrary, analyzing standardized values is appropriate. However, if the units are meaningful or if the researcher intends to compare across groups or with other published results, standardized values are not appropriate (see Blalock's discussion within the context of path analysis [7]; Blalock [11]; Wiley [48]) and unstandardized values will be more informative.

The initial order of the measured variables is not important when entering the data for use in the program. However, when one specifies the causal model, the measured indicators for endogenous dimensions will be assumed to be the first p number of indicators within the program, while the indicators of the exogenous concepts will be the last $t - p = q$ variables (where t is the total number of measured variables in the estimated model). The program allows the user to reorder the measured variables provided to the program and to select from a larger set of measured variables within a data set. However, the reordering must eventually list the indicators of endogenous concepts first, followed by the indicators of exogenous concepts. For the first illustration, the number of observations is 500, the matrix to be analyzed is a covariance matrix, the number of variables in the data is 15, and their order is $c_1, c_2 \ldots , f_2, f_3, a_1, a_2 \ldots , b_2,$ and b_3.

In the second stage of the program the researcher formulates the matrices shown in equations (14.10)–(14.16) which form the core of the program. In these matrices, the user communicates the following information: (1) the number of measured indicators of endogenous and exogenous dimensions and the number of exogenous and endogenous dimensions in the model; (2) which matrices will be used in the model; (3) their shape (e.g., diagonal or symmetric); and (4) which parameters within them are free or constrained.

The program uses four conventions to define the matrices and the elements within them. The shape of the matrix is defined—the shapes are either full, diagonal, or symmetric. The former implies that each element or cell in the matrix is relevant, the second states that only the diagonal of the matrix is relevant, and the last implies that the diagonal and the subdiagonal elements are relevant. Symmetric is used because the above-diagonal elements and subdiagonal elements are equal, as in a correlation matrix.

The matrices are also defined as either being "free" or "fixed." This specifies whether all the elements of the matrix are to be estimated or that all the elements are fixed at some value (i.e., 0.0). The elements freed or fixed are restricted by the shape of the matrix, that is, if a matrix has been defined as a "diagonal matrix and free," only the elements of the diagonal constitute and are estimated in the matrix. Specifying that a matrix is free or fixed is often a matter of convenience, since in most cases the matrix will have a mixture of these types of parameters.

Once the user specifies the matrices to be used, their shape, and general content, the program allows for freeing and fixing individual elements. This can be done either by a "pattern" matrix whose elements are 1 for parameters to be estimated and 0 for constrained parameters, or by specifying the coordinates of the element that is free or fixed by row and column, that is, fix lambda$_y$ (2,3) where 2 refers to the second row and 3 to the third column of the lambda$_y$ matrix.

Finally, particular values can be assigned to particular elements in a matrix or set equal to some other element in the same or different matrix — an equality constraint. Unless otherwise specified, all fixed values are assumed to be zero by the program; at times, however, values other than zero may be appropriate. For example, a parameter may be fixed at a non-zero value obtained from prior research in order to achieve identification. Equality constraints can be used as specific hypotheses to test, that is, the epistemic relations between two different indicators of the same dimension may be set equal to each other or such a constraint may be used to achieve identification. Such constraints are specified by defining the row and column of the two or more parameters constrained to be equal.

In the present example, the researcher begins the second stage by stating the number of indicators for endogenous dimensions ($9 - c_1, c_2, c_3, d_1, d_2, d_3, f_1, f_2, f_3$) and exogenous dimensions ($6 - a_1, a_2, a_3, b_1, b_2, b_3$) and the number of dimensions — 2 ksi (exogenous dimensions) and 3 eta (endogenous dimensions). The matrices are also defined as having certain properties as discussed above. For Fig. 14.4, the properties are: beta = full,free; gamma = full,free; lambda$_x$ = full,free; lambda$_y$ = full,free; phi = symmetric,free; psi = diagonal,free; theta epsilon = diagonal,free; and theta delta = diagonal,free.

The "pattern" matrices that represent each of the matrices described above — the beta, gamma, lambda, psi, phi, theta delta, and theta epsilon — are used to specify the free and fixed parameters.[13] The fixed and constrained parameters are given their value by: stating the value the parameter is equal to, the matrix in which it is located, and the row and column relevant to the constraint. In this particular example, there are fixed values but no equality constraints. In Table 14.2, the necessary pattern matrices and the constraints are given for each matrix in the specification of the model represented in Fig. 14.4.

A pattern matrix for phi, psi, theta epsilon, and theta delta would be redundant given that phi was defined as symmetric free and the other three as diagonal free. In this circumstance, all the elements in these four matrices are estimated.

For the structural model, 1's in the pattern matrices of gamma and beta and the free parameters implicit in the definition of phi and psi correspond to the non-zero values entered in the matrices given in equations 14.10 – 14.16; while the zeros are in the same position in these matrices as before. Recall that an entry of "1" here denotes a value to be estimated (a "free" parameter), while a zero entry denotes a "fixed" parameter. For the measurement part of the model, entries of "1" also appear where previously

---

[13] The LISREL program provides alternative means of fixing and freeing parameters and stating their initial values not shown in these examples. The "pattern" matrices are used because they are similar to the matrix formulation introduced earlier.

Table 14.2. Pattern matrices for estimating model Figure 14.4

Pattern matrices for the structural equations

	beta				gamma	
	C	D	F		A	B
C	0	0	0	C	1	1
D	1	0	0	D	0	0
F	1	0	0	F	0	0

Pattern matrices for the measurement equations

	lambda$_y$				lambda$_x$	
	C	D	F		A	B
c_1	1	0	0	a_1	0	0
c_2	0	0	0	a_2	1	0
c_3	1	0	0	a_3	1	0
d_1	0	0	0	b_1	0	0
d_2	0	1	0	b_2	0	1
d_3	0	1	0	b_3	0	1
f_1	0	0	0			
f_2	0	0	1			
f_3	0	0	1			
Start 1.0	$\lambda_y(2,1)$	$\lambda_y(4,2)$	$\lambda_y(7,3)$			
Start 1.0	$\lambda_x(1,1)$	$\lambda_x(4,2)$				

(i.e., in matrices shown in equations 14.10–14.16) there were parameters to be estimated. Again, the theta delta and theta epsilon matrices have been defined with free parameters (i.e., "1"'s) along the diagonal and zeros in the off-diagonal elements. However, there are five exceptions.

For each column of the two lambda matrices, one previously estimated parameter has been set at a fixed value (i.e., designated by a zero in the "pattern" matrix). In this case, the fixed value is 1.0 as stated in the "start" statement below the matrices (the fixed indicators are: a_1, b_1, c_2, d_1, and f_1). When analyzing covariance matrices, one indicator for each dimension must be set to some value (generally 1.0) to give the dimension a unit of measure. The unmeasured dimension has no inherent scale or units, therefore, setting one "reference" indicator per dimension scales the dimension to the units of the reference indicator. For example, if a_1 were the income level of an individual's family in dollars, setting the lambda$_x$ path equal to 1.0 places the unit of measure of the dimension in dollars. Likewise for dimension C, setting the lambda$_y$ path for c_1, measured in years of schooling, to 1.0 means that dimension C has years as its unit of measure. Therefore, the path from dimension A to C (the gamma path) is expressed in years change per dollar change (i.e., an increase in "x" dollars leads to an increase of "y" years). Setting the parameter linking an indicator to a dimension at 1.0, however, does not imply a perfect relationship between indicator and dimension. A perfect relation is indicated if the error term (or error variance) of the indicator is zero. Since the respective theta epsilon and theta delta terms for each of indicators fixed at 1.0 (a_1, b_2, c_1, d_1, and f_2) are not fixed at zero, there is no claim of a perfect relationship here. The choice of which indicator to fix at 1.0 is arbitrary, although it is convenient to choose one that provides easy interpretation and discussion.

The phi matrix (covariances among the exogenous dimensions, A and B) has been defined as symmetric and free. This implies that the exogenous dimensions (two in this instance) are correlated, as shown in Fig. 14.4 by a double-headed arrow. The error covariance matrices — psi, theta epsilon, and theta delta — were all defined as being diagonal and free (i.e., having no off-diagonal terms). They are defined in this manner since there are no correlated errors postulated in Fig. 14.4 between either the endogenous dimensions or between the pairs of measured indicators within the set of indicators for exogenous dimensions or within the set of indicators for endogenous dimensions. Each of the diagonal elements are free to assume an estimated value. Again, a fixed zero in either the theta delta or theta epsilon matrix would imply that the indicator associated with that error variance is a perfect indicator of the dimension. A zero error term in the diagonal of the psi matrix is unlikely as it implies that the eta term (the endogenous dimension) it is associated with is fully explained by the other terms in the model that "cause" that dimension, an unlikely event and one that should be estimated rather than fixed in a causal model.

The content of the output of the program is determined by the user. A variety of results can be obtained. Among these are: standard errors of the estimates, "t-values" of the estimates (defined as the parameter estimate divided by twice its standard error), factor scores, the sigma matrix, the "residual" matrices, as well as other diagnostics. For this example, the basic output is obtained which includes: a listing of the parameters that will be estimated, the entries in the data matrix analyzed, the parameter estimates and goodness of fit measures. For a thorough description of the program output options, see Jöreskog and Sörbom [29].

Having set the units of measure for each dimension and checked the necessary condition for identification, the model is estimated — refer to equation (14.19); there are 40 unknowns in the model which is less than $(1/2)$ $(9 + 6 + 1)$ $(9 + 6)$, where 9 and 6 are the number of indicators for endogenous and exogenous dimensions, respectively. Table 14.3 provides the estimates for the model as they would appear in the matrix equations. Figure 14.6 illustrates how these estimates would appear in a diagram. The

Table 14.3. Parameter estimates for estimating model Figure 14.4 (standard errors)

	lambda$_y$				lambda$_x$	
	C	D	F		A	B
c_1	.80 (.03)	0.00	0.00	a_1	1.00*	0.00
c_2	1.00*	0.00	0.00	a_2	1.12 (.05)	0.00
c_3	2.10 (.08)	0.00	0.00	a_3	1.63 (.07)	0.00
d_1	0.00	1.00*	0.00	b_1	0.00	1.00*
d_2	0.00	1.34 (.05)	0.00	b_2	0.00	2.43 (.11)
d_3	0.00	2.73 (.09)	0.00	b_3	0.00	.83 (.04)
f_1	0.00	0.00	1.00*			
f_2	0.00	0.00	2.32 (.04)			
f_3	0.00	0.00	1.31 (.03)			

	beta				gamma	
	C	D	F		A	B
C	0.00	0.00	0.00	C	.43 (.06)	.91 (.08)
D	.41 (.04)	0.00	0.00	D	0.00	0.00
F	1.09 (.05)	0.00	0.00	F	0.00	0.00

	phi			psi		
	A	B		C	D	F
A	3.05 (.27)			2.66 (.26)	2.70 (.23)	1.92 (.17)
			R^2	.54	.26	.78
B	.94 (.14)	2.12 (.22)				

	theta epsilon								
	c_1	c_2	c_3	d_1	d_2	d_3	f_1	f_2	f_3
	1.20 (.09)	2.90 (.20)	2.09 (.32)	1.36 (.11)	1.96 (.17)	1.90 (.47)	.95 (.07)	1.39 (.23)	1.18 (.10)
R^2	.75	.66	.92	.73	.77	.94	.90	.97	.93

	theta delta					
	a_1	a_2	a_3	b_1	b_2	b_3
	1.25 (.11)	1.53 (.14)	1.97 (.24)	1.48 (.11)	1.43 (.33)	.86 (.07)
R^2	.71	.71	.80	.59	.89	.63

Chi square = 23.13 with 85 df; probability level = .99.
* Fixed value.

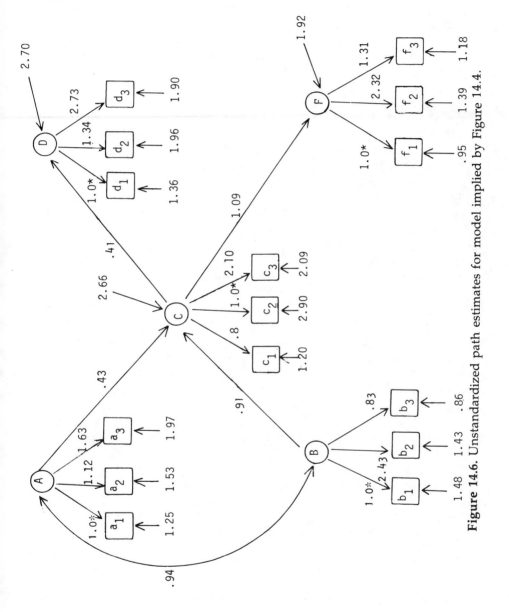

Figure 14.6. Unstandardized path estimates for model implied by Figure 14.4.

unstandardized path estimates correspond to the values that appear in the matrices in Table 14.3. The LISREL program lists the estimates much as they appear in the table. In Fig. 14.6, there are no "path" estimates for the arrows representing the error terms for the endogenous dimensions and the indicators. Listed at the end of each error path is the "error variance" for these measured variables (theta delta and theta epsilon) and dimensions (psi).

In addition to these estimates, the proportion of variation explained for each of the indicators as well as for each endogenous dimension is provided in the program output. The individual R^2 values have the same interpretation as in multiple regression. Each is calculated by subtracting from 1.0 the error sum of squares divided by the total sum of squares. In addition, we are given two values that suggest the amount of variation explained in the indicators as a set — the coefficient of determination. This assesses the reliability of the set of indicators for the endogenous dimensions and for the exogenous dimensions; the greater the coefficient, the greater the reliability of the indicators as a set. The reliability of each indicator will be guaged by its R^2. The reliabilities of individual indicators are generally high in this simulation, and this is reflected in high coefficients of determination (.9) for both sets of indicators. The explained variance in the three endogenous dimensions are .537, .263, and .781 for dimensions C, D, and F, respectively.

The chi square and the indices of fit suggest that the model estimated is a very good approximation of the "true" model that generated the observed covariance structure, S. The chi square is 23.13 ($df = 85$) with a 1.0 probability level; the probability level has been rounded. The null hypothesis that the implied covariances differ from the observed covariances only because of random sampling variation cannot be rejected, suggesting that the "estimated" model adequately represents the "true" causal process. In this case, minor misspecifications were deliberately created so that there would be a certain amount of discrepancy between the implied and observed covariances; had this not been done or had these minor misspecifications been corrected in the estimated model, the chi-square value would have been exactly 0.0 and the probability level exactly 1.0. Such a small value of chi-square using "real" data would generally indicate that the model is not so restrictive as it could be (i.e., too many paths have been included and some of these are probably zero). Inspecting the normalized residuals matrix, not shown here, shows only minor deviations from zero — none greater than 2.0 and, therefore, none are significant deviations from the observed covariance matrix, S.

Example 2: A Confirmatory Factor Analysis

Often the researcher needs to assess and respecify the measurement specification of a causal model (see Herting and Costner [23]). Or the researcher may simply be interested in the factor structure and the measure-

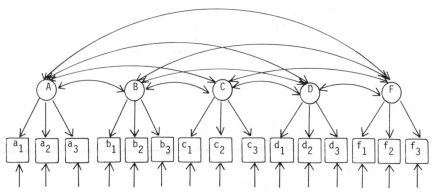

Figure 14.7. Confirmatory factor analysis model implied by Figure 14.4.

ment relations rather than the causal structure among dimensions (see Bielby and Hauser [4]; Bridges [12] for applications of confirmatory factor analysis). This next example posits a measurement model in which merely "covariation" between dimensions is specified, rather than "cause." We illustrate by specifying the "measurement model" implicit in Fig. 14.4 (see Fig. 14.7).

The model illustrated in Fig. 14.7 is quite simple. No exogenous or endogenous distinctions are imposed upon the dimensions. Given this, we need not specify a beta matrix or a gamma matrix — since these represent the relationships among the endogenous and exogenous dimensions. Neither is a psi matrix required, nor are the lambda$_y$ and theta epsilon matrices since there are only exogenous dimensions. There is no structural equation specified in a confirmatory factor analysis. The measurement equation can be formulated as equation (14.5) as previously developed above; a phi matrix is included whose element are the estimates of the covariation among the dimensions. The theta delta matrix represents the error covariances of the measurement equations; the diagonal elements are the error variances of each indicator, the off-diagonal elements are the covariation or correlated error between pairs of indicators.

The following matrices are defined and the "pattern" matrices established as shown below. (In this formulation, we are still estimating unstandardized values; the input information for the first stage of the program is the same.)

The second stage would consist of the following matrix definitions: lambda$_x$ = full,free; theta delta = diagonal,free; phi = symmetric,free (implicitly the beta, gamma, psi, lambda$_y$, and theta epsilon are set to zero). The pattern matrices and fixed values are given in Table 14.4.

The pattern matrix for lambda$_x$ has 15 rows representing the measured variables and 5 columns representing the dimensions. The order of the variables is the same as in the first example. The first 3 columns (C, D, and F)

Table 14.4. The lambda$_x$ pattern matrix for the
confirmatory factor analysis
represented in Figure 14.7

	lambda$_x$				
	C	D	F	A	B
c_1	1	0	0	0	0
c_2	0	0	0	0	0
c_3	1	0	0	0	0
d_1	0	0	0	0	0
d_2	0	1	0	0	0
d_3	0	1	0	0	0
f_1	0	0	0	0	0
f_2	0	0	1	0	0
f_3	0	0	1	0	0
a_1	0	0	0	0	0
a_2	0	0	0	1	0
a_3	0	0	0	1	0
b_1	0	0	0	0	0
b_2	0	0	0	0	1
b_3	0	0	0	0	1
Start 1.0	$\lambda_x(2,1)$	$\lambda_x(4,2)$	$\lambda_x(7,3)$	$\lambda_x(10,4)$	$\lambda_x(13,5)$

represent the previous endogenous dimensions, the second two are A and B as before. The pattern matrix is constituted by linking dimension (cause) to indicator (result) as previously discussed. As before, one indicator per dimension has been fixed equal to 1.0 to provide identification and units of measure for each dimension; these indicators are the same as in the first example. The error covariance matrix, theta delta, and the phi matrix do not require pattern matrices since the first was defined as diagonal and free and the second as symmetric and free. The first is diagonal because no correlated errors occur among indicators have been specified. The latter is symmetric, with the variances of the dimensions on the diagonal and the covariances between pairs of dimensions as off-diagonal elements.

The estimates of the lambda and phi parameters are given in Table 14.5. The goodness of fit results and other values are the same as in the first example. Had there been misspecifications in the structural model for Fig. 14.4, for example, an omitted path between A and F, such a specification error in the structural part of the model would have been irrelevant in the corresponding confirmatory factor analysis model which considers only the covariances between dimensions. The confirmatory factor analysis model would then have had a better fit than the LISREL model. Since there was, in this instance, no misspecification in the structural part of the model, the goodness of fit is identical in the confirmatory factor analysis model and the LISREL model (see Herting and Costner [23]). Only one coefficient of deter-

Table 14.5. The parameter estimates for the confirmatory factor analysis represented in Figure 14.7

	\multicolumn{5}{c}{lambda$_x$}				
	C	D	F	A	B
c_1	.80	0.0	0.0	0.0	0.0
c_2	1.00*	0.0	0.0	0.0	0.0
c_3	2.10	0.0	0.0	0.0	0.0
d_1	0.0	1.00*	0.0	0.0	0.0
d_2	0.0	1.34	0.0	0.0	0.0
d_3	0.0	2.73	0.0	0.0	0.0
f_1	0.0	0.0	1.00*	0.0	0.0
f_2	0.0	0.0	2.32	0.0	0.0
f_3	0.0	0.0	1.31	0.0	0.0
a_1	0.0	0.0	0.0	1.00*	0.0
a_2	0.0	0.0	0.0	1.12	0.0
a_3	0.0	0.0	0.0	1.63	0.0
b_1	0.0	0.0	0.0	0.0	1.00*
b_2	0.0	0.0	0.0	0.0	2.43
b_3	0.0	0.0	0.0	0.0	.83

phi

	C	D	F	A	B
C	5.74				
D	2.35	3.66			
F	6.27	2.56	8.79		
A	2.18	.89	2.38	3.05	
B	2.34	.96	2.57	.94	2.12

Chi square = 23.13; 80 *df*
* Fixed values.

mination is produced for the confirmatory factor analysis model since the measured variables are all taken as a set of indicators of exogenous dimensions. The coefficient of determination for this example is .96.

Example 3: A More Complicated Measurement Model

Figure 14.7 is useful in demonstrating the general confirmatory factor analysis model, but it does not demonstrate a number of issues that will generally appear in a causal model with multiple indicators. More often, the researcher will face, among other problems, measured variables that (1) have a spurious source of covariation with some other indicator (i.e., correlated error); (2) will be a result of two or more dimensions in the model (i.e., a "shared" indicator); or (3) may be the only indicator of a construct available in the data, that is, multiple indicators for this dimension are not available or the dimension is fully represented by the measured variable. Figure 14.8 is introduced as an example of a confirmatory factor analysis with the three features listed above. In this example, we also analyze a correlation matrix to demonstrate how to obtain standardized results directly.

As before, we first define the data and form of analysis desired. For this simulation, the number of cases is 750, the data entered is a covariance matrix; however, we designate that we want to analyze a correlation matrix. Because of this statement, the LISREL program will convert the covariance matrix to a correlation matrix and proceed to estimate the parameters of the model on the basis of that correlation matrix (i.e., in the estimation procedure, S will be standardized). The results will be a standardized Σ (the model-implied covariance matrix) and standardized parameters (beta weights and correlations).

In the model, as represented in Fig. 14.8, two dimensions only have one indicator each (V and Z). Indicator v_1 is assumed to have some measurement error shown by including an arrow indicating error, while z_1 is assumed to be equivalent to the construct Z, shown by encircling the indicator z_1 in the figure. For example, in a model of earnings, the measured variable, "dollars earned," is the variable of interest and, therefore, not a reflector of some latent unmeasured variable. However, if one were modeling "total income" rather than earnings, "dollars earned" would be a reflector of the total income dimension and would be treated similarly to indicator v_1 in this example (i.e., some measurement error would have to be assumed for the single indicator). Correlated error is specified between indicators y_1 and y_2. Such error may be due to both indicators being affected by a response bias (i.e., social desirability). The indicator, u_3, is "shared" by dimension U and V; it "reflects" both dimensions. These added relations, plus the fact that we are analyzing a correlation matrix, will change the specification of the model in the second stage of the program. As in the second example, the only relevant matrices are the lambda$_x$, phi, and theta delta. The model has 13 measured variables and 6 unmeasured variables. The matrices are defined as: lambda$_x$ = full,free; theta delta = symmetric,fixed; and phi = symmetric,free. The pattern matrices and fixed values are given in Table 14.6.

The second stage of the program for this example differs from the previous example (Fig. 14.7) because of the additional relations among indicators and because we have specified that a correlation matrix be analyzed. First, the phi matrix has been specified as symmetric and free, but the diagonal of the phi matrix has been fixed and each element of the diagonal set equal to one — the zeros in the diagonal of the phi "pattern" matrix and the start statement for the phi matrix accomplish this. Because we are analyzing a correlation matrix and the elements of phi represent the correlations among pairs of dimensions fixing 1's in the diagonal is appropriate. The 1's in the diagonal of phi also fix the units of the dimensions to standard units,

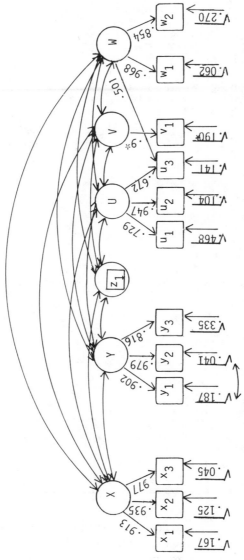

Figure 14.8. Confirmatory factor analysis with standardized path estimates.

Table 14.6. The pattern matrices for the confirmatory factor analysis represented in Figure 14.8

	lambda$_x$							phi						
	X	Y	Z	U	V	W		X	Y	Z	U	V	W
x_1	1	0	0	0	0	0	X	0					
x_2	1	0	0	0	0	0	Y	1	0				
x_3	1	0	0	0	0	0	Z	1	1	0			
y_1	0	1	0	0	0	0	U	1	1	1	0		
y_2	0	1	0	0	0	0	V	1	1	1	1	0	
y_3	0	1	0	0	0	0	W	1	1	1	1	1	0
z_1	0	0	0	0	0	0							
u_1	0	0	0	1	0	0							
u_2	0	0	0	1	0	0							
u_3	0	0	0	1	0	1							
v_1	0	0	0	0	0	0							
w_1	0	0	0	0	0	1							
w_2	0	0	0	0	0	1							

Start 1.0 $\phi(1,1)$ $\phi(2,2)$ $\phi(3,3)$
$\phi(4,4)$ $\phi(5,5)$ $\phi(6,6)$

Start 1.0 λ_x (7,3)
Start .9 λ_y (11,5)

theta delta

	x_1	x_2	x_3	y_1	y_2	y_3	z_1	u_1	u_2	u_3	v_1	w_1	w_2
x_1	1												
x_2	0	1											
x_3	0	0	1										
y_1	0	0	0	1									
y_2	0	0	0	1	1								
y_3	0	0	0	0	0	1							
z_1	0	0	0	0	0	0	0						
u_1	0	0	0	0	0	0	0	1					
u_2	0	0	0	0	0	0	0	0	1				
u_3	0	0	0	0	0	0	0	0	0	1			
v_1	0	0	0	0	0	0	0	0	0	0	0		
w_1	0	0	0	0	0	0	0	0	0	0	0	1	
w_2	0	0	0	0	0	0	0	0	0	0	0	0	1

Start .19 θ_δ (11,11)

so correspondingly there is no need to fix one indicator per dimension in the lambda$_x$ matrix as in the previous confirmatory factor analysis analyzing a covariance matrix. In this example, the elements of the lambda$_x$ pattern matrix are 1's for all indicators associated with a given dimension and, therefore, are estimated. Two exceptions are evident: The lambda$_x$ element for indicators z_1 and v_1 have both been fixed since they are the only indicators of their respective dimensions. The relation between dimension V and indicator v_1 has been assumed to be less than perfect and, therefore, fixed at

.9 (the epistemic correlation of .9 may or may not be accurate; the model is technically misspecified if the parameter is not truly .9). The error variance for this indicator is set to .19 since the error variance is equal to $1 - .9^2$ in the standardized case (see below). The epistemic correlation between dimension Z and z_1 has been set at 1.0 (and its error variance set to zero in the theta delta matrix) reflecting that z_1, as measured, is the variable of interest.

The theta delta matrix is defined as symmetric and fixed because of the correlated error specified between y_1 and y_2 in the figure. The pattern matrix indicates that all the diagonal elements of the matrix are to be estimated, except that the error variance of z_1 is fixed and set to zero, and the error variance for v_1 is set at .19 as discussed above. The off-diagonal elements are all fixed except for the element representing the correlated error between y_1 and y_2 (the program sets all "fixed" elements at zero unless otherwise specified, so no statements setting these values to zero are required for the off-diagonal elements or for the error variance of indicator z_1).

The parameter estimates are given in Fig. 14.8; we meet the necessary condition for identification, and there is no evidence from the program suggesting a problem with identification. The fit of the confirmatory factor analysis model to the data is quite good; chi-square 24.86 with 52 degrees of freedom is not significant. The goodness of fit index is .969, and the mean square residual is quite small. The normalized residual matrix and other specific measures of fit also indicate that the model is an adequate representation of the data.

In this example, the elements of the lambda matrix are estimates of the epistemic correlations between the dimensions and indicators (i.e., they are the square roots of the reliabilities of the indicators in this model). In this simulated illustration, most are quite high. Having analyzed a standardized matrix, the error paths shown in Figure 14.8 can be calculated for the indicators by taking the square root of the error variance of the indicator; this is the same as calculating the square root of $1 - R^2$ as is typically done in path analysis. Given that we fixed the epistemic correlation of indicator v_1 to .9, the error variance, .19, was derived by taking one minus the square of the epistemic correlation. The error path, therefore, is the square root of .19. In a circumstance where an unstandardized error variance must be fixed to some *a priori* determined value for v_1, determining that value is more problematic.

Example 4: Nested Models and a Standardized Causal Model

Having fit the measurement model in Figure 14.8, we can attempt to estimate a causal model for the same observed covariances. Figure 14.9 illustrates the researcher's specification of the model, where the solid lines represent the initial specification (H_0) and the two dotted paths (U to Y and V to Y) represent a sequence of additional models to be estimated. Adding the path from U to Y constitutes the second model to be estimated (H_1); adding

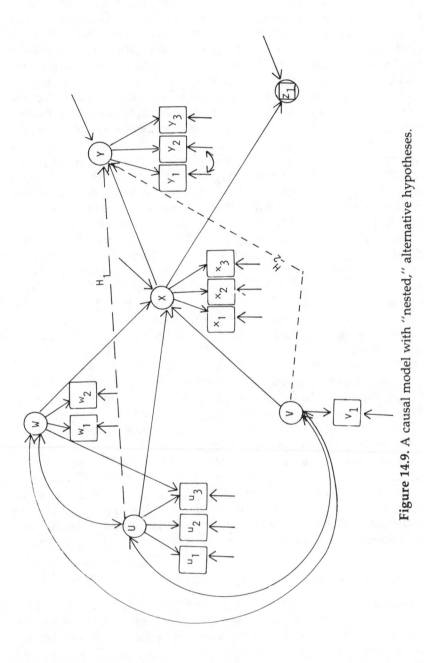

Figure 14.9. A causal model with "nested," alternative hypotheses.

the path V to Y to this specification is the third model (H_2). The tests are nested: H_0 is a submodel of H_1 and H_2; H_1 is a submodel of H_2.

The first stage of the program is the same as in the confirmatory factor analysis in Example 3. We are analyzing a standardized covariance matrix with the same sample size and same number of variables.

The second stage is similar, though there are differences from both Example 1 of estimating a causal model and from the confirmatory factor analysis in Example 3 just completed. The number of indicators of endogenous dimensions is 7, the number of indicators for exogenous dimensions is 6. There are three eta terms and three ksi terms. The shape and general content of the matrices for the specification H_0 are defined as: beta = full,fixed; gamma = full,free; phi = symmetric,free; psi = diagonal,free; lambda$_y$ = full,free; lambda$_x$ = full,free; theta epsilon = symmetric,fixed; theta delta = diagonal,free. Table 14.7 gives necessary pattern matrices defining the elements as free and fixed.

The pattern matrices and fixed values for the structural portion of the model are similar to the Example 1, except that the pattern matrix for phi is fixed down the diagonal at 1.0, thus standardizing the covariances among pairs of the exogenous dimensions and fixing the unit of measure for these dimensions to standard unit variance (e.g., 1.0). The fixed and free elements in the gamma and beta pattern matrices are associated with each path in the first model (H_0). The psi matrix is diagonal and free; all diagonal elements are estimated.

The pattern for the lambda$_x$ matrix appears as it did in the lower right-hand part of the previous confirmatory factor analysis. It is not necessary to fix any of these paths to 1.0 because the 1.0's in the diagonal of the phi matrix have set the units for the exogenous dimensions; the lambda$_x$ path associated with v_1 has been fixed at .9 as before, as has its error variance been fixed to .19 in the theta delta matrix. The lambda$_y$ pattern matrix for the indicators of the endogenous dimensions appears as it does when analyzing a covariance matrix. A fixed element for each dimension must be included to set the units for the endogenous dimensions. In this example, x_1, y_1, and z_1 are fixed, the z_1 indicator is fixed to 1.0 as before. The lambda$_y$ parameters representing the paths of dimension X to indicator x_1 and dimension Y to indicator y_1, however, have been fixed to the square roots of their reliabilities as estimated in the confirmatory factor analysis previously completed, .913 and .909, respectively.[14]

Setting these two parameters at the square root of the reliability of their respective indicators sets these two endogenous dimensions (X and Y) to unit variance, that is, the dimension is set to the same scale as in the confir-

---

[14] The user does have the option of fixing these parameters to 1.0 instead of the square roots of their reliabilities. The lambda coefficients, in this case, would be interpreted as beta coefficients (as in multiple regression), which may exceed 1.0.

Table 14.7. The pattern matrices for estimating model Figure 14.9

Pattern matrices for the structural model

	beta				gamma				phi		
	X	Y	Z		U	V	W		U	V	W
X	0	0	0	X	1	1	1	U	0		
Y	1	0	0	Y	0	0	0	V	1	0	
Z	1	0	0	Z	0	0	0	W	1	1	0

Start 1.0 $\phi(1,1)$ $\phi(2,2)$ $\phi(3,3)$

Pattern matrices for the measurement model

	lambda$_y$				lambda$_x$		
	X	Y	Z		U	V	W
x_1	0	0	0	u_1	1	0	0
x_2	1	0	0	u_2	1	0	0
x_3	1	0	0	u_3	1	0	1
y_1	0	0	0	v_1	0	0	0
y_2	0	1	0	w_1	0	0	1
y_3	0	1	0	w_2	0	0	1
z_1	0	0	0				

Start .935 $\lambda_y (1,1)$
Start .909 $\lambda_y (4,2)$
Start 1.00 $\lambda_y (7,3)$
Start .900 $\lambda_x (4,2)$

	theta epsilon								theta delta					
	x_1	x_2	x_3	y_1	y_2	y_3	z_1		u_1	u_2	u_3	v_1	w_1	w_2
x_1	0							u_1	1					
x_2	0	1						u_2		1				
x_3	0	0	1					u_3			1			
y_1	0	0	0	0				v_1				0		
y_2	0	0	0	1	1			w_1					1	
y_3	0	0	0	0	0	1		w_2						1
z_1	0	0	0	0	0	0	0							

Start .168 $\theta_\epsilon (1,1)$ Start .19 $\theta_\delta (4,4)$
Start .207 $\theta_\epsilon (4,4)$

matory factor analysis in Example 3 that had generated these two estimates of the lambda$_y$ coefficients. The theta epsilon includes the correlated error between y_1 and y_2 and the zero error variance for z_1. The error variances for x_1 and y_1 are fixed to the values previously estimated in the confirmatory factor analysis in Example 3.

We estimate the model as specified in H_0. The parameter estimates for λ_x, λ_y, θ_ϵ and θ_δ for the measurement part of the model are all equal to the

Table 14.8. Estimates of structural parameters for
the "nested" models in Figure 14.9

For hypothesis			
	H_0	H_1	H_2
β_{YX}	.712 (.029)	.614 (.033)	.602 (.040)
β_{ZX}	.547 (.030)	.547 (.032)	.547 (.032)
γ_{XU}	.231 (.028)	.224 (.028)	.224 (.028)
γ_{XV}	.371 (.030)	.371 (.030)	.371 (.030)
γ_{XW}	.479 (.031)	.481 (.030)	.482 (.031)
γ_{YU}	0.0	.211 (.031)	.209 (.031)
γ_{YV}	0.0	0.0	.019 (.038)
	$\chi^2 = 89.72$	$\chi^2 = 44.11$	$\chi^2 = 43.86$
	df 58	df 57	df 56

previous confirmatory factor analysis. The structural parameters and their
standard errors are given in Table 14.8. The chi square for this model (H_0) is
89.72, with 58 df and a probability of .005. Based on this chi-square value,
we would reject the model as an adequate representation of the data. The
other measures (the normalized residuals, the goodness of fit index, and the
root mean square residual) also indicate that the model does not fit the data
well.

The alternative model, H_1, is estimated. We add the path from dimension
U to Y. This changes the gamma pattern matrix only:

$$
\begin{array}{c c c c}
 & U & V & W \\
X & 1 & 1 & 1 \\
Y & 1 & 0 & 0 \\
Z & 0 & 0 & 0
\end{array}
$$

The parameters in this model are estimated; the results for the structural
level are given in the second column of Table 14.8. The chi square is 44.11
with 57 df. The probability level suggests that the model as specified by H_1
fits the data. The chi-square test between the two alternative models
($89.72 - 44.11 = 45.61$) shows that the additional decrease in chi square by
adding the path from dimension U to Y is significant (45.61 with 1 df). These
results suggest that the model as specified by H_1 fits the data significantly
better than H_0. We now test the hypothesized model H_2 against H_1.

For model H_2, we change the specification by adding the paths from U to
Y and V to Y shown by the dotted lines in Figure 14.9. The gamma pattern
matrix is changed accordingly (the path V to Y is added to the gamma matrix
shown for H_1):

$$
\begin{array}{c c c c}
 & U & V & W \\
X & 1 & 1 & 1 \\
Y & 1 & 1 & 0 \\
Z & 0 & 0 & 0
\end{array}
$$

The rest of the specification remains the same. The results are given in the third column of Table 14.8. The chi square is 43.86 with 56 df. The model fits the data, but there is only a decrease of .25 in the chi-square. This decrease is not significant, hence, model H_2 is not an improvement upon H_1. We would accept H_1 or H_2 as an adequate representation of the data; however, the path from V to Y is superfluous in achieving a good fit as shown by the nested chi-square test. The standard error of the path also suggests that it is not different from zero; the parameter estimate for the path is less than twice its standard error. H_1 is the preferred model.

Having posed the hypotheses (H_0, H_1, and H_2) *a priori* to estimation, we can treat the chi-square statistic as testing a null hypothesis. As indicated above, we may also utilize this same type of procedure for ferreting out which paths may or may not improve an original specification of a given model. In this circumstance, we do not directly test null hypotheses but rather guage the importance of a path by the resulting reduction per degree of freedom in chi-square when it is included in the model. Large reductions indicate a given path should be included.

Example 5: Analysis of Causal Models across Samples

The final example demonstrates how the general LISREL model can be extended to an analysis across samples. In this circumstance, the researcher may ask whether the same causal model applies to two or more groups or samples, for example, whether a status attainment model is the same for females and males or whether the process leading to innovation in organizations is the same for French, Japanese, and United States firms. The samples must be independently drawn and the subgroups must be mutually exclusive. Such model comparisons are prevalent in the sociological literature, although using multiple indicator models in such analyses is less evident.

In the general causal model discussed earlier, the elements of Σ depend on Λ_y, Λ_x, β, Γ, Φ, Ψ, θ_δ and θ_ϵ. When the researcher has different samples or can divide a sample into distinct subgroups (e.g., treatment-control; ethnic groups) it is possible to test whether differences exist between elements of these matrices for one group compared to another. We may be interested in, for example, whether $\Lambda_y^{(1)} = \Lambda_y^{(2)} = \ldots = \Lambda_y^{(k)}$ or $\Gamma^{(1)} = \Gamma^{(2)} =, \ldots, = \Gamma^{(k)}$, where the superscript represents a distinct matrix for each sample or group. The researcher may impose various restrictions on the similarity of parameters across groups. For example, one might require that all elements in each of these matrices be equal across groups. Less restrictive would be the requirement that selected elements in these matrices be equal for all groups while other elements are free to vary from one group to another. The least restrictive model would allow all the parameters to vary across the samples. The procedure also allows for some paths to be fixed at zero in one sample, while in other samples the path is estimated. It is also permissible to have a different number of dimensions or measured variables in each sample, and one may specify different measured variables for particular dimensions in different samples. For example, reading skills may be assessed in an entirely different manner for one age group from that for another. However, for comparison purposes, the models will generally be similar, and the measures for the dimensions will have similar content though not necessarily equivalent forms or be derived from similar research instruments (e.g., questionnaires). The procedure is outlined below using the model illustrated in Fig. 14.7 including the path from U to Y, shown in the last example to be significant.

The analysis is similar to what has been discussed earlier. It is assumed that the model for each group is defined by the following equations:

$$\eta^g = \beta^g \eta^g + \Gamma^g \xi^g + \zeta^g \tag{14.21}$$

$$x^g = \Lambda_x^g \xi^g + \delta^g \tag{14.22}$$

$$y^g = \Lambda_y^g \eta^g + \epsilon^g \tag{14.23}$$

where "g" $= 1, 2, \ldots , k$, with k equal to the number of groups. If k equals 1, then this is the same formulation given in equations (14.4) and (14.6) for the general LISREL model. Here, each sample, indicated by some value for "g," has a matrix in equations (14.21)–(14.23) which may or may not contain different elements from the matrices of other samples. Again, the free and constrained elements within these matrices define the model of interest. The constraints in the matrices may be within and/or across the samples. The fitting function and logic of estimation is similar to what has been described.[15] Furthermore, a chi-square test is still available to test the adequacy of the model to fit the data (i.e., to fit the two or more observed covariance matrices). A chi-square test is calculated as before with the degrees of freedom equal to

$$[(1/2)(k)(p + q)(p + q + 1) - u]$$

where p is the number of indicators of endogenous dimensions, q the number of indicators of exogenous dimensions, k is the number of groups, and u

---

[15] The fitting function F changes slightly when analyzing more than one sample:

$$F = \sum_{g=1}^{G} N^{(g)}/N \, F^{(g)} \, (\Sigma^{(g)} S^{(g)})$$

where

$$F^{(g)} = \log |\Sigma^{(g)}| + \text{tr}(S^{(g)}\Sigma^{(g)-1}) - \log |S^{(g)}| - t$$

where $g = 1, 2, \ldots , k$; N is the total N; and $N^{(k)} = $ the number in group k.

the number of independently estimated parameters. The output provides the chi-square value, degrees of freedom, and the probability level of the chi-square value.

The parameter estimates are neither averages across the groups nor based upon separate calculations for each group. Rather, the estimation procedure simultaneously calculates values for the elements that minimize the fitting function. Meaning the estimation procedure calculates the elements of the matrices — Λ_y, Λ_x, β, Γ, Φ, Ψ, θ_δ, and θ_ϵ — that will make the estimated Σ's (i.e., Σ^1, Σ^2, . . . , Σ^k) as nearly identical to their respective observed S's (i.e., S^1, S^2, . . . , S^k) as they can possibly be, given the free and constrained elements of the model as specified by the researcher — those elements in the matrices set to be estimated, or equal across samples, or set to some fixed value such as zero.

The program basically has the same structure as before. There are three stages for each group. Each group or sample follows in sequence: Stages 1, 2, and 3 are defined for Group 1, then Stages 1, 2, and 3 for Group 2 and so forth. For ease of presentation, each stage is defined together for the two samples in this example. The major focus will be on the second stage.

In the first stage of the program, the researcher defines the data and analysis as before, though when analyzing more than one group, the user also states the number of samples to be analyzed. In this illustration, there are two samples. The user also specifies that a covariance matrix will be analyzed since we consider inappropriate the comparison of standardized parameters across two samples. The number in each sample is 750 and 350, respectively.

The second stage is similar to Example 4. First, we specify the model for the first group. The matrices are defined in the same manner as in Example 4 (i.e., the shape and definition of the theta epsilon matrix is the same: symmetric and free). The pattern matrices are changed for the analysis of covariance (Table 14.9 lists the pattern matrices for the first sample); recall when analyzing a covariance matrix we fix one indicator for each dimension to 1.0 to set the units of measure of the dimension. In this case, we fix the paths for indicators x_1, y_1, u_1, and w_2. Also we fix the error term for the v_1 indicator (theta delta 4,4) to some reasonable value other than zero to indicate that the measured indicator of dimension V is not perfect. For the second group, the user alters the specifications in accord with the hypothesized model across the samples. These specifications define which parameters are constrained specifically in the second sample and which are constrained across samples. For this example, we allow the measurement part of the model to be basically unconstrained across the groups, though for each dimension with only one indicator (V and Z in this model) the values for the error terms are equal across the groups. The structural relations are initially constrained to be equal across the samples. Simply stated, the model is specified such that the

Table 14.9. Pattern matrices for the first sample in Example 5

	gamma				beta		
	U	V	W		X	Y	Z
X	1	1	1	X	0	0	0
Y	1	0	0	Y	1	0	0
Z	0	0	0	Z	1	0	0

	lambda$_x$				lambda$_y$		
	U	V	W		X	Y	Z
u_1	0	0	0	x_1	0	0	0
u_2	1	0	0	x_2	1	0	0
u_3	1	0	1	x_3	1	0	0
v_1	0	0	0	y_1	0	0	0
w_1	0	0	1	y_2	0	1	0
w_2	0	0	0	y_3	0	1	0
				z_1	0	0	0

	theta delta							theta epsilon						
u_1	1						x_1	1						
u_2	0	1					x_2	0	1					
u_3	0	0	1				x_3	0	0	1				
v_1	0	0	0	0			y_1	0	0	0	1			
w_1	0	0	0	0	1		y_2	0	0	0	1	1		
w_2	0	0	0	0	0	1	y_3	0	0	0	0	0	1	
							z_1	0	0	0	0	0	0	0

Start 1.0 $\lambda_x(1,1)$ $\lambda_x(4,2)$ $\lambda_x(6,3)$
Start 1.1 $\lambda_y(1,1)$ $\lambda_y(4,2)$ $\lambda_y(7,3)$
Start 1.1 $\theta_\delta(4,4)$
Start 0.0 $\theta_\epsilon(7,7)$

measurement model differs across the groups, that is, the reliabilities and relations between dimensions and their indicators are not the same across the samples, while the structural relations (the causal relations among the dimensions) are constrained to be equal. Figure 14.10 depicts the model being estimated; the paths designated with EQ are those set to be equivalent across the samples.

For the second group, the matrices in the second stage are defined as follows: lambda$_y$ = full,free; lambda$_x$ = full,free; theta epsilon = symmetric,fixed; theta delta = diagonal,free; beta = invariant; gamma = invariant; phi = invariant; and psi = invariant. Here "invariant" implies that the matrix is defined exactly as in the first sample/group. Table 14.10

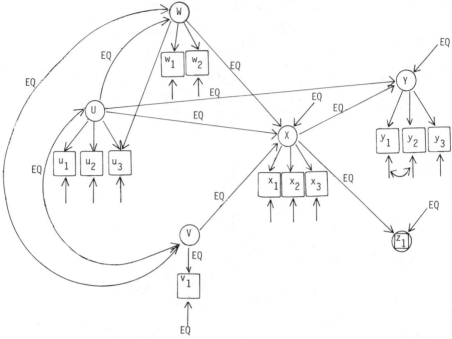

Figure 14.10. A causal model with specified equality constraints across two
samples.

shows the pattern matrices for the second sample; the gamma, phi, beta,
and psi are not shown as they are invariant and, thus, are defined as in the
first sample (see Table 14.9). The start statements below set the units of
measure for each of the dimensions. The same indicators are fixed to one in
the respective samples; if this were not the case, the invariance on the
structural level may be rejected inappropriately (i.e., because one parameter
is measured in years while another is measured in months). The equality
statement, like the invariant statement, specifies that two parameters are set
equal across the samples. In this case, we specify that when there is but a
single indicator of a given dimension, the error term for that indicator will
have the same value across the two samples. The equality constraint designates the two parameters that are set equal by specifying the coordinates in
the following order: sample first, row second, and column third; in this
example, the theta delta entry for indicator v_1 is fixed to the same value
across the samples (i.e., theta delta [1,4,4]).

Table 14.10. Pattern matrices for second sample
Example 5

	lambda$_x$				lambda$_y$		
	U	V	W		X	Y	Z
u_1	0	0	0	x_1	0	0	0
u_2	1	0	0	x_2	1	0	0
u_3	1	0	1	x_3	1	0	0
v_1	0	0	0	y_1	0	0	0
w_1	0	0	0	y_2	0	1	0
w_2	0	0	1	y_3	0	1	0
				z_1	0	0	0

	theta delta								theta epsilon						
u_1	1						x_1	1						
u_2	0	1					x_2	0	1					
u_3	0	0	1				x_3	0	0	1				
v_1	0	0	0	0			y_1	0	0	0	1			
w_1	0	0	0	0	1		y_2	0	0	0	1	1		
w_2	0	0	0	0	0	1	y_3	0	0	0	0	0	1	
							z_1	0	0	0	0	0	0	0

Start 1.0 $\lambda_x(1,1)$ $\lambda_x(4,2)$ $\lambda_x(6,3)$
Start 1.0 $\lambda_y(1,1)$ $\lambda_y(4,2)$ $\lambda_y(7,3)$
Equality $\theta_\delta(1,4,4)$
Equality $\theta_\epsilon(1,7,7)$

The results of the analysis are given in Table 14.11. The chi square for this formulation is 178.48 with 127 df, suggesting that the model does not fit the data; the probability level is .002. The goodness of fit indices are .911 and .820 for the first and second samples, respectively. These indicate a moderate to good fit of the data, and the higher mean square residual for the second sample (2.30) suggests that the model is more accurate for the first sample than for the second sample. The R^2 values for the endogenous dimensions are equal for the two samples. Since the elements of the psi matrix were constrained to be equal across the samples, the error variances for the endogenous dimensions must be equal. However, the measurement model for the two groups shows distinct differences. For example, the covariance between the Y dimension and each of its indicators differs between the two groups. The reliabilities, based on the R^2 for each indicator, are similar, although some indicators have slightly lower reliabilities in the second group than for the first group. Were the estimation procedure not

Table 14.11. Parameter estimates for estimating model Figure 14.10

Parameter	Sample 1	Sample 2	Parameter	Sample 1	Sample 2
$\lambda_{x_1 X}$	1.00*	1.00*	$\theta_\epsilon x_1$	2.21	2.19
$\lambda_{x_2 X}$	1.68	1.65	$\theta_\epsilon x_2$	4.32	4.25
$\lambda_{x_3 X}$	2.74	2.04	$\theta_\epsilon x_3$	3.68	3.78
$\lambda_{y_1 Y}$	1.00*	1.00*	$\theta_\epsilon y_1$	1.81	2.47
$\lambda_{y_2 Y}$	2.19	1.95	$\theta_\epsilon y_2$	3.45	4.21
$\lambda_{y_3 Y}$	.87	.65	$\theta_\epsilon y_3$	2.98	2.83
$\lambda_{z_1 Z}$	1.00*	1.00*	$\theta_\epsilon y_1 y_2$	.49	1.30
$\lambda_{u_1 U}$	1.00*	1.00*	$\theta_\epsilon z_1$	0.00*	0.00**
$\lambda_{u_2 U}$	2.93	2.18	$\theta_\delta u_1$	2.72	2.72
$\lambda_{u_3 U}$	1.52	1.49	$\theta_\delta u_2$	2.94	3.30
$\lambda_{v_1 V}$	1.00*	1.00*	$\theta_\delta u_3$	2.25	2.20
$\lambda_{w_1 W}$	2.40	1.78	$\theta_\delta v_1$	1.10*	1.10**
$\lambda_{w_2 W}$	1.00*	1.00*	$\theta_\delta w_1$	1.83	2.10
$\lambda_{u_3 W}$	.91	1.07	$\theta_\delta w_2$	1.82	1.83
β_{YX}**	.61	.61			
β_{ZX}**	.40	.40			
γ_{XU}**	.30	.30			
γ_{XV}**	1.25	1.25			
γ_{XW}**	.64	.64			
γ_{YU}**	.29	.29			

phi**			psi**		
	U	V	W	X	Y	Z

	U	V	W	X	Y	Z
U	3.08			3.10	3.44	4.38
V	.90	1.71				
W	.93	1.16	4.88			

Chi square 178.48; 127 *df*; probability level .002.
Goodness of fit index: Sample 1 = .911; Sample 2 = .820.
Root mean square residual: Sample 1 = 1.303; Sample 2 = 2.297.
* Fixed value.
** Equality constraint.

able to allow for such differences in the measurement model, the fit of the model to the data would not have been so good. In single indicator models, where differences in measurement models are not recognized or allowed for, parameter estimates might suggest substantial structural differences when in fact the differences are due to differentials in the reliabilities of the measures from sample to sample, or in varying patterns of differential biases among the indicators from sample to sample, or both.

A test for differences at the structural level between the two samples is

Table 14.12. Structural parameter estimates for estimating model Figure 14.10 with gamma and psi unconstrained in the second sample

Parameter	Sample 1	Sample 2
β_{YX}**	.54	.54
β_{ZX}**	.40	.40
γ_{XU}	.34	.19
γ_{XV}	1.11	1.70
γ_{XZ}	.65	.56
γ_{YU}	.31	.28
γ_{YW}	.00*	.55

	phi**			psi					
				Sample 1			Sample 2		
	U	V	W	X	Y	Z	X	Y	Z
U	3.08			3.58	3.44	4.38	1.95	3.02	5.00
V	.90	1.71							
W	.91	1.16	4.88						

Chi square 81.12; 119 *df*; probability level .98.
Goodness of fit index: Sample 1 = .941; Sample 2 = .906.
Root mean square residual: Sample 1 = .518; Sample 2 = .489.
* Fixed value.
** Equality constraint.

straightforward. For example, in Fig. 14.10, the researcher may assume differential effects of the exogenous dimensions across samples but speculate that the relation between the intervening dimension, X, and the endogenous dimensions are the same across the groups. We test this alternative model by altering the definitions and pattern matrices for gamma and psi in the second stage of the program for the second group. For this illustration, we assume that the exogenous dimensions have differential effects on X across the groups and for the second group, W has a direct effect on Y. We also assume that the error variance in the endogenous dimensions are no longer equal across groups due to these differential effects.

Previously, we had defined the gamma and psi matrices as invariant for Group 2. The user now sets the gamma matrix as full and free and the psi as diagonal and free. The rest of the matrices for Sample 2 are defined as before. The pattern matrix for gamma for the second group adds a path from W to Y, and since the matrix is no longer invariant, each parameter in gamma will be estimated and not be constrained to be equal across the 2 groups. The pattern matrices for the first sample are unchanged. The pattern matrix for gamma is:

	U	V	W
X	1	1	1
Y	1	0	1
Z	0	0	0

The estimates for the structural parameters are given in Table 14.12. The chi square is 81.12 with 119 degrees of freedom and a probability value near 1.0. The model fits the data quite well. We test whether freeing the elements in the gamma and psi matrices for the second sample had a significant effect on fit by comparing the chi square in the first model, where the gamma and psi matrices were invariant to the second model, where the gamma and psi matrices were unconstrained across the samples and gamma included the direct effect of W on Y (the models are nested). The difference in chi square is 97.36 (178.48 − 81.12) distributed with 8 df (127 df − 119 df). This suggests that freeing the parameters in the second sample represents a significant improvement in fit of model and data; the difference of 97.36 is significant (for 8 df) at the .001 level. The implications are that the structural parameters from the exogenous dimensions to the endogenous dimensions differ across the two groups.

CONCLUSION

All possible problems that may confront the user of LISREL have not been covered in the above examples nor have all possible ways of using LISREL to explore questions of substantive interest. Longitudinal models and models examining mean structure have been omitted. Longitudinal models can be developed from the general models given in the examples above; Sörbom [44], Dwyer [19] and Kenny [31] illustrate the use of multiple indicators in this context. For analyses involving structural means, see Sörbom [43]; and Schoenberg [41]. Problems in goodness of fit, specification and correction of models are developed further in Herting and Costner [23]. The importance of using multiple indicator models, however, should be clear in the context of causal models and the ability of the LISREL program to facilitate a variety of such analyses should be evident.

REFERENCES

[1] Bagozzi, Richard P. *Causal Models in Marketing.* New York: John Wiley & Sons, 1980.
[2] Bentler, P. M. "Multivariate Analysis with Latent Variables: Causal Modeling." *Annual Review of Psychology* 31 (1980): 419–456.
[3] Bentler, P. M., and Bonett, Douglas G. "Significance Tests and Goodness of Fit in the Analysis of Covariance Structures." *Psychological Bulletin* 88 (1980): 588–606.

[4] Bielby, William T., and Hauser, Robert M. "Response Error in Earnings Functions for Nonblack Males." *Sociological Methods and Research* 6 (1977): 241–280.

[5] Bielby, William T., and Hauser, Robert, M. "Structural Equation Models." *Annual Review of Sociology* 3 (1977): 137–161.

[6] Blalock, H. M., Jr. *Causal Inferences in Nonexperimental Research*. North Carolina: Chapel Hill, 1961.

[7] Blalock, H. M., Jr. "Path Coefficients versus Regression Coefficients." *American Journal of Sociology* 72 (1967): 675–676.

[8] Blalock, H. M., Jr. "The Measurement Problem: A Gap between the Languages of Theory and Research." In *Methodology in Social Research*, edited by H. M. Blalock and Ann B. Blalock. New York: McGraw-Hill, 1968.

[9] Blalock, H. M., Jr. "Multiple Indicators and the Causal Approach to Measurement Error." *American Journal of Sociology* 75 (1969): 264–272.

[10] Blalock, H. M., Jr. "Measurement and Conceptualization Problems: The Major Obstacle to Integrating Theory and Research." *American Sociological Review.* 44 (1979): 881–894.

[11] Blalock, H. M., Jr. *Conceptualization and Measurement in the Social Sciences* Beverly Hills: Sage, 1982.

[12] Bridges, George S. "Estimating the Effects of Response Errors in Self-Reports of Crime." In *Methods in Quantitative Criminology*, edited by James Alan Fox. New York: Academic Press, 1981.

[13] Browne, M. W. "Generalized Least-Squares estimators in the Analysis of Covariance Structures." *South African Statistical Journal* 8 (1974): 1–24.

[14] Burt, Ronald S. "Interpretational Confounding of Unobserved Variables in Structural Equation Models." *Sociological Methods and Research* 5 (1976): 3–52.

[15] Clogg, Clifford C. "New Developments in Latent Structure Analysis." In *Factor Analysis and Measurement in Sociological Research*, edited by David J. Jackson and Edgar F. Borgatta. London: Sage, 1981.

[16] Costner, Herbert L. "Theory, Deduction and Rules of Correspondence." *American Journal of Sociology.* 75 (1969): 245–263.

[17] Costner, Herbert L., and Schoenberg, Ronald. "Diagnosing Indicator Ills in Multiple Indicator Models." In *Structural Equation Models in the Social Sciences.* New York: Seminar Press, 1973.

[18] Duncan, O. D. *Introduction to Structural Equation Models.* New York: Academic Press, 1975.

[19] Dwyer, James H. *Statistical Models for the Social and Behavioral Sciences.* New York: Oxford, 1983.

[20] Goodman, L. A. "A General Model for the Analysis of Surveys." *American Journal of Sociology* 77 (1972): 1035–1086.

[21] Gruvaeus, G. T., and Jöreskog, Karl G. "A Computer Program for Minimizing a Function of Several Variables." *Research Bulletin* 70–14. Princeton: Educational Testing Service, 1970.

[22] Hauser, R. M., and Goldberger, A. S. "The Treatment of Unobservable Variables in Path Analysis." In *Sociological Methodology: 1971.* San Francisco: Jossey-Bass, 1971.

[23] Herting, Jerald R., and Costner, Herbert L. "Respecification in Multiple Indicator Models." In *Causal Models in the Social Sciences* Vol. 1, edited by H. M. Blalock, Jr. Chicago: Aldine, 1984.

[24] Hoelter, Jon W. "The Analysis of Covariance Structures: Goodness of Fit Indices." *Sociological Methods and Research* 11 (1983): 325–344.

[25] Jöreskog, Karl G. "A General Method for Analysis of Covariance Structures." *Biometrika* 57 (1970): 239–251.

[26] Jöreskog, Karl G. "Structural Equation Models in the Social Sciences: Specification, Estimation, and Testing." In *Applications of Statistics,* edited by P. R. Krishnaiah. Amsterdam: North-Holland, 1977.

[27] Jöreskog, Karl G. "Analysis of Covariance Structures." *Scandinavian Journal of Statistics* 8 (1981): 65–92.

[28] Jöreskog, Karl G. "Introduction to LISREL V." *DATA* 1 (1981): 1–7.

[29] Jöreskog, Karl G., and Sörbom, Dag. *LISREL V: Analysis of Linear Structural Relationships by the Method of Maximum Likelihood; User's Guide.* University of Uppsala, Uppsala, 1981.

[30] Jöreskog, Karl G., and van Thillo, M. "LISREL: A General Computer Program for Estimating a Linear Structural Equation System Involving Multiple Indicators of Unmeasured Variables." *Research Bulletin* 71–1. Princeton: Educational Testing Service, 1972.

[31] Kenny, David A. *Correlation and Causality.* New York: John Wiley & Sons, 1979.

[32] Kmenta, Jan. *Elements of Econometrics.* New York: Macmillan, 1971.

[33] Land, Kenneth C. "Path Coefficients for Unmeasured Variables." *Social Forces* 48 (1970): 506–511.

[34] Lawley, D. N., and Maxwell, A. E. *Factor Analysis as a Statistical Method.* New York: American Elsevier, 1971.

[35] Lee, S. Y., and Jennrich, R. I. "A Study of Algorithms for Covariance Structure Analysis with Specific Comparisons using Factor Analysis." *Psychometrika* 44 (1979): 99–113.

[36] Long, J. Scott. "Estimation and Hypothesis Testing in Linear Models Containing Measurement Error." *Sociological Methods and Research* 5 (1976): 157–206.

[37] Long, J. Scott. *Confirmatory Factor Analysis: A Preface to LISREL.* Beverly Hills: Sage, 1983.

[38] Muthen, B. "Contributions to Factor Analysis of Dichotomous Variables." *Psychometrika* 43 (1978): 551–60.

[39] Pindyck, Robert S., and Rubinfeld, Daniel L. *Econometric Models and Economic Forecasts.* New York: McGraw-Hill, 1981.

[40] Rao, Potluri, and Miller, Roger LeRoy. *Applied Econometrics.* Belmont, California: Wadsworth, 1971.

[41] Schoenberg, Ronald. "Multiple Indicator Models: Estimation of Unconstrained Construct Means and Their Standard Errors." *Sociological Methods and Research* 10 (1982): 421–433.

[42] Siegal, P. M., and Hodge, R. W. "A Causal Approach to the Study of Measurement Error." In *Measurement in the Social Sciences,* edited by H. M. Blalock and A. B. Blalock. Chicago: Aldine, 1968.

[43] Sörbom, Dag. "A General Method for Studying Differences in Factor Means and Factor Structures between Groups." *British Journal of Mathematical and Statistical Psychology* 27 (1974): 138–151.

[44] Sörbom, Dag. "Detection of Correlated Errors in Longitudinal Data." *British Journal of Mathematical and Statistical Psychology* 28 (1975): 138–151.

[45] Sullivan, John L. "Multiple Indicators and Complex Causal Models." In *Causal Models in the Social Sciences,* edited by H. M. Blalock, Jr. Chicago: Aldine, 1971.

[46] Van Valey, Thomas L. "On the Evaluation of Simple Models Containing Multiple Indicators of Unmeasured Variables." In *Causal Models in the Social Sciences,* edited by H. M. Blalock, Jr. Chicago: Aldine, 1971.

[47] Werts, C. E., Linn, R. L., and Jöreskog, K. G. "Estimating the Parameters of Path Models Involving Unmeasured Variables." In *Causal Models in the Social Sciences,* edited by H. M. Blalock, Jr. Chicago: Aldine, 1971.

[48] Wiley, David E. "The Identification Problem for Structural Equation Models with Unmeasured Variables." In *Structural Equation Models in the Social Sciences,* edited by A. S. Goldberger and O. D. Duncan. New York: Seminar Press, 1973.

Respecification in Multiple Indicator Models

Jerald R. Herting

Herbert L. Costner

15

The wide use of structural equation models with unmeasured variables, and the widespread adoption of the computer programs developed by Jöreskog ([5]; Jöreskog and Sörbom [7]; Jöreskog and van Thillo [8]) to estimate and assess such models have introduced a new vocabulary into the analysis of data by sociologists. A decade ago, a conversation about "specification," "identification," "goodness of fit," "residuals," and "correlated error" would have been incomprehensible to most sociologists. Methodological and pedagogical papers, especially the review by Long [11] and numerous publications reporting on the analysis of substantively interesting "multiple indicator models" have now made such terms a part of the recognition vocabulary of almost all quantitative sociologists. Furthermore, a substantial number of sociologists now have a reasonably sophisticated comprehension of the reasoning underlying structural equation models with unmeasured variables, of the available techniques for estimating the parameters of such models, and of some of the indices for assessing the fit between model and data.

A disappointingly common experience in assessing fit between model and data is to conclude that the fit is not satisfactory, suggesting some unknown specification errors. When this occurs, we have limited understanding of how to proceed to remedy the poor fit. Published papers on this topic (Byron [2]; Costner and Schoenberg [3]; Kalleberg [9]; Sörbom [13]; Saris, Pijper, and Zegwaart [12]) provide little guidance in the use of the "diagnostic" print-out of the latest version of the Jöreskog computer program (LISREL V), and the program documentation is scant and inadequate on these features. These "diagnostics" introduce still another new vocabulary to sociologists (e.g., "normalized residuals" and "modification indices"), and few sociologists are currently skilled in comprehending and using this vocabulary wisely. But if multiple indicator models are to realize their potential for aiding in the creative development of more adequate conceptualizations of basic processes underlying social phenomena, we must be able to go beyond parameter estimation and assessing the goodness of fit. We must be able to respecify a poorly fitting model so that it represents more accurately the "way things work," including the way our imperfect indicators are contaminated by measurement flaws as well as the way the underlying phenomena of interest are affected by their causal antecedents. In this chapter, we do not offer a formula or a fixed set of rules for respecifying the "correct" model. We can offer only some insights into the diagnostic tools now available and some guidelines for their use.

We have three purposes in this chapter:

1. To describe the broad types of possible specification errors in structural equation models with unmeasured variables. Such a description of types will serve as a reminder of possible ways of respecifying a model. Through the systematic exploration of simulated data in conjunction with models deliberately misspecified in known ways, we will show that the diagnostic tools now available are better suited for discovering certain types of specification error than for discovering other types.

2. To clarify the nature and interpretation of the diagnostic indices included in the print-out for LISREL V. The concepts underlying these indices are quite abstract, and their meaning is not always evident from the program documentation. Our intent is to describe the diagnostic indices in ways that will facilitate their interpretation.

3. To suggest how best to use the diagnostic print-out of LISREL V in respecifying a model. In doing this, we will draw on the results of simulated data that differ in specified ways from the estimating model. Useful as the diagnostic indices are, the effective respecification of a model remains a matter of substantive judgment, not a task that is readily achieved by "mechanically" following the diagnostics.

TYPES OF SPECIFICATION ERROR IN CAUSAL MODELS WITH UNMEASURED VARIABLES

A structural equation model is "specified" by describing the causal dependencies or other sources of covariation between each pair of variables in the model. A specification error occurs when the "true" dependencies are inaccurately described or when some sources of covariation are omitted. Structural equation models with unmeasured variables (i.e., "multiple indicator models") are composed of two components: the *structural model* describing the dependencies and other sources of covariation between the theoretical variables for which there are no empirical measures, and the *measurement model* describing the links between theoretical variables and their empirical indicators or between the empirical indicators themselves. There may be specification errors in the structural model, in the measurement model, or in both. We will describe six broad types of specification errors, each of which may appear in either the structural model or the measurement model, and a seventh type of specification error that is simultaneously in both the structural model and the measurement model.

It is common practice to limit discussion of structural equation models to dependencies that are additive and linear,[1] although nothing in the basic concept of a structural equation model is limited to such dependencies. We point out that assuming additive and/or linear dependencies, when the "true" dependencies are not as assumed, is to make a specification error. Hence, our first two types of specification error are:

1. *Additivity specification error:* assuming the effects of two or more variables are additive when the effects are actually nonadditive, that is, when the effect of a change in one variable depends on the level of another variable.

2. *Linearity specification error:* assuming linear dependencies when the actual dependencies are curvilinear.

When an *additivity specification error* occurs in the *structural model,* we are simplifying the way things work by omitting interaction effects as they actually occur in the phenomena of interest. Alternatively stated, we describe effects as if they were uniform for all cases when in fact those effects are conditional on some other variable or variables. It should be evident that conditional effects complicate generalization from one sample or population to another. But conditional or nonadditive effects are not simply complications that make generalization difficult; they may also be of real substantive or practical import, as when the effectiveness of a teaching technique depends on the initial skill level of the students, or when the effect of education on income depends on the employing industry.

When an *additivity specification error* occurs in the *measurement model,* we have what might be called "conditional indicators," i.e., empirical indicators that are strongly or weakly related to an underlying concept depending

---

[1] We follow common practice and refer to "linear" dependencies in contrast to "curvilinear" dependencies. The more etymologically correct distinction is between "rectilinear" and "curvilinear" dependencies or effects.

on other conditions. For example, voting may be a good indicator of political interest in the United States but not in Australia where eligible voters are required by law to go to the polls. Conditional indicators may lead one to assume that there are conditional effects in the underlying processes when no such conditional effects exist. They may also lead to other puzzling results in data analysis, as when relationships between variables are attenuated in some populations and not in others as a result of the differential adequacy of the indicators in those populations.

A *linearity specification error* in the *structural model* means that the effect of some variable is dependent on its own level, so that a curve rather than a straight line best describes the relationship between cause and effect. When linearity is assumed but curvilinearity occurs, the linear relationship described will fit the data less adequately than a curvilinear function, and if the curvilinearity is extreme, a strong effect may be estimated to be very weak or nonexistent. Curvilinear relationships, like nonadditive relationships, complicate the generalization of conclusions since the relationship evidently differs for different segments along the range of the independent variable. But curvilinear relationships may also be of substantive import, for example, the effect of size on structural features of complex organizations may gradually "peter out" as organizations become very large.

A *linearity specification error* in the *measurement model* means that an indicator has a curvilinear relationship to its underlying concept rather than the linear relationship assumed. Thus, for some ranges of variation in the underlying concept, the indicator may not vary at all, while for other ranges the indicator may closely reflect variation in the underlying concept. For example, when a simple arithmetic test distinguishes between poor students and good students in the third grade but shows little variation and no relationship to performance among physics graduate students, we readily infer that this indicator (i.e., scores on this test) reflects variation in ability at the lower ranges but not at the higher ranges. The relationship between ability and test score is curvilinear because of a "ceiling" on the test scores. In other instances of different results for different populations, it is not easy to distinguish such a curvilinear relationship from a conditional relationship, nor is it always easy to know whether the curvilinearity or conditionality is in the relationship between indicators and their concepts or in the underlying processes.

Currently available techniques for estimating the parameters of structural equation models with unmeasured variables do not allow for the incorporation of nonlinearities and nonadditive effects. The transformation of a variable (e.g., to logarithms) may linearize monotonic nonlinearities, and large samples may be subdivided into subgroupings to accommodate (approximately) to a very few nonadditive effects (see Stolzenberg, 1974). But these devices are limited in application and hence even if additivity specification errors and linearity specification errors are discerned or sus-

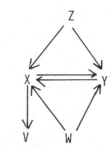

Figure 15.1. "True" model.

pected, a model respecified to correct for them cannot be estimated without recourse to such limited devices as those mentioned above. Relatively slight departures from additivity and linearity may not lead to seriously misleading substantive conclusions but may be responsible for some intractable lack of good fit in structural equation models. Substantial departures from additivity and linearity in data analyzed by additive linear models may lead to grossly misleading conclusions. Linearizing transformations and sample subdivision into subsets approximately homogenous on the values of selected variables will be necessary for such analyses until new techniques that accomodate to these complications have been developed. Because the data simulations and model estimation and assessment techniques used in the remainder of this chapter are not suitable for the discovery of or correction for additivity specification errors or linearity specification errors, they will not be further discussed in this chapter.

Restricting our discussion henceforth to additive and linear dependencies, we proceed to discuss other types of specification errors, first concentrating on several types that entail omitted sources of covariation between variables in the model. Our discussion will be facilitated by reference to Figs. 15.1 and 15.2. Assume that Fig. 15.1 represents the "true" nature of things, the way things really work. Figure 15.2 describes our model of the way things work. Since the models differ, Fig. 15.2 must include some specification errors and, indeed, it includes an illustration of four different types of specification error.

We focus first on the dependency between V and X. The estimating model claims that X is dependent on V. The "true" model makes the reverse

Figure 15.2. Estimating model.

claim, that is, that V is dependent on X. Clearly, the estimating model includes a specification error in that the dependency claimed is in the wrong direction. We therefore describe a third type of specification error:

3. *Direction of effect specification error:* assuming a dependency between two variables in the direction opposite to the true dependency.

If a *direction of effect specification error* occurs in the *structural model*, we are misinterpreting cause as effect, and vice versa, in the underlying process. For example, if engaging in a particular behavior leads to a positive attitude toward that behavior, but we assume that a positive attitude leads one to engage in the behavior, we have made a direction of effect specification error. The problem of distinguishing between cause and effect is ancient and puzzling in nonexperimental research. Of course, an effect may occur in both directions (see below), but it may be unidirectional, and our model may claim the wrong direction. We then have a specification error of this type in the structural model and may draw misleading conclusions unless we recognize the error.

If a *direction of effect specification error* occurs in the *measurement* model, we are treating an indicator as an effect of the underlying concept when it is actually a cause of variation in that concept (or much more rarely, we treat an indicator as a cause when it is actually an effect). The common assumption in "multiple indicator models" is that indicators are effects of the underlying concepts. For example, years of school completed may be treated as an indicator of social class and may be represented in a structural equation model as an effect of social class, that is, as a "reflector indicator" of class. But this conceptualization, suggesting that one achieves a certain social class level and then acquires a certain educational level as a result, is not realistic (for a single generation case). A more realistic conception would treat education as a cause of social class rather than as an effect.[2] A model that treats years of school completed as a "reflector indicator" of social class has thus incorporated a direction of effect specification error in the measurement model.

We shift attention next to the dependency of Y on W in the "true" model (Fig. 15.1) and the absence of any direct dependence of Y on W in the estimating model. This omission illustrates our fourth type of specification error:

4. *Omitted path specification error:* assuming no direct dependence of one variable on another when such a direct dependence should have been included.

---

[2] The true model is probably more complicated than these comments suggest. "True" education may be a cause of social class, but measured education is an effect of "true" education. Hence, measured education is a correlate of social class but is neither a cause nor an effect of social class.

When an *omitted path specification error* occurs in the *structural model*, we have failed to include in the estimating model all of the ways in which one variable has an effect on another and hence have misrepresented "how things work." As illustrated in Fig. 15.2, the claim in this estimating model is that W has an effect on Y, but that effect is represented as being mediated by X. In contrast, the "true" model shows an effect of W on Y mediated by X and *also* a "direct" effect of W on Y, indicating that the effect of W on Y is not entirely mediated by X but is also mediated by some variable or variables omitted from the model. One of the goals of research on causal processes is to identify the variables through which effects are mediated, that is, the "intervening variables" that "transmit" effects. That is one of the fundamental ways of understanding a basic process, and substantively detailed models will therefore frequently claim indirect effects through specified mediating variables and no direct effects. Such models assert, in effect, that we have found all of the mediating variables and hence we understand how an effect is transmitted from one variable to another. One of the important substantive points at issue is whether this claim is correct, that is, whether a "direct" effect should have been included, or, alternatively stated, whether an omitted path specification error has been made. We may recognize, for example, that the education and occupation of fathers affect the occupational attainment of sons and that this effect is mediated, at least in part, through the educational attainment of the sons. But it is important to know whether that effect is entirely or only partially mediated by the educational attainment of sons. Are there or are there not other mechanisms through which parental status affects the status of offspring?

When an *omitted path specification error* occurs in the *measurement model* it means that an indicator has been treated as a reflector of one underlying concept, but it should also have been treated as a reflector of other concepts as well. This might be referred to as a "shared indicator," but the sharing is unrecognized if this type of specification error occurs. In the language of factor analysis, one might say that a given indicator has a loading on more than one factor, which is evidently a common occurrence even though it is not commonly represented in "multiple indicator models." When shared indicators are not represented as such (i.e., when loadings on other factors have been omitted), we have misrepresented the connections between those indicators and the underlying structural model. For example, number of citations may be treated as a reflector of scholarly productivity and properly so. But if number of citations is also a reflector of centrality in an "invisible college" of scholars working in a particular specialty, and this concept is also included in the underlying structural model, the failure to include this additional "loading" (of citations on centrality in an invisible college) constitutes a specification error — and a potential source of bias and misleading conclusions unless it is identified as an error.

We now shift attention in Figs. 15.1 and 15.2 to the dependence between

X and Y. The estimating model (Fig. 15.2) claims that Y is dependent on X but not the reverse (i.e., X is not dependent on Y). In the "true" model, however, X is dependent on Y as well as Y being dependent on X. Thus, the true relationship between X and Y is one of reciprocal causation, rather than unidirectional cause as claimed in the estimating model. Alternatively stated, the estimating model claims a recursive process, whereas the true model shows a nonrecursive process to be operating. This brings us to our fifth type of specification error:

5. *Recursive assumption specification error:* assuming a unidirectional effect when the true effect is reciprocal.

A *recursive assumption specification error* in the *structural model* is probably a common type of specification error because it greatly simplifies identification and parameter estimation. But simplified or not, such assumptions may misrepresent the "way things work" and lead to distorted parameter estimates. For example, it has long been claimed that association with delinquents leads to delinquency. But it is also claimed that being delinquent leads one to associate with others who behave similarly. Both may be true — and they probably are — indicating that a nonrecursive process is operating. But identification and estimation problems commonly lead investigators to assume a unidirectional effect between these two variables and hence to a specification error in the conceptual model.

Although a *recursive assumption specification error* may occur in a *measurement model,* it is not intuitively evident that such errors are common. We typically conceive of indicators as "reflectors" of underlying concepts (i.e., as effects of those concepts), or as "producers" of underlying concepts (i.e., as causes of variation in underlying concepts, as in experimental manipulation), or as correlates of underlying concepts (i.e., as being dependent on the same sources of variation as are the underlying concepts). But we do not ordinarily think of the relationship between concept and indicator as one of reciprocal causation, and there seems to be little intuitive reason to consider such relationships in respecifying a model.

We now consider the dependence of both X and Y in Fig. 15.1 (the "true" model) on a common source of variation, Z. This common source of variation in X and Y, and hence a source of covariation between X and Y, is not included in the estimating model. This type of specification error constitutes our sixth type:

6. *Spurious association specification error:* omitting a common source of variation between two variables.

If a *spurious association specification error* occurs in the *structural model,* this means that a source of spurious covariation has been omitted, but since the covariation is there in the data, it will be "reassigned" in parameter estimation (e.g., it will augment the estimated effect of one variable on the other). This is the classical problem of mistaking dependence on a common cause for evidence of an effect and is the source of the aphorism: "Correlation does not imply causation."

If a *spurious association specification error* occurs in the *measurement model*, this means that a source of covariation between two (or more) indicators has not been represented in the model. For example, if two personality tests are treated as indicators of different concepts but both have high loadings on the "acquiescence factor," and if this factor (or if one or both of these loadings) is omitted from the model, we have a spurious association specification error in the measurement model, and our estimate of the covariation between the underlying concepts will be distorted by that specification error.

The three last-named types of specification error (omitted path, recursive assumption, spurious association) may be represented in a more generic form without specifying the type in greater detail. To illustrate, we assume that the initial estimating model (Fig. 15.2) did not demonstrate a satisfactory fit between model and data, and so it is respecified. We now refer to Fig. 15.3, showing a respecified model. It is a relatively simple model, asserting that W has an effect on X, that X has an effect on V and on Y, but that there are other unspecified sources of covariation between X and Y (represented by the curved, double-headed arrow connecting these two variables). This is a "correctly" respecified model in some narrow sense, that is, the true model (Fig. 15.1) also shows these same effects and other sources of covariation between X and Y. But Fig. 15.3 shows an inadequately respecified model, even if it is "correct," in that it provides very little information and fails to describe the other sources of covariation that were omitted in the original estimating model. Alternatively stated, those other sources of covariation have been "located" but not "specified." In accord with established practice, we will refer to a general claim of other sources of covariation between two variables (i.e., in a particular "location") as *correlated error*. In our subsequent discussion, we will find that the diagnostic indices available as print-out for LISREL V sometimes (but not always!) help "locate" correlated error (i.e., suggest other sources of covariation probably exist between a specific pair of variables), although they sometimes mislocate correlated error also. Even when they locate a correlated error correctly, they do not

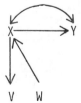

Figure 15.3. Respecified model.

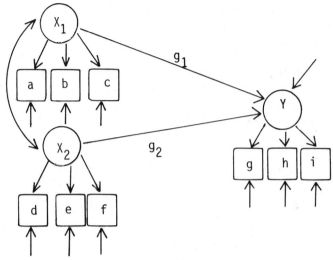

Figure 15.4. "True" model.

necessarily provide useful clues to the nature of the omitted sources of covariation. Some nonobvious clues, plus substantive insights, will always be useful in going beyond the "location" of correlated errors to an understanding of the underlying processes.

Finally, we recognize the possibility of a specification error that consists of treating as one theoretical dimension what would more properly be treated as two or more distinct dimensions. This type of specification error is illustrated in Figs. 15.4 and 15.5, showing an estimating model that has "fused" (i.e., failed to distinguish between) two structural dimensions. We define this type of specification error as follows:

7. *Conceptual fusion specification error:* assuming that a set of indicators reflect one and only one underlying dimension in the structural model when they actually reflect two or more underlying dimensions.

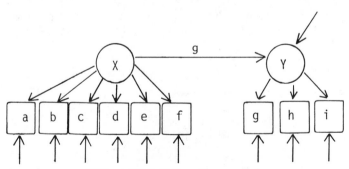

Figure 15.5. Estimating model.

This type of specification error, unlike other types listed above, does not have one variant in the structural model and another variant in the measurement model. This type intrudes into both the structural model and the measurement model simultaneously. Such a specification error might occur, for example, if several indicators of "anomie" were treated as reflecting a single dimension instead of two different dimensions that have distinct causes and effects. Referring to Figs. 15.4 and 15.5, it should be clear that if $g_1 = g_2 = g$, the distinction between X_1 and X_2 is inconsequential for this model. In general, the omission from a model of all conceivable conceptual distinctions does not necessarily imply that a conceptual fusion specification error has occurred. Such a specification error will have occurred if and only if the parameters describing the links to at least one cause or at least one effect are different for the distinct concepts, indicating that such a parameter will inevitably be inaccurately estimated for the "fused" concept.

GOODNESS OF FIT MEASURES IN LISREL V

The "goodness of fit" between model and data refers to the accuracy with which the model with its parameter estimates can reproduce the covariances between observed variables. Estimation in LISREL proceeds by finding the parameters that fit best, that is, those parameters that minimize a function of the discrepancies between observed covariances and those implied by the model. When that function is as small as possible, LISREL provides us with the parameter estimates that achieve that degree of good fit. But even the best fitting parameters may not fit the data very well, suggesting that there is some flaw in the model, possibly a specification error, but possibly some other flaw such as a departure from the multivariate normal distribution assumed for maximum likelihood parameter estimates in LISREL. The print-out for LISREL V includes several kinds of information designed to assist the user in assessing the goodness of fit and to help locate the sources of a lack of good fit.

Following Jöreskog [6, p. 4], we distinguish between two kinds of measures of fit: (1) overall measures of fit for the entire model and (2) measures that focus on the fit exhibited in specific features of the model, that is, specific relationships between observed variables or specific parameters. We refer to the latter set of measures as "focused" measures in contrast to "overall" measures.

"Overall" Measures of Goodness of Fit

THE CHI-SQUARE TEST. The chi-square value provided in the LISREL print-out (for maximum likelihood estimates only) is sensitive to all sources of departure from good fit, that is, all discrepancies between the empirical covariances and those implied by the model and also departures from the distribution assumptions (i.e., a multivariate normal distribution of the observed variables). Like all other chi-square measures of fit, it is also sensitive to sample size, so that with a large sample, very small discrepancies will yield very large values of chi square. Formally, the chi-square test is a test of the hypothesis that the observed discrepancies are due to sampling variation alone rather than being due to some misspecification in the model or some other departure from the underlying assumptions. Jöreskog and Sörbom [7, p. 39] have suggested that the problem is commonly not one of testing this hypothesis so much as deciding whether the fit achieved is reasonably adequate for one's purposes.

THE GOODNESS OF FIT INDEX. This index should range from 0 to 1.0, although it is possible for it to assume negative values, in which case the model is very seriously flawed. Larger values of the index indicate better fit. This index depends on a comparison between the observed covariances and the residuals (i.e., the discrepancies between the implied covariances and the observed covariances). An intuitive understanding of the goodness of fit index may be achieved by conceiving of the sum of squares of the observed covariances as the starting point of the estimation process and the sum of squares of the residuals as the ending point of that process, which is designed to reduce all residuals to values as close to zero as possible. The goodness of fit index measures the reduction achieved from the starting point to the ending point, expressed as a proportion of the possible reduction that could have been achieved (i.e., the magnitude of the starting point).

The goodness of fit index is designed to summarize the fit, not to test a null hypothesis as is the chi-square test. Hence, the goodness of fit index is not dependent on sample size, as is chi square, and Jöreskog and Sörbom report that the goodness of fit index is "relatively robust against departures from normality" [7, p. 41].

THE ROOT MEAN SQUARE RESIDUAL. Small residuals imply a good fit; hence, it seems reasonable to have an index of fit that is dependent in a relatively simple way on the magnitude of the residuals. The root mean square residual is such a measure. It is a modification of a simple mean of the residuals that gives extra weight to larger residuals by squaring them. As the name suggests, the root mean square residual is the square root of the mean of the squared residuals.

Unlike the goodness of fit index, the root mean square residual is not normed to vary (normally) between 0 and 1.0, and a single value of the root mean square residual, standing in isolation, has no evident interpretation (except, of course, when its value is zero, in which case the fit is perfect). This measure may be used, however, to compare the fit of two (or more) different models for the same data (including nonnested models), since the units of measure and the variances and covariances would be held constant in the comparison. The model with the smallest root mean square residual has the better fit.

THE Q-PLOT. This mode of representing graphically the overall fit is based on the "normalized residuals" discussed below. The Q-plot is described following the discussion of normalized residuals.

UNREASONABLE ESTIMATES AS INDICATORS OF POOR FIT. Negative variances, estimated correlations larger than one, squared multiple correlations below zero, or (available for maximum likelihood estimates only) excessively large standard errors are examples of unreasonable outcomes that indicate that the model has flaws and possibly specification errors. These quantities are not ordinarily used as measures of good fit, of course. But when they assume abnormal values, they suggest specification errors and, in that sense, serve as overall measures of the fit between model and data.[3]

"Focused" Measures of Goodness of Fit

THE RESIDUALS AND NORMALIZED RESIDUALS. The residuals (discrepancies between the covariances implied by the estimated model and the corresponding covariances as observed in the data) may be examined one by one to glean clues about the location of specification errors. Large residual covariances suggest the possibility of a specification error in that aspect of the model that links (or might link) those two variables, but this clue is sometimes misleading (see Costner and Schoenberg [3]).

Normalized residuals are designed to help identify residuals that are larger than would be expected as a result of sampling variation alone. If the model is perfect, the expected value of each residual is zero, with departures from zero due to sampling variation expected to be normally distributed with a standard deviation that can be estimated from the sample size and the variances and covariances of the specific variables involved. Hence, a residual divided by this estimated standard deviation yields a "normalized residual," that is, a quantity that can be interpreted like a "standard score" or "z score." Hence, a discrepancy greater than 2.0 in absolute value suggests (at approximately the .05 level) a residual that is too large to be reasonably attributed to sampling variation alone.

The quantile plot (Q-plot) of normalized residuals is a visual presentation that allows one to see in the general configuration of points the overall fit of the model, and to see, by the location of the larger residuals, how seriously these residuals challenge the assumption that they arise simply from random sampling variation. As indicated above, the normalized residuals are expected to be normally distributed around zero. Given 40 normalized residuals, for example, we should not be surprised to find that two of these exceed 2.0 in absolute value since approximately 5% of such normalized variables would be more than 2.0 standard deviations away from the mean.

---

[3] Unreasonable values may arise for other reasons also. For example, large standard errors may result from lack of identification in the model.

In an analogous way (i.e., using a table of areas under the normal curve), we would be able to specify the interval in which each of 40 residuals would be expected to fall if they were distributed as nearly in accord with normality as 40 values can possibly be. For a set of normalized residuals thus distributed, a plot of these expected intervals (quantiles) on the ordinate (vertical axis) against the actual normalized residuals on the abscissa (horizontal axis) would yield a straight line at a 45-degree angle. This would be the expected pattern given no flaws in the model and residuals greater than zero simply because of random sampling variation. In contrast, if the larger residuals on the positive side fall clearly below the idealized 45-degree line, or if the larger residuals on the negative side fall clearly above it, or both, then some of the residuals are larger than would be expected as a result of random variation, and the residuals most clearly "out of line" suggest possible locations for a specification error. On the other hand, if the larger residuals on the positive side fall above the 45-degree line and the larger residuals on the negative side fall below it, then some of the residuals are not even as large as would be expected simply because of sampling variation, and a good fit of the model to the data is suggested. If at the extreme, all residuals are zero, the points in this plot will be in a vertical line at zero, suggesting abnormally low variation of the residuals around zero (i.e., the fit is "suspiciously" close).

THE MODIFICATION INDEX. Whereas the residuals focus attention on those aspects of the model that provide the connections between observed variables, one pair at a time, the modification indices focus attention on the fixed parameters of the model, typically "nonconnections" (i.e., fixed at zero) between structural variables, between a structural variable and an observed variable, or between observed variables. The modification index for a given fixed parameter tells how much the chi-square value would decrease if we reestimated the model with that parameter "free" (i.e., to be estimated) instead of "fixed" (at a specific value, frequently zero), and all previously estimated parameters were held fixed at their estimated value. A relatively large modification index for a particular fixed parameter thus indicates that we could improve the fit by freeing that parameter. The largest of the modification indices indicates that we could achieve a greater improvement in the goodness of fit by freeing its associated parameter than by freeing any other. Unless there are clear substantive reasons for retaining that parameter as a fixed parameter, we could then free it, reestimate, and again examine the modification indexes for an indication of still another fixed parameter that might be freed. This process may be continued until a satisfactory fit is achieved, until all identified parameters have been freed, or until substantive concerns argue against freeing additional parameters.

The modification index is a complex quantity, being a function of both the first and second derivatives of the fitting function with respect to each fixed parameter. LISREL IV included a print-out option for the first derivatives, but they were difficult to interpret. The square of the first derivative with respect to a given fixed parameter, expressed as a ratio to the second derivative, can be modified in a relatively simple way to provide the minimum decrease in chi-square that would be achieved by freeing that parameter. This quantity, the modification index, can thus be interpreted without computing standard errors.

Freeing a previously fixed parameter entails the loss of one degree of freedom for chi square so that very small reductions in chi square represent no real improvement in fit. The significance of the reduction (i.e., the significance of the modification index) can be assessed by locating it in a chi-square distribution with one degree of freedom. Hence a modification index smaller than about 4.0 provides an insignificant improvement in fit.[4]

USING MEASURES OF FIT TO RESPECIFY MODELS ESTIMATED WITH SIMULATED DATA

In this section, we examine the performance of the diagnostic indices described above in assessing goodness of fit and in pointing to needed respecifications for models that have been deliberately misspecified in particular ways. The central questions to be explored here are: What types of misspecifications are accurately identified and located by what diagnostic indices? What types of misspecifications are likely to go "unnoticed" by any existing indices of fit? Under what circumstances are we in danger of being misled by respecifying in accord with the suggestions implicit in the diagnostic indices, for example, respecifying with correlated errors between indicators when the actual specification error is the omission of a loading for one of those indicators?

Before proceeding to the discussion of results of our exploration of simulated data bearing on these questions, we first describe briefly the simulation procedure we have used and the two-stage analysis procedure that we consider helpful.

The Simulation Procedure

Real data are generated by unknown processes, and the purpose of analysis is to achieve a clearer understanding of those underlying processes. Simulated data can be generated by known processes, and the purpose of

---

[4] As Sörbom [13] notes, the information provided by the first derivatives (and hence by the modification indices) assumes the model is correctly specified. To the degree that the original model was misspecified, the information from the modification indices is therefore suspect. Hence, the interpretation of the modification indices should be guided by experience, such as the simulations discussed later in this chapter.

analysis is commonly to assess the strengths and shortcomings of our methodology in discerning those known processes. Our purpose here is to assess certain aspects of our methodology, and hence we find it useful to explore data generated by known processes, (i.e., simulated data).

To generate our simulated data, we first specify a causal model and the values of its parameters. The model and its parameters are then used to create the covariance matrix for a set of (simulated) empirical indicators. The causal model used in generating this covariance matrix is called the "true" model. We then proceed to specify an "estimating model" that differs from the "true" model in one or more ways, thus building in a known specification error. For example, the "true" model may have a given indicator loaded on two structural dimensions, whereas the "estimating model" omits one of those loadings. We then use the data generated by the "true" model to estimate the parameters of the deliberately misspecified "estimating model." On the basis of these estimates, the various goodness of fit measures, both "overall" and "focused," are calculated, allowing us to assess the goodness of fit of our (misspecified) "estimating model" to the data generated by the "true" model and to examine the focused measures of fit to see what respecifications, if any, are suggested. If we find a lack of good fit and if the respecifications suggested recreate the "true model," then the diagnostic indices are performing successfully. On the other hand, if we are led to conclude that the fit is good even though we know that the model is misspecified, or if we conclude that the fit is poor but due to misspecifications other than those deliberately built in, then we must conclude that the diagnostic indices are not performing successfully for that particular kind of misspecification.

In the simulated data examined here, we have included illustrations of several of the types of misspecification described earlier in this chapter, and for some types of misspecification several illustrations have been examined so as to be able to discern the effects of changes in "context" (i.e., other features of the model) and parameter magnitudes on the performance of the diagnostic indices. We have not systemmatically explored the performance of the diagnostic indices in highly complex models with multiple misspecification errors of different types. The complexities of relatively simple models with single misspecifications in each will be our major focus of attention here.

A Two-Stage Analysis Procedure

A reasonable investigator presumably specifies first a model that is believed to be theoretically sound and then proceeds to assess its fit to the data. If the fit is adequate, this does not demonstrate that alternative specifications would fail to fit equally well, and one may wish to explore the fit of selected alternatives to see if they are acceptable or are rejected in favor of the original specification. However, if the model does not fit the observed covariance structure, the problem becomes one of discovering the misspecifications causing the lack of fit. The difficulty in this task is compounded by the high probability that more than one misspecification is involved. To minimize the complexity, it seems best to consider some form of the model that will highlight certain misspecifications (if present) while suppressing the impact of others on the goodness of fit. This may be accomplished by specifying only covariations (rather than causal relations) on the structural level, thereby focusing on the measurement model as the only possible locus of specification errors. This implies proceeding with a confirmatory factor analysis of the data. A confirmatory factor analysis model specifies which variables "load" on which structural dimensions, but while these dimensions are allowed to correlate with each other, there is no specification of the causal structure underlying those correlations. If there were *structural* misspecifications in the original model, they will have been suppressed in such a confirmatory factor analysis model. Thus, any failure of fit in the confirmatory factor analysis model must be due to measurement misspecifications. Furthermore, whereas LISREL does not permit a given indicator to "load" on, or have correlated error with, both an exogenous and an endogenous variable, such restrictions are absent in the confirmatory factor analysis model in which there is no distinction between exogenous and endogenous factors. This feature permits one to explore sources of poor fit in a confirmatory factor analysis model that are impossible to explore within the general causal model as presently developed in LISREL.

With the confirmatory factor analysis model it is possible to utilize the diagnostic indices, such as the modification indices or normalized residuals, to help locate a specification error in the measurement model. After respecification to correct for that error, the same indices may be used again and again. When the measurement model has been respecified, step by step, as suggested by the focused measures of fit and by appropriate theoretical reasoning to achieve a satisfactory fit, one can then move back to the original model with the revised measurement model and assess the fit of the model as a whole, including the specification of causal structure among the structural dimensions. If this model fits well, it is, again, not necessarily the only model that would fit equally well. If this model does not fit satisfactorily, one is again obliged to search for possible misspecifications. At this point, however, we are aided by having eliminated misspecifications in the measurement model (assuming a satisfactory fit was achieved with the confirmatory factor analysis model), and we can concentrate our attention now on respecifications in the causal structure linking the structural dimensions. Again, proceeding by making changes one at a time, we may continue this process until a satisfactory fit has been achieved.

This two-stage procedure will be alluded to as we proceed with the analysis of simulated data below, although in most instances one stage or the other may be trivial because our examples are contrived to include only one specification error in most instances.

EXAMPLE: A MODERATELY COMPLEX MISSPECIFICATION. With even moderately complex models of the LISREL type, it is not unusual to find the overall measures of fit indicating that the model fails to fit the data, while the residuals and other focused measures of fit present ambiguous clues as to the nature and location of the misspecifications responsible for the lack of good fit. To illustrate this common circumstance, we have estimated the model pictured in Fig. 15.6 using data generated by a "true" model that differed in various ways from the estimating model. The "true" model is not revealed just yet so that the reader will lack that crucial information just as one ordinarily does in research, that is, the "true" model is unknown, and the purpose of the analysis is to come to some conclusion about what the "true" model is.

We are forced to conclude that the estimating model is not the "true" model, or even very close to it. The chi-square value of 276.71 with 50 degrees of freedom is so large that we would not reasonably assume that the discrepancies between implied and observed covariances are a result of random variation alone. Furthermore, the adjusted goodness of fit index is relatively low at .703, and the root mean square residual (.561) is relatively large compared to some of the entries in the matrix of observed covariances. The Q-plot (Fig. 15.7) points to the same conclusions; the slope is less than one and the nonlinear shape of the plot suggests that the model is misspecified. Thus, all of the overall goodness of fit measures point to an incorrectly specified model — which is, of course, the proper conclusion since the model used to generate the data was deliberately different from the estimating model. The challenge is not to draw this obvious conclusion of poor

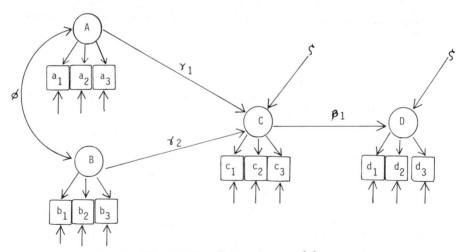

Figure 15.6. Estimating model.

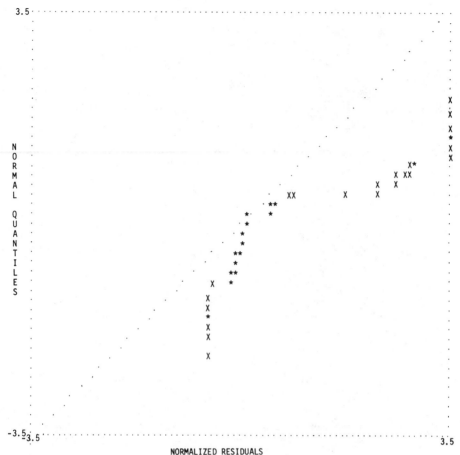

Figure 15.7. Q-plot of normalized residuals for estimating model of Figure 15.6.

fit but to respecify the model to approximate as closely as possible the "true" model.

   To help with this task, we turn to the focused measures of fit, that is, the residuals, normalized residuals, and the modification indices. But we approach these measures with some caution for a number of reasons. First, we know that some misspecifications have little or no impact on parameter estimates and hence on the goodness of fit. To take a simple example, a causal chain from left to right will yield the same parameter estimates as a causal chain from right to left through the same variables. Hence, if one of these is the "true" model and the other is the estimating model, the misspecification would never be discovered in clues entirely dependent on the compatibility between parameter estimates and empirical covariances. Second, the "full information" estimation procedure used to derive parameter estimates in LISREL is such that a misspecification in one segment of the model may have ramifications in other parts of the model (see Burt [1]). Thus, each misspecification may "mess up" the fit in other well-specified parts of the model in ways that are not intuitively obvious. Third, some types of misspecifications are known from prior simulation studies to produce patterns in the residuals matrix that do not point clearly to the misspecification responsible (Costner and Schoenberg [3]). Despite these recognized difficulties, the focused measures of fit are the best clues to misspecification we have available, so we proceed to use them as best we can.

Table 15.1 shows the residuals and normalized residuals for the example described above. A comparison of corresponding values in these two matrices will demonstrate that the normalized residuals convey different information from the "raw" residuals. The normalized residuals are "normed" to the variation expected on the basis of sampling variation and hence are more comparable from one to another than are the "raw" residuals. Focusing on the normalized residuals, we see that they point to several possible points of misspecification (i.e., a number of the normalized residuals are greater than 2.0). But the most appropriate respecifications, given the pattern of normalized residuals, are not immediately evident. For example, the covariances between variable b_3 and all indicators of dimension A and dimension D seem to be too large to be attributed to random variation alone. But what does this mean? Does it mean that the model should be respecified to include correlated error between this variable and each of these other indicators? Does it mean that this variable has a non-zero loading on each of these two additional dimensions and that the model should be respecified to so state? Is it some combination of these two types of misspecifications that should be tested for improvement in fit? Or is there some misspecification that does not entail this variable at all but which has permeated many parts of the model because of the "full information" nature of the estimation procedure?

The modification indices presented in Table 15.2 show a similar variety of potential respecifications. Freeing any one of a number of parameters would seem likely to decrease the magnitude of the chi-square statistic by a significant amount. Should we respecify to include correlated error between b_1 and a_2, or should a structural path be included between dimensions A and D, or both simultaneously?

The disembodied nature of our illustration deprives us of one potentially rich source of insight into such questions. Clearly, some potential respecifications can be ruled out on the basis of their theoretical implausibility. But rarely will theory suffice to narrow down the potential respecification to one only. More typically, the researcher is faced with a model in which a variety of respecifications seem acceptable on theoretical and intuitive grounds,

Table 15.1. Residuals matrix and normalized residuals matrix for estimating model of Figure 15.6

Residuals

	c_1	c_2	c_3	d_1	d_2	d_3	a_1	a_2	a_3	b_1	b_2	b_3
c_1	-.000											
c_2	.132	-.000										
c_3	-.083	-.049	-.000									
d_1	-.023	-.016	.395	.000								
d_2	-.172	-.199	.034	-.024	.000							
d_3	-.227	-.261	.097	-.003	.009	.000						
a_1	-.075	-.087	-.004	.900	1.027	1.461	.000					
a_2	-.101	-.119	-.062	.977	1.113	1.584	-.014	.000				
a_3	-.115	-.133	.007	1.443	1.648	2.344	.025	-.015	.000			
b_1	-.119	-.137	.055	.102	.026	.047	-.260	.296	-.411	-.000		
b_2	-.082	-.075	.714	.370	.231	.351	-.498	-.588	-.786	.279	-.000	
b_3	-.135	-.155	.058	.625	.642	.921	.439	.463	.708	-.122	-.072	-.000

Normalized residuals

	c_1	c_2	c_3	d_1	d_2	d_3	a_1	a_2	a_3	b_1	b_2	b_3
c_1	-.000											
c_2	.443	-.000										
c_3	-.116	-.056	-.000									
d_1	-.090	-.051	.544	.000								
d_2	-.558	-.530	.038	-.058	.000							
d_3	-.558	-.529	.082	-.005	.014	.000						
a_1	-.454	-.437	-.009	4.797	4.475	4.903	.000					
a_2	-.540	-.522	-.114	4.562	4.248	4.657	-.082	.000				
a_3	-.458	-.438	.010	5.058	4.720	5.167	.110	-.057	.000			
b_1	-.588	-.555	.094	.463	.096	.135	-1.717	1.718	-1.779	-.000		
b_2	-.201	-.151	.601	.854	.434	.504	-1.667	-1.728	-1.722	.738	-.000	
b_3	-.634	-.599	.094	2.744	2.296	2.519	2.799	2.590	2.950	-.616	-.180	-.000

Table 15.2. Significant modification indices for estimating model of Figure 15.6

Parameter	Modification index
$A \rightarrow D$	50.80
$A \rightarrow b_3$	45.88
$a_2 \leftrightarrow b_1$	38.93
$A \rightarrow b_2$	25.45
$a_3 \leftrightarrow b_3$	18.48
$a_3 \leftrightarrow b_1$	17.09
$b_1 \leftrightarrow b_2$	14.65
$a_2 \leftrightarrow b_2$	14.56
$c_1 \leftrightarrow c_2$	12.48
$a_1 \leftrightarrow b_3$	12.35
$a_1 \leftrightarrow b_1$	11.67
$C \leftrightarrow D$	10.70
$b_1 \leftrightarrow b_3$	6.92
$D \leftarrow C$	5.45

and several of these may be suggested by the focused measures of fit. The choice among several plausible alternatives is ordinarily not going to be obvious.

Before proceeding to respecify this initial example in a step-by-step procedure, we turn our attention to models deliberately misspecified in known and relatively simple ways. These simpler problems will allow a more detailed examination of the utility of the diagnostic clues available and will help us discern when they provide reliable clues to respecification and when they do not. We begin with misspecifications in the measurement model only and subsequently turn to misspecifications in the structural part of the model. Within each of these broad categories of misspecification, we have grouped specific misspecifications into types to facilitate the discussion.

In the illustrations to follow, Fig. 15.6 remains the estimating model. But we consider a variety of data sets, each generated by a "true" model that differs from Fig. 15.6 in a particular way. Hence, in each instance, the estimating model (Fig. 15.6) is misspecified in a particular and relatively simple way. After the effects of these misspecifications on the goodness of fit measures, and especially on the focused fit measures, have been explored, we will return to the initial example described above and apply our "educated judgment" to the problem of respecifying in the light of the diagnostics shown in Tables 15.1 and 15.2, which seem, on initial examination, to be pointing in many directions at once.

An Omitted Common Source of Variation Between Indicators; Correlated Error in the Measurement Model

As illustrated in Fig. 15.6, pairs of indicators may have a common source of variation because they are indicators of correlated dimensions (e.g., a_1 and c_1). But any two indicators may, in addition, have a common source of variation that has been omitted entirely from the estimating model. If so, we should conclude that there is correlated error between those two indicators, respecify the model to include that correlated error, and thereby achieve a better fit. Will the diagnostic measures — the normalized residuals and the modification indices — correctly inform us when our estimating model includes this kind of misspecification? No less important, will the diagnostic measures point to this kind of misspecification *only* if that kind of misspecification has occurred? We will see in this section that the answer to the first of these questions is typically "yes," but in subsequent sections we will discover that the answer to the second question is, unhappily, "no."

We consider first the case of an omitted common source of variation between two indicators of two different endogenous dimensions. To do so, we have generated data for a model like that of Fig. 15.6, except for correlated error between c_3 and d_1, but have proceeded to use Fig. 15.6 as the estimating model. The resulting parameter estimates are somewhat biased, but if we did not know the "true" model and its parameters (and normally we would not), these biases would not be evident, that is, none of the parameter estimates is beyond the range of plausibility and there are no negative error variances. The chi-square value (115.65, with 50 degrees of freedom) would lead us to conclude that the fit is poor, but the other overall measures of fit suggest a fit that is at least "fairly good." The adjusted goodness of fit index is reasonably high (.941), the root mean square residual, .032, is small relative to the observed covariances, and the Q-plot exhibits only one extreme outlier. Nonetheless, we conclude that the model is in need of respecification and therefore initiate the two-stage strategy described above, first transforming the model into a confirmatory factor analysis model (Fig. 15.8). Because there are no specification errors in the structural part of this model, the goodness of fit results are identical to those discussed above. The matrix of normalized residuals, Table 15.3, shows that the normalized residual between c_3 and d_1 is greater than 2.0 and evidently larger than any of the other normalized residuals. Furthermore, the modification indices (Table 15.4) indicate that freeing the parameter for the correlated error between c_3 and d_1 would reduce the chi-square value by a sizable and significant amount. We, therefore, conclude that the diagnostic measures perform quite satisfactorily in pinpointing the specification error in this instance. Furthermore, similar results occur if the correlated error is between indicators of two different exogenous dimensions.

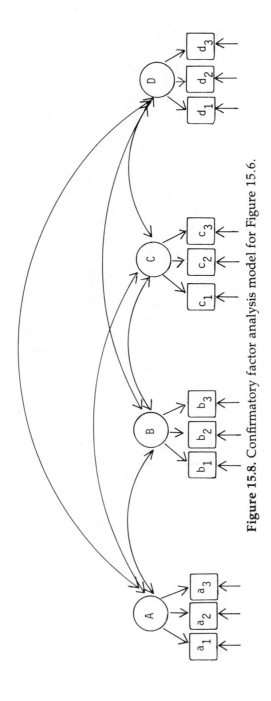

Figure 15.8. Confirmatory factor analysis model for Figure 15.6.

Table 15.3. Normalized residuals matrix for estimating model of Figure 15.8 in presence of correlated error between d_1 and c_3

	c_1	c_2	c_3	d_1	d_2	d_3	a_1	a_2	a_3	b_1	b_2	b_3
c_1	.000											
c_2	.153	.000										
c_3	−.142	−.069	.000									
d_1	−.604	−.575	2.354	.000								
d_2	−.222	−.175	−.370	−.029	.000							
d_3	−.256	−.214	−.384	−.111	.306	−.000						
a_1	.016	.079	−.149	−.111	.153	.084	.000					
a_2	.015	.069	−.136	−.098	.132	.069	.001	.000				
a_3	.012	.068	−.112	−.083	.119	.069	−.001	−.000	.000			
b_1	.014	.066	−.121	−.088	.124	.068	−.000	−.000	−.001	.000		
b_2	.016	.071	−.137	−.099	.136	.072	.000	.001	−.000	−.000	.000	
b_3	.013	.055	−.109	−.076	.102	.051	.000	.000	−.000	−.000	.000	.000

Table 15.4. Significant modification indices for estimating model of Figure 15.8 in presence of correlated error between d_1 and c_3

Parameter	Modification index
$c_3 \leftrightarrow d_1$	103.66
$c_3 \leftrightarrow d_2$	31.63
$c_2 \leftrightarrow d_1$	21.55
$c_3 \leftrightarrow d_3$	11.45
$c_1 \leftrightarrow d_1$	9.75
$D \rightarrow c_3$	8.21
$c_2 \leftrightarrow d_2$	5.89

The second stage of the two-stage procedure would, of course, be to respecify until the fit is satisfactory for the confirmatory factor analysis model (hence, for the measurement part of the original model) and then proceed with a LISREL analysis to determine fit and needed respecifications in the structural part of the model. In these overly simple simulations, the second stage is evidently redundant since the fit is perfect in the LISREL analysis when it is perfect in the confirmatory factor analysis model, there being no specification errors in the structural part of the model.

We turn now to the case of a common source of variation between an indicator of an exogenous dimension and an indicator of an endogenous dimension. To explore this case, we have generated data for a model like that of Fig. 15.6 except for correlated error between a_3 and c_1, but have used Fig. 15.6 as the estimating model. Again, some of the resulting parameter estimates are somewhat biased, but not in ways that would be evident if we did not know the true model and its parameters. The overall measures of fit again suggest a "fairly good" fit, with the exception of the chi-square value (177.5, with 50 degrees of freedom) which suggests that the model is misspecified. We conclude, as above, that the model is in need of respecification and proceed with the consideration of the focused measures of fit for the corresponding confirmatory factor analysis model. We should point out here that without shifting to the corresponding confirmatory factor analysis model, we would never be able to discover the specification error in this instance because the general LISREL model assumes no correlated errors of this type, that is, between two indicators, one of which is an indicator of an exogenous dimension and the other of which is an indicator of an endogenous dimension. The results with the corresponding confirmatory factor analysis model (in which the distinction between exogenous and endogenous dimensions is not relevant) point quite clearly to the misspecification responsible for the departure from perfect fit in a manner parallel to the case described above.

Having located the specification error with the confirmatory factor analysis model and having found that it is a correlated error between indicators of an exogenous and an endogenous dimension, respectively, we face a problem. We cannot proceed with a LISREL analysis incorporating the correlated error that we have concluded is there; LISREL does not allow for such a correlated error. We can only suggest that one of the "contaminated" indicators be dropped in the LISREL analysis. The LISREL analysis is redundant for an assessment of fit in this oversimplified simulation, of course, since there are no specification errors in the structural part of the model.

Adding more than one correlated error between indicators does not alter these results substantially. It does serve to illustrate, however, that respecifications should be made one parameter at a time. Consider a "true" model having correlated error between c_3 and d_1 and between c_3 and d_2, with Figure 15.6 again used as the estimating model. The estimated model does not fit the data (chi square = 140.4, 50 df; adjusted goodness of fit index = .883). Turning to the focused measures of fit, the normalized residuals show two main outliers, suggesting problems between the correct pairs of variables, as shown in Table 15.5. It is evident that the affected indicators stand out, although there is a tendency for discrepancies for other pairs of indicators, especially those involving indicators of dimension C, to be larger than when there was only one correlated error. Examining the modification indices (Table 15.6), we find the largest index to be that for c_3 and d_2 (one of the correct pairs), but the second largest modification index is not for c_3 and d_1, as one might suppose it would be. Only after freeing the parameter for correlated error between c_3 and d_2 and reestimating the model as thus respecified do the modification indices suggest freeing the parameter for correlated error between c_3 and d_1. Thus, whereas the normalized residuals pointed correctly to the two pairs of indicators with correlated error, the modification indices do so only if one proceeds one step at a time.

When there is a correlated error, or when there are multiple correlated errors, between indicators of a single dimension, the focused measures of fit cannot be depended on to ferret out the omission of those correlated errors from the estimating model. Table 15.7 shows the normalized residuals when correlated errors exist between d_1 and d_2 and between d_2 and d_3; a fourth indicator of dimension D, d_4, has been added to Fig. 15.6 in this simulation. The normalized residuals suggest a problem with the nonaffected indicator, d_4. The modification indices do not do much better, as shown in Table 15.8. They indicate the correlated error but also suggest

Table 15.5. Normalized residuals matrix for estimating model of Figure 15.8 in presence of correlated error between d_1 and c_3 and between d_2 and c_3

	c_1	c_2	c_3	d_1	d_2	d_3	a_1	a_2	a_3	b_1	b_2	b_3
c_1	.000											
c_2	.356	.000										
c_3	−.262	−.138	−.000									
d_1	−.831	−.770	1.895	.000								
d_2	−.900	−.842	2.600	−.088	−.001							
d_3	−.593	−.532	−.888	.172	.057	−.000						
a_1	.070	.178	−.301	.014	−.051	.080	.000					
a_2	.063	.159	−.272	.013	−.049	.065	.000	.000				
a_3	.053	.140	−.229	.010	−.033	.066	−.001	−.000	.000			
b_1	.058	.147	−.246	.013	−.040	.065	−.000	.000	−.001	.000		
b_2	.065	.162	−.276	.015	−.048	.068	−.000	.001	−.000	−.000	.000	
b_3	.051	.126	−.217	.011	−.039	.049	−.000	.000	−.000	−.000	.000	.000

Table 15.6. Significant modification indices for estimating model of Figure 15.8 in presence of correlated error between d_1 and c_3 and between d_2 and c_3

Parameter	Modification index
$c_3 \rightleftharpoons d_2$	46.42
$c_3 \rightleftharpoons d_3$	36.26
$D \rightarrow c_3$	22.72
$c_3 \rightleftharpoons d_1$	12.09
$c_1 \rightleftharpoons c_2$	8.44
$D \rightarrow c_2$	6.74
$c_2 \rightleftharpoons d_3$	5.66
$c_1 \rightleftharpoons d_2$	5.61
$D \rightarrow c_1$	4.14

freeing parameters involving the nonaffected indicator, d_4, including suggesting an omitted path from dimension C to d_4.

In summary, the focused goodness of fit measures perform quite adequately in locating the misspecification when there is correlated error between indicators of two different dimensions. When there are correlated errors between indicators of the same dimension, the focused measures of fit provide misleading information. In practice, an investigator may not know when the diagnostic indices are misleading and when they are not but will discover that (sometimes!) when the model is respecified in accord with the suggestions of the diagnostic indices and reestimated, the fit is not substantially improved. Hence, we caution that the suggestions provided by the focused measures of fit may not be substantiated in further analysis, and the investigator will need to keep in mind the nonintuitive specification errors that may give rise to misleading first-stage suggestions. For example, when the focused measures of fit suggest a problem with a particular indicator and a respecification in accord with that suggestion does not result in substantially improved fit, one should keep in mind the possibility that the actual specification problem is correlated error between other indicators of the same dimension as the suggested "problematic" indicator. Clearly, the use of the normalized residuals and modification indices is not always going to be so straightforward as one would like.

Table 15.7. Normalized residuals matrix for estimating model of Figure 15.8 (including d_4) in the presence of correlated error between d_1 and d_2, d_2 and d_3

	c_1	c_2	c_3	d_1	d_2	d_3	d_4	a_1	a_2	a_3	b_1	b_2	b_3
c_1	.000												
c_2	.000	−.000											
c_3	−.001	−.000	.000										
d_1	.380	.392	.377	−.000									
d_2	−.252	−.262	−.251	.055	−.000								
d_3	.253	.266	.250	−1.845	.216	−.000							
d_4	2.444	2.536	2.428	.622	−.186	.457	−.000						
a_1	−.000	−.000	.001	.177	−.120	.117	1.120	−.000					
a_2	−.000	−.001	.005	.155	−.107	.100	.986	−.000	−.000				
a_3	−.000	.001	−.005	.130	−.086	.089	.818	.000	.000	.000			
b_1	−.000	−.000	.001	.251	−.170	.167	1.600	−.000	−.000	−.000	−.000		
b_2	.000	−.001	.001	.292	−.197	.195	1.862	−.000	−.000	−.000	.000	−.000	
b_3	−.000	−.001	.005	.250	−.170	.163	1.589	−.000	−.000	−.000	−.000	.000	.000

Table 15.8. Significant modification indices for estimating model of Figure 15.8 (including d_4) in the presence of correlated error between d_1 and d_2, d_2 and d_3

Parameter	Modification index
$d_1 \rightleftharpoons d_3$	82.10
$d_2 \rightleftharpoons d_3$	26.17
$C \rightarrow d_4$	25.63
$d_2 \rightleftharpoons d_4$	22.37
$C \rightarrow d_2$	13.03
$d_1 \rightleftharpoons d_4$	11.11

Omitted Paths in the Measurement Model: Indicators "Shared" by Two or More Dimensions

In exploratory factor analyses, it is common to conclude that a given variable has a non-zero loading on two or more factors. In contrast, the usual beginning assumption in LISREL models is that each indicator has a non-zero loading on one and only one dimension. This assumption is always problematic (but perhaps less problematic than the typical finding of multiple loadings in exploratory factor analyses would suggest), and our only clues to multiple loadings or "shared" indicators are the lack of good fit and the clues in the focused measures of fit. How well will the focused measures of fit perform in pointing to such a misspecification when it has been deliberately built into the data?

Before proceeding to consider the focused measures of fit for models deliberately misspecified by omitting loadings, we should point out that such a misspecification might be expected intuitively to result in focused measures of fit that suggest a particular pattern of correlated errors between indicators. In Figure 15.6, for example, if indicator d_1 has a non-zero loading on dimension C as well as on dimension D, we should expect to find that the covariances between d_1 and each of the indicators of C are higher than would be suggested by the model of Fig. 15.6. This is expected because there is an additional source of covariation between these pairs of indicators that has been omitted from the estimating model. If the covariances implied by the estimating model were based on the correct parameters, the "excess" component of these covariances should be unambiguously evident. But given the misspecification, the estimated parameters will not be the correct parameters, since the estimates will represent an attempt to fit the model as given (and as misspecified) as closely as possible to the empirical covariances, which include the "excess" component. Hence, the estimation procedure will, in effect, "absorb" the "excess" component of these covariances to the extent possible by making inaccurate parameter estimates. To the extent that the estimation procedure is successful in thus "absorbing" the "excess" component of the covariances, we will be deprived of clues that would lead us to conclude that the model is misspecified.

But if the omitted path representing a loading of d_1 on dimension C is likely to appear in the form of suggested correlated error between d_1 and each of the indicators of C, why would we prefer to recognize it as an omitted path rather than as a set of correlated errors? This preference is, first, a preference for a clear understanding of how things actually work. But it is also a preference for conserving degrees of freedom, and this we do by adding a single parameter to be estimated (i.e., the one omitted "loading") as compared to adding three parameters to be estimated (i.e., the correlated errors between d_1 and each of the indicators of C).

Turning now to simulations representing omitted loadings, we consider first the case of such an omission entirely in the part of the model associated with exogenous dimensions. Otherwise stated, an indicator of an exogenous dimension also has a non-zero loading on another exogenous dimension, but this additional loading has not been included in the estimating model. Referring back to Fig. 15.6, our specific illustration is the omission of a loading for b_1 on dimension A in the estimating model (Fig. 15.6) even though that loading was included in the "true" model used to generate the data. The result of this misspecification is pervasive in the parameter estimates, that is, the estimates for loadings (and especially the loadings of the indicators of dimension B, which includes the shared indicator) are distorted, and the estimates for the structural parameters between exogenous dimensions as well as those between exogenous and endogenous dimensions are biased downward. Despite these distortions in parameter estimates, resulting from the fact that the estimation procedure is designed to maximize the fit between the model (here misspecified) and the empirical covariances, the fit is poor (chi square = 323.99, 50 df; adjusted goodness of fit index = .79; Q-plot shows clear departure from linearity). The focused measures of fit for the corresponding confirmatory factor analysis model are displayed in Tables 15.9 and 15.10. There is one dominant entry in the matrix of normalized residuals, and it suggests a correlated error between b_2 and b_3, neither of which is directly involved in the misspecification. Nothing in the matrix of normalized residuals points toward correlated error between b_1 and the three indicators of dimension A, even though that is what we would intuitively expect. In fact, for all pairs of indicators involving the shared indicator, b_1, the normalized residuals suggest that there is nothing

Table 15.9. Normalized residuals matrix for estimating model of Figure 15.8 in presence of omitted path A to b_1

	c_1	c_2	c_3	d_1	d_2	d_3	a_1	a_2	a_3	b_1	b_2	b_3
c_1	.000											
c_2	.000	.000										
c_3	.000	−.000	.000									
d_1	−.001	−.002	.002	.000								
d_2	−.000	.001	.001	−.000	.000							
d_3	−.003	.001	−.003	.001	−.000	.000						
a_1	.000	.000	.000	−.001	−.000	.001	.000					
a_2	.000	.001	−.002	−.002	.001	.000	.000	.000				
a_3	.000	.001	−.004	−.001	.002	.002	.000	.000	.000			
b_1	−.008	−.008	−.008	−.006	−.005	−.004	.013	.014	.012	−.000		
b_2	.808	.825	.798	.453	.521	.354	−1.414	−1.522	−1.243	−.021	−.000	
b_3	.891	.910	.885	.505	.582	.398	−1.523	−1.632	−1.345	−.010	6.431	−.000

353

Table 15.10. Significant modification indices for estimating model of Figure 15.8 in presence of omitted path A to b_1

Parameter	Modification index
$b_2 \leftrightarrow b_3$	214.99
$a_2 \leftrightarrow b_3$	14.46
$b_1 \leftrightarrow b_2$	14.16
$a_2 \leftrightarrow b_2$	9.66
$a_2 \leftrightarrow b_1$	7.23
$a_1 \leftrightarrow b_3$	5.06

wrong at all, and there is not even the slightest hint that there might be an omitted loading of b_1 on dimension A. On the other hand, there is a clear indication that there is correlated error between b_2 and b_3. The modification indices suggest the same conclusion. Freeing the parameter for correlated error between b_2 and b_3 should, the modification index suggests, result in a sizable and significant reduction in the chi-square value. Some other modification indices, although much smaller than that for the correlated error between b_2 and b_3, suggest other parameters that might be freed to achieve a moderately small decrease in the chi-square value. But surprisingly, the modification index for the loading of b_1 on dimension A — the actual misspecification — suggests that freeing this parameter would result in a trivial reduction in chi square (1.9). Thus, an investigator would be led to the conclusion that the model should be respecified to include a correlated error between b_2 and b_3, and that there is probably no other respecification required.

If we respecify the model to allow correlated error between b_2 and b_3, as suggested, we achieve an improved fit (chi square $= 17.03$). But this respecification also yields an impossible result: a negative error variance for indicator b_1. Ironically, then, following a false lead has led us to the correct conclusion, that is, respecifying the model as suggested by the focused measures of fit yields a negative error variance for the shared indicator, suggesting that it is problematic. More generally, we suggest that this sequence (i.e., respecification in accord with the suggestions of the normalized residuals and modification indices followed by finding a negative error variance for a given indicator) suggests that the indicator thus pinpointed has an omitted loading on some other dimension. This other dimension seems always to be the dimension that had the highest residuals with the co-indicators of the shared indicator (in this instance, dimension A).

We next focus on an omitted loading entirely in the part of the model associated with endogenous dimensions. Our specific illustration, referring again to Fig. 15.6, is an omitted loading of d_1 on dimension C. The general pattern in the measures of fit parallels the pattern for the case considered above. The impact on the parameter estimates is similar. The focused measures of fit exhibit the same pattern, with the largest normalized residual occurring between the two co-indicators of the shared indicator. The modification indices, as before, suggest correlated error between the same two indicators, d_2 and d_3, and this index is large enough to suggest a respecification of that kind. This respecification, however, leads to a negative error variance for d_1, pointing once again to the problematic indicator. For each of the co-indicators of this problematic indicator (i.e., d_2 and d_3), the normalized residuals were highest when these were paired with indicators of dimension C, thus suggesting that d_1 has a non-zero loading on that dimension. This is, of course, the "correct" respecification.

Two other forms of this type of misspecification were considered. In one of these, one indicator of an endogenous dimension was shared with an exogenous dimension, and in the other, an indicator of an exogenous dimension was also an indicator of an endogenous dimension. In both cases, the results were similar to those described above, suggesting that following the above strategy should be expected to pinpoint the omitted loading. The problem in both of these illustrations is that there is no easy way of respecifying the model to include such shared indicators between endogenous and exogenous dimensions in a LISREL model. But the problematic indicator may be omitted in such a model.

To check the consistency of these findings, the strength of the omitted path was varied. It was found that the diagnostic indices shift depending on the strength of the omitted loading. When the omitted loading is strong, compared to the loading for that indicator that is included in the estimating model, the patterns described above emerge. When the omitted path to the shared indicator is relatively weak, a different pattern emerges, and the diagnostic indices point in a straightforward way to the omitted path as a specification error. In the first example, we decreased the relationship of dimension A to indicator b_1 (Fig. 15.6) and then reestimated the model omitting this specification. With this weaker coefficient for the omitted path, the other parameter estimates were less distorted, and the goodness of fit measures point to a better fit between model and data. Even so, the estimated model can be deemed misspecified on the basis of the chi-square value (75.45) and on the basis of the magnitude of the normalized residuals. The two focused measures (normalized residuals and modification indices) in this instance point directly to b_1 as a troublesome indicator. The residuals are high with the indicators of A, and the modification index specifies that the path from dimension A to indicator b_1 should be freed. The focused measures thus identify the correct respecification. This holds whether the shared indicator is between exogenous dimensions, between endogenous dimensions, or between an exogenous and an endogenous dimension.

In summary, for omitted paths in the measurement model, the diagnostics perform adequately, although somewhat indirectly under some circumstances. When the omitted path is relatively strong, respecification as suggested by the diagnostic indices and then reestimation of the new model may yield a negative error variance. In that event, one should suspect that the indicator with the negative error variance has an additional loading, and that that loading is a relatively strong connection to the dimension whose indicators have the highest normalized residuals with the co-indicators of the problematic (i.e., shared) indicator. When the modification indices and pattern of normalized residuals suggest freeing a specific loading originally fixed at zero, this may be the proper respecification, and the omitted path should be expected to be relatively weak. But recall that the same pattern in the focused measures of fit may emerge when there are correlated errors between indicators of a single dimension (see the discussion above).

Direction of Effect Specification Errors in the Measurement Model: "Producer Indicators"

One generally assumes that indicators "reflect" variation in "their" underlying dimension as opposed to creating variation in the dimension. However, the causal direction may at times be reversed, in which case it is more appropriate to refer to a "producer indicator" rather than a "reflector indicator." We now consider simulations incorporating specification errors of this type to check the performance of the diagnostic indices in discerning the correct respecification.

Figure 15.9 shows a "true" model in which a_3, one of the indicators of exogenous dimension A, is a producer indicator even though the estimating model (Fig. 15.6) treats it as a reflector indicator. The resulting parameter estimates will typically be distorted since there is no implied correlation between a_3 and, for example, b_1, in the true model while such a correlation is implied by the estimating model. The overall measures of fit indicate that some respecification is needed. The chi square is 93.69, and the adjusted goodness of fit index is .813.

The normalized residuals (Table 15.11) do not follow a readily interpretable pattern. The producer indicator, a_3, has larger than expected covariances with all other indicators in the model except for its own co-indicators, a_1 and a_2. (These residuals will depend, in part, on the arbitrarily set covariances between a_3 and other variables, set at zero in the "true" model here.) Although the correct respecification could not be achieved by freeing any parameter in the estimating model, the modification indices show large values for the path between dimension B and a_3 and between indicators b_2

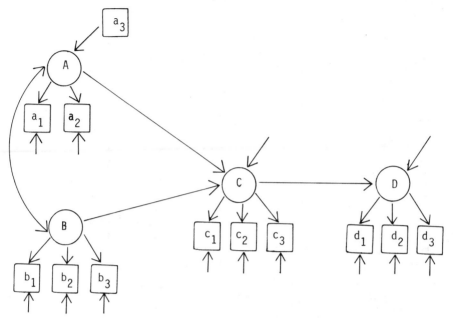

Figure 15.9. One "producer indicator" for dimension A.

and a_3. A reasonable inference from this pattern might be that a_3 has an omitted loading on dimension B. Reestimating this respecified model seems appropriate, and it fits the data exactly. The loading of a_3 on B is negative, and the other estimates approach their "true" value. But this is an incorrect respecification, despite the good fit.

Very similar results occur when a producer indicator of an endogenous dimension is misspecified as a reflector indicator.

The general conclusion from these simulations is that the diagnostic indices give misleading clues to respecification when a producer indicator is misspecified as a reflector. The diagnostic clues point to the inclusion of an added loading for the misspecified indicator. Ironically, when the actual misspecification is an omitted loading, the diagnostic clues give this kind of clue only when the omitted loading is relatively weak. Our results suggest that when (*a*) a single indicator has larger than expected covariances with all other indicators in the model except for its own co-indicators, and (*b*) the modification indices suggest an additional loading for that same indicator, and, on reestimation, that additional loading improves the fit and has a negative parameter, the actual misspecification may be the misrepresentation of a producer indicator as a reflector indicator. But it seems evident that the diagnostic indices do not point very clearly or very directly to this type of misspecification.

Table 15.11. Normalized residuals matrix for estimating model of Figure 15.8 in presence of producer indicator a_3 to A

	c_1	c_2	c_3	d_1	d_2	d_3	a_1	a_2	a_3	b_1	b_2	b_3
c_1	.000											
c_2	.000	-.000										
c_3	-.001	.000	-.000									
d_1	-.002	-.003	-.001	-.000								
d_2	.001	.004	.005	-.000	-.000							
d_3	-.002	-.004	.004	.002	-.002	-.000						
a_1	.519	.545	.515	.362	.346	.292	.000					
a_2	.564	.587	.559	.382	.375	.310	-.318	-.000				
a_3	-2.646	-2.786	-2.612	-1.793	-1.717	-1.436	.377	.431	.000			
b_1	.000	.001	-.002	.000	.004	.003	.932	.951	-4.535	.000		
b_2	.003	-.002	-.000	-.004	.006	-.006	1.053	1.077	-5.142	.000	.000	
b_3	.003	-.001	.001	-.006	-.002	-.001	.804	.824	-3.903	-.000	.000	.000

An Omitted Common Source of Variation between Structural Dimensions: Correlated Error in the Structural Model

The model represented in Fig. 15.10 includes a dimension, F, which is a common source of variation in dimensions D and E. We may also represent this causal structure (although somewhat less completely) without including dimension F, that is, without naming it or measuring it or even suggesting that there is only one F that is a common source of variation in D and E, as shown in Fig. 15.11. Here the curved, double-headed arrow states simply that there is some source of covariation between D and E that is not otherwise represented or accounted for in the model. We refer to such covariation which is not otherwise accounted for by the model as *correlated error.*

It is common in structural equation models to begin with the assumption that the causal structure specified accounts for all of the covariation between each pair of variables included in the model. Otherwise stated, one commonly begins by assuming no correlated error. Thus, even if the model described in Fig. 15.11 were the "true" model, one might (in ignorance of that "true" model) begin by using Fig. 15.11 *without* the curved, double-headed arrow as the estimating model. For that reason, that is, to distinguish the "true" model from the estimating model, we have made the curved arrow in Fig. 15.11 a broken line rather than a solid line. Thus, the entire model as shown (including the broken line) is the "true" model, whereas the model without the broken line is the estimating model.

We have assigned specific values to each of the parameters in the "true" model of Fig. 15.11 and thereby generated a covariance matrix. With that matrix, we have proceeded to estimate the parameters of the estimating model (i.e., omitting that part represented by a broken line). The true parameter values and the estimates for the misspecified estimating model are somewhat distorted, and the chi-square value (225.13, $df = 85$) alerts us to a lack of good fit and a likely misspecification, although the adjusted goodness of fit index (.911) suggests a reasonably good fit. Table 15.12 shows the normalized residuals for this example. Large positive deviations are found for the indicators of dimension D when paired with indicators of dimension E, whereas all other normalized residuals are insignificantly small. The largest modification index (Table 15.13) is for the parameter representing correlated error between dimensions D and E (187.47), and the other significant modification indices suggest an effect of D on E (97.05) or E on D (35.94). Thus, both focused measures of fit suggest a problem between D and E dimensions, and the modification index appropriately suggests that the correlated error between D and E should be estimated. Estimating the model respecified in accord with this suggestion (i.e., including the correlated error between D and E) provides a perfect fit, of course.

The focused measures of fit seem typically to perform well when the estimating model is misspecified by omitting a correlated error between

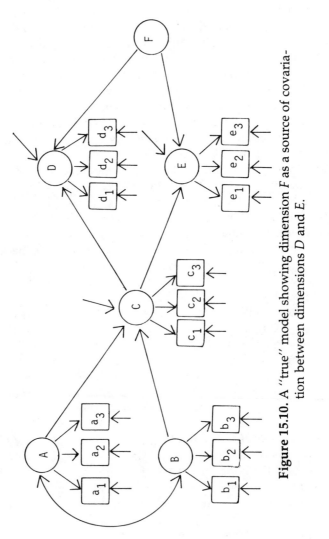

Figure 15.10. A "true" model showing dimension *F* as a source of covariation between dimensions *D* and *E*.

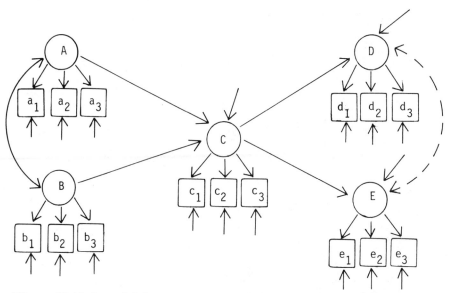

Figure 15.11. A model showing a source of covariation between dimensions
D and E that has been omitted from the estimating model.

structural dimensions. We consider a second simulation, as represented in
Fig. 15.12. Here the "true" model includes correlated error between dimensions C and D, while the estimating model omits this feature. The results
indicate that the parameter estimates are not strikingly distorted; true parameters and estimated parameters (in parentheses) shown in Fig. 15.12.
But the chi-square value (169.34, $df = 85$) is significant, implying a lack of
good fit between model and data. In contrast, the other overall measures of
fit are reassuring about the adequacy of the estimating model. The adjusted
goodness of fit index is .945, the root mean square residual is .337, which is
relatively small in comparison to the observed covariances, and the Q-plot
shows no major deviations from linearity. The focused measures of fit,
replicating the implication of the chi-square value, suggest a less than adequate fit of model and data. As shown in Table 15.14, the normalized
residuals are relatively large for indicators of the exogenous dimensions
when these are paired with indicators of dimension D, although these are
rather far removed from the actual specification error. The modification
indices (Table 15.15) suggest that a large reduction in the chi-square value
might be achieved by freeing the parameter for correlated error between
dimensions C and D, or, secondarily, by freeing the structural paths from A
to D, or from B to D, or from D to C. The researcher is faced with a choice of
(1) specifying direct effects of one or both exogenous variables on D, as
suggested by the normalized residuals but not by the largest of the modifica-

Table 15.12. Normalized residuals matrix for estimating model of Figure 15.11

	c_1	c_2	c_3	d_1	d_2	d_3	e_1	e_2	e_3	a_1	a_2	a_3	b_1	b_2	b_3
c_1	−.000														
c_2	−.028	−.000													
c_3	.046	.045	.000												
d_1	−.372	−.385	−.344	−.000											
d_2	−.380	−.391	−.349	−.000	−.000										
d_3	−.386	−.396	−.355	−.000	−.000	−.000									
e_1	−.143	−.148	−.081	3.796	3.849	3.921	.000								
e_2	−.146	−.152	−.084	3.948	4.006	4.079	.000	−.000							
e_3	−.145	−.150	−.082	3.908	3.965	4.039	−.000	−.000	−.000						
a_1	−.016	−.018	.032	−.227	−.231	−.236	−.091	−.095	−.093	.000					
a_2	−.017	−.019	.034	−.246	−.250	−.253	−.098	−.102	−.101	.000	.000				
a_3	−.018	−.020	.035	−.255	−.260	−.265	−.101	−.105	−.103	−.000	−.000	.000			
b_1	−.021	−.023	.040	−.298	−.302	−.308	−.116	−.121	−.119	.000	−.000	−.000	.000		
b_2	−.023	−.025	.044	−.337	−.342	−.348	−.130	−.135	−.132	.000	.000	−.000	−.000	−.000	
b_3	−.021	−.023	.041	−.304	−.309	−.315	−.119	−.122	−.123	.000	.000	−.000	.000	.000	.000

Table 15.13. Significant modification indices for estimating model of Figure 15.11

Parameter	Modification index
$D \leftrightarrow E$	187.47
$D \rightarrow E$	97.05
$D \leftarrow E$	35.94

tion indices or (2) including correlated error between C and D, as suggested by the largest of the modification indices but not by the normalized residuals.

Evidently, if the researcher chose the second of these alternatives and estimated the model respecified in this way, the resulting fit would be perfect. But what if the researcher made the wrong choice? Freeing either of the structural paths in alternative (1) would yield an improved fit, and freeing both of these parameters would provide an excellent fit, even though that is not the correct respecification. But the largest of the modification indices, not the normalized residuals, pointed to the correct respecification. Thus, when the largest of the modification indices suggests freeing the parameter for correlated error between dimensions, that is a reasonable

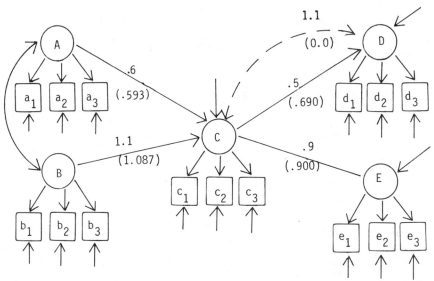

Figure 15.12 An omitted source of covariation between dimensions C and D. (Estimated parameters in parentheses.)

Table 15.14. Normalized residuals matrix for estimating model of Figure 15.12

	c_1	c_2	c_3	d_1	d_2	d_3	e_1	e_2	e_3	a_1	a_2	a_3	b_1	b_2	b_3
c_1	.000														
c_2	.069	.000													
c_3	−.033	−.030	.000												
d_1	.217	.223	.141	.000											
d_2	.217	.225	.142	.000	−.000										
d_3	.220	.229	.144	.000	.000	.000									
e_1	.059	.063	−.032	.201	.201	.203	.000								
e_2	.063	.066	−.033	.207	.209	.212	.000	.000							
e_3	.062	.065	−.032	.205	.206	.210	−.000	−.000	.000						
a_1	.106	.111	.047	−2.000	−2.026	−2.058	.097	.101	.100	.000					
a_2	.114	.119	.051	−2.154	−2.182	−2.214	.104	.108	.107	.000	.000				
a_3	.110	.123	.052	−2.237	−2.265	−2.300	.108	.118	.112	−.000	−.000	−.000			
b_1	.135	.140	.060	−2.591	−2.621	−2.659	.124	.129	.128	.000	.000	.000	.000		
b_2	.150	.156	.066	−2.915	−2.947	−2.986	.139	.141	.143	.000	.000	.000	.000	.000	
b_3	.137	.143	.061	−2.644	−2.675	−2.713	.126	.132	.129	.000	.000	−.000	.000	.000	.000

Table 15.15. Significant
modification
indices for
estimating model
of Figure 15.12

Parameter	Modification index
$C \leftrightarrow D$	100.10
$C \leftarrow D$	31.62
$B \rightarrow D$	21.95
$A \rightarrow D$	13.70

respecification to try. As we will show in the next section, however, the modification indices may indicate freeing the parameter for a correlated error between dimensions when that is not the correct respecification. In that event, other information is available to correct this misleading clue.

Omitted Paths in the Structural Model

An omitted path in the structural model is a misrepresentation of "how things work." The model may propose, for example, that the effect of one dimension on another is mediated entirely by "intervening" dimensions represented in the model when, in fact, a part of that effect is "direct" (i.e., mediated by dimensions not included at all). In such instances, will the focused measures of fit accurately lead the researcher to include a direct path that was initially fixed at zero?

Figure 15.13 depicts the "true" model, including a direct path from A to D. Again, we distinguish between the "true" and the estimating models by use of a broken line. Thus, the estimating model in this instance differs from the "true" model in omitting the path represented by the broken line, that is, the path from A to D. The true parameters and the estimated parameters are also shown in Fig. 15.13, with the estimates in parentheses. The inflated estimates of the paths from A to C and from C to D illustrate the attempt of the estimating procedure to adapt to misspecification by "absorbing" into the estimates of free parameters any covariation that would, in a correctly specified model, be assigned to parameters fixed at zero in the estimating model.

The fit of model and data is not very good. The chi square is 138.74 ($df = 50$), and the adjusted goodness of fit index is .850. The Q-plot shows a cluster of deviations and a major departure from linearity. Hence, we reject the model as misspecified. The normalized residuals (Table 15.16) show positive deviations between indicators of dimension D when these are paired with indicators of A. Smaller negative deviations exist between indi-

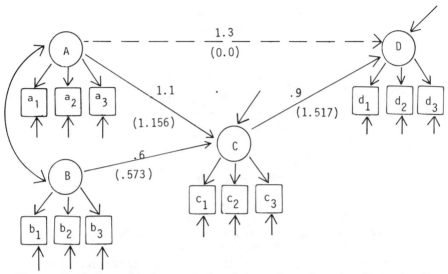

Figure 15.13. An omitted structural path from dimension A to dimension D. (Estimated parameters in parentheses.)

cators of B paired with indicators of D. These latter deviations are all less than 1.0 and would not be considered significant deviations by the usual criteria. The significantly large modification indices (Table 15.17) are for the parameters estimating correlated error between dimensions C and D (45.89), for the direct path from A to D (32.76), for the path from B to D (7.85), and for the direct path from D to C (5.84). Given the ability of the modification index to diagnose correlated error accurately (as discussed above) and the outcome here with the largest of the modification indices pointing toward a correlated error, it might seem reasonable and appropriate to try this respecification first. Respecifying in accord with that suggestion, we are obliged to recognize our error because of the resulting estimate of a negative error variance for dimension C in the reestimated model. The second highest modification index (and the respecification suggested by the pattern of the largest normalized residuals) points us in the correct direction (i.e., a respecification to include the direct path from A to D).

Figures 15.14 and 15.15 illustrate two models simulated to further explore the above result. In both cases, the estimated models (with the broken-line paths in the figure omitted) showed an inadequate fit of the estimated model to the data. The patterns of normalized residuals were similar to those found when correlated error between dimensions had been omitted in the estimating model. The residuals between the indicators of the two dimensions with the omitted path exhibit larger residuals than are found for indicators of other pairs of dimensions. In both of these cases, the largest modification index suggests a correlated error between two endogenous

Table 15.16. Normalized residuals matrix for estimating model of Figure 15.13

	c_1	c_2	c_3	d_1	d_2	d_3	a_1	a_2	a_3	b_1	b_2	b_3
c_1	.000											
c_2	−.027	−.000										
c_3	.088	.087	.000									
d_1	−.193	−.199	−.092	.000								
d_2	−.195	−.200	−.092	.000	.000							
d_3	−.195	−.201	−.092	.000	.000	.000						
a_1	−.340	−.351	−.272	1.999	2.007	2.016	.000					
a_2	−.359	−.373	−.288	2.127	2.133	2.142	−.000	.000				
a_3	−.373	−.384	−.296	2.192	2.200	2.208	.000	−.000	.000			
b_1	.068	.068	.154	−.841	−.845	−.849	.000	−.000	.000	−.000		
b_2	.076	.077	.173	−.948	−.951	−.956	.000	−.000	.000	−.000	.000	
b_3	.069	.070	.157	−.859	−.863	−.867	.000	−.000	.000	−.000	.000	.000

Table 15.17. Significant
modification
indices for
estimating model
of Figure 15.13

Parameter	Modification index
$C \leftrightarrow D$	45.89
$A \rightarrow D$	32.76
$B \rightarrow D$	7.85
$C \leftarrow D$	5.84

dimensions. In Fig. 15.14, the largest modification index suggests correlated error between D and E, and for Fig. 15.15 the largest modification index suggests correlated error between C and D. Freeing the parameter for the correlated error in each instance produces a negative error variance for one of the endogenous dimensions, indicating that this specification is incorrect. Freeing the direct path between the two dimensions with the largest normalized residual (and as suggested by the second highest modification index) provides the correct respecification of the model.

When the estimating model omits a path at the measurement level, we showed above that the accuracy of the focused measures in pointing to the

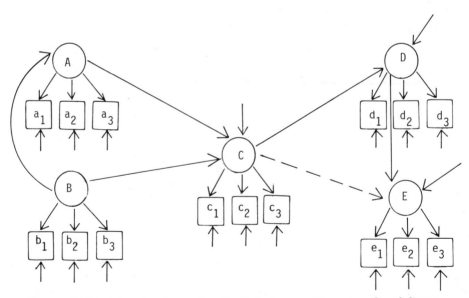

Figure 15.14. An omitted structural path between dimension C and dimension E.

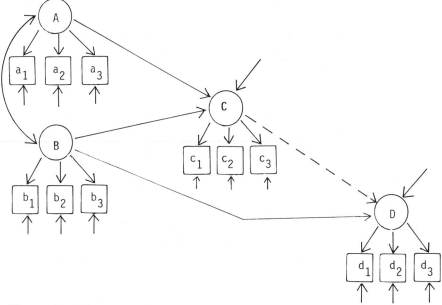

Figure 15.15 An omitted structural path between dimension *C* and dimension *D*.

correct respecification depended on the strength of the relationship omitted. Similarly, when the estimating model omits a path at the structural level, the diagnostic information differs according to whether the omitted path is strong or weak relative to other paths in the model being estimated. In either situation, however, the focused measures of fit perform adequately in that either (1) the modification index and normalized residuals both point directly to the correct respecification (when the omitted path is weak), or (2) the diagnostic measures perform as discussed above, that is, they give a misleading clue that is discovered subsequently when reestimation yields a negative error variance.

Direction of Effect Specification Error in the Structural Model

Our causal reasoning, which is made explicit in a causal model, may propose an effect in one direction when it actually runs in reverse of that assumption or runs in both directions. We explore first the performance of the focused measures of fit in identifying correctly a reversal of direction (i.e., although we assume *X* causes *Y*, in actuality *Y* causes *X*). At the outset, it is difficult to anticipate the ability of the diagnostic measures to point to this form of misspecification. The modification indices may not be helpful because they are designed to call our attention to a fixed parameter that should be freed. But here the misspecification entails a free parameter that should be fixed.

Figure 15.16 illustrates the model used to simulate the first covariance structure. The estimated model is represented in Fig. 15.17. In the estimated model, dimension D is assumed to be a dependent or endogenous dimension, while in the "true" model it is an exogenous cause of dimension C. The fit of the estimated model to the data is moderately good as measured by the adjusted goodness of fit index (.935), while the chi square (128.06, $df = 50$) suggests an unsatisfactory fit. Figure 15.16 displays the true parameters, while Figure 15.17 displays the parameter estimates generated by the misspecified model. The effect of D on C has been largely "credited" to its closest correlate, A. The normalized residuals (Table 15.18) are large between the indicators of D when paired with indicators of B but are otherwise small. The modification indices (Table 15.19) indicate that a significant reduction in chi square would be achieved by freeing the parameter representing correlated error between C and D, or by freeing the direct path from dimension B to D, or by including a reciprocal relation between dimension D and dimension C. The largest modification index points to correlated error

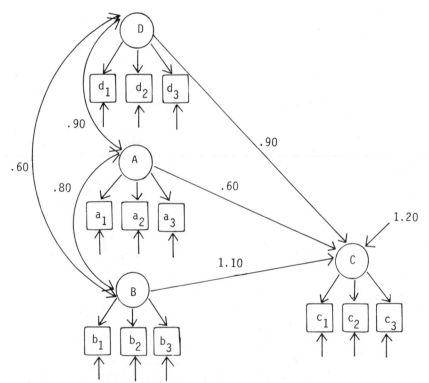

Figure 15.16. "True" model in which D is a cause of C.

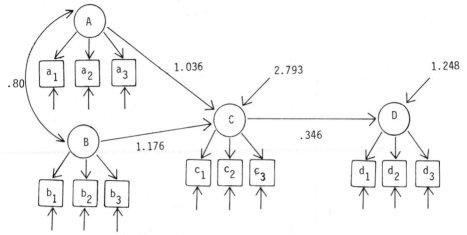

Figure 15.17. Estimating model in which *D* is represented as an effect of *C*. (Parameters estimated on the basis of the covariance structure generated by the model of Figure 15.16.)

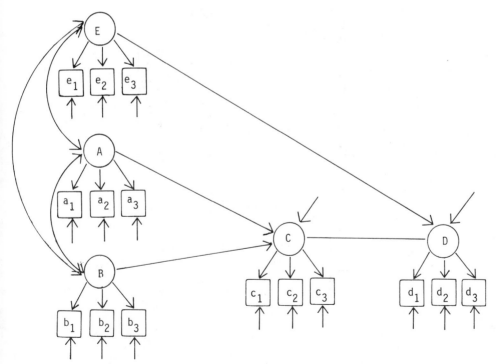

Figure 15.18. "True" model in which *E* is a cause of *D*.

Table 15.18. Normalized residuals matrix for estimating model of Figure 15.17

	c_1	c_2	c_3	d_1	d_2	d_3	a_1	a_2	a_3	b_1	b_2	b_3
c_1	.000											
c_2	.037	.000										
c_3	-.010	-.008	.000									
d_1	.107	.111	.078	-.000								
d_2	.110	.113	.079	-.000	-.000							
d_3	.112	.115	.080	-.000	-.000	-.000						
a_1	.027	.029	-.003	-.267	-.273	-.280	-.000					
a_2	.029	.031	-.003	-.288	-.294	-.302	-.000	.000				
a_3	.029	.032	-.003	-.299	-.305	-.314	.000	-.000	-.000			
b_1	.078	.081	.047	-3.817	-3.897	-4.002	.000	-.000	.000	.000		
b_2	.087	.090	.052	-4.343	-4.431	-4.545	.000	-.000	.000	-.000	.000	
b_3	.080	.082	.048	-3.902	-3.984	-4.091	.000	-.000	.000	.000	.000	-.000

Table 15.19. Significant
modification
indices for
estimating model
of Figure 15.17

Parameter	Modification index
$C \leftrightharpoons D$	54.23
$B \rightarrow D$	42.50
$C \leftarrow D$	22.35

between D and C, while the pattern of normalized residuals suggests the inclusion of an omitted path between B and D. Neither is the correct respecification, of course, but will there be clues calling that error to our attention if we respecify and reestimate?

Freeing the parameter for correlated error between C and D provides a much improved fit and does not lead to any estimates of negative error variance. We are led to accept the model with correlated error, even though it is wrong. The parameter estimates in this specification do not exhibit unusual values. Although the correlated error between C and D is negative while all other parameters are positive, there is no *a priori* reason to rule out a negative correlated error between the two dimensions.

Figure 15.18 represents the model used to simulate a second covariance structure, and Fig. 15.19 shows the corresponding estimated model. We estimate the model using E as a dependent dimension rather than in its true role as a cause of D. The estimated model is difficult to reject based on the overall goodness of fit measures. The chi-square value is small, and the adjusted goodness of fit index is moderately high. However, there are numerous negative and positive deviations of about 2.1 among the normalized residuals (Table 15.20), suggesting something less than a good fit. The departures from linearity in the Q-plot (Fig. 15.20) also suggest that the model is misspecified. The modification indices (Table 15.21) suggest that there are numerous parameters which, if freed, would result in moderate decreases in the chi-square value, the largest being the correlated error parameter between dimensions C and E. Thus, the fit of model and data can be seriously questioned, but the focused diagnostics are unclear as to where the misspecification may be located. As a first step, the correlated error parameter is freed, but examining the normalized residuals and the Q-plot (not shown) suggest that this respecification is not accurate. Major deviations remain in the Q-plot, and there are sizable deviations in the normalized residuals matrix, which are more pervasive than in the original. None of the diagnostic indices points to the correct specification, and having respecified in accord with the suggestions of the focused measures, we have

Table 15.20. Normalized residuals matrix for estimating model of Figure 15.19

	c_1	c_2	c_3	d_1	d_2	d_3	e_1	e_2	e_3	a_1	a_2	a_3	b_1	b_2	b_3
c_1	−.000														
c_2	−.015	.000													
c_3	.009	.009	−.000												
d_1	.002	−.001	.019	−.000											
d_2	.001	−.001	.019	−.003	.000										
d_3	−.002	−.002	.017	−.006	−.007	.000									
e_1	−2.094	−2.160	−2.262	.057	.058	.056	.000								
e_2	−2.306	−2.376	−2.485	.062	.062	.061	.000	.000							
e_3	−2.247	−2.315	−2.423	.061	.062	.060	−.000	.000	.000						
a_1	−.065	−.068	−.054	2.153	2.173	2.198	2.405	2.672	2.596	.000					
a_2	−.070	−.073	−.058	2.324	2.345	2.371	2.608	2.895	2.813	−.000	−.000				
a_3	−.072	−.075	−.060	2.416	2.438	2.465	2.718	3.016	2.932	−.000	.000	−.000			
b_1	−.010	−.011	.010	−.112	−.113	−.116	−1.816	−2.010	−1.955	−.000	.000	−.000	.000		
b_2	−.011	−.012	.011	−.126	−.128	−.131	−2.085	−2.303	−2.241	−.000	−.000	−.000	−.000	.000	
b_3	−.010	−.011	.010	−.114	−.117	−.119	−1.859	−2.057	−2.001	.000	.000	.000	.000	.000	−.000

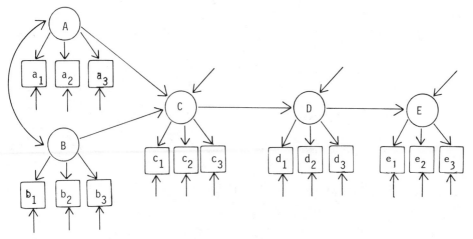

Figure 15.19. Estimating model in which E is represented as an effect of D.

failed to find either a good fitting model or additional clues that point to the correct respecification.

A third example of a direction of effect misspecification is illustrated in Fig. 15.21 (the "true" model) and Fig. 15.22 (the estimating model). Here, the misspecification is entirely in the endogenous part of the model. The "true" model shows that E causes C, whereas the estimated model reverses this direction of effect. The chi-square value (170.71) indicates that the estimated model does not fit the data well, although the other overall measures of fit suggest at least a moderately good fit. The adjusted goodness of fit index is .933, and nothing in the Q-plot (not shown) would alert us to a specification error. The focused measures suggest a lack of good fit, thus

Table 15.21. Significant
modification
indices for
estimating model
of Figure 15.19

Parameter	Modification index
$C \leftrightarrow E$	17.58
$C \leftarrow E$	15.22
$C \rightarrow E$	13.12
$B \rightarrow E$	12.83
$A \rightarrow D$	12.25
$D \leftrightarrow E$	9.99
$C \leftrightarrow D$	4.74

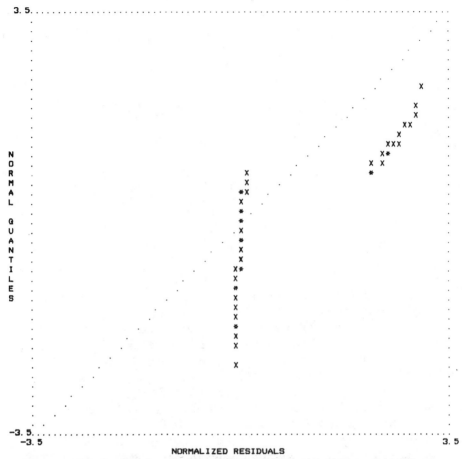

Figure 15.20. Q-plot of normalized residuals for estimating model of Figure 15.17.

supporting the conclusion based on chi square. The indicators for dimension *E*, paired with those for dimension *B*, have large negative normalized residuals, as shown in Table 15.22. The modification indices (Table 15.23) have large values for the parameters for the direct path from *B* to *E* and for the correlated error between *C* and *E*. Including a direct path from *B* to *E* provides a substantial reduction in chi square, and this respecified model would be judged by the usual criteria as providing an adequate fit to the data, even though it does not duplicate the "true" model and yields distorted parameter estimates.

When other covariance structures were generated and estimated with a direction of effect specification error, the results were generally similar to those reported above in that none of the diagnostics gave correct clues as to the correct respecification of the estimated models. Some of the misspecifications were difficult to reject on the basis of the overall goodness of fit measures, although the focused measures usually pointed to significant problems. It is further dismaying to find that freeing the parameter suggested by the focused measures of fit sometimes (but not always) leads to tolerable fits of the model and data—even though the respecified model was not the correct one. Our only clues that this respecification may be wrong are that some major deviations may still be present among the normalized residuals, but even these clues are not always present. We are forced to conclude that at this time, we lack reliable clues to a direction of effect specification error.

Recursive Assumption Misspecification in the Structural Model

This type of specification error means that the model fails to specify that two variables cause each other or that some feedback occurs within the system. Frequently, models that include such feedback will have parameters that are not identified, and this creates a twofold problem. First, the modification indices will be zero whenever a particular parameter is not identified (Jöreskog [6]). We are likely to assume that a modification index of zero means that freeing that parameter would not improve the fit rather than interpreting it as pointing to an underidentified parameter. Second, underidentification will limit our ability to reestimate a respecified model, even if the correct respecification has been ascertained.

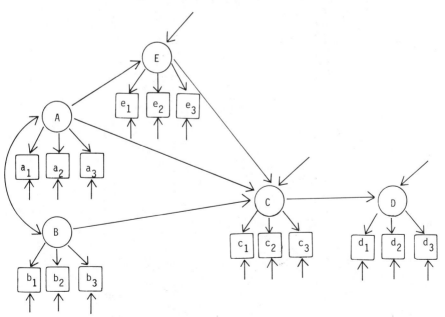

Figure 15.21. "True" model in which E is a cause of C.

Table 15.22. Normalized residuals matrix for estimating model of Figure 15.22

	c_1	c_2	c_3	d_1	d_2	d_3	e_1	e_2	e_3	a_1	a_2	a_3	b_1	b_2	b_3
c_1	.000														
c_2	.021	.000													
c_3	−.007	−.006	.000												
d_1	.016	.016	−.007	.000											
d_2	.016	.017	−.007	.000	.000										
d_3	.015	.017	−.007	−.000	.000	.000									
e_1	.067	.068	.047	.055	.055	.057	.000								
e_2	.071	.073	.051	.061	.059	.062	.000	−.000							
e_3	.072	.073	.050	.061	.060	.061	−.000	−.000	−.000						
a_1	.003	.006	−.015	.003	.005	.004	−.191	−.206	−.201	−.000					
a_2	.009	.010	−.013	.007	.007	.005	−.201	−.217	−.211	.014	.000				
a_3	.026	.028	.005	.021	.022	.021	−.192	−.206	−.203	.033	.038	.000			
b_1	.070	.071	.053	.056	.055	.056	−3.536	−3.847	−3.760	.182	.199	.216	−.000		
b_2	.076	.079	.059	.064	.063	.065	−4.022	−4.358	−4.265	.209	.228	.247	−.000	−.000	
b_3	.070	.071	.054	.060	.059	.060	−3.615	−3.929	−3.841	.187	.203	.221	.000	−.000	−.000

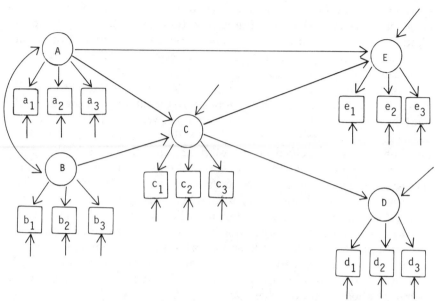

Figure 15.22. Estimating model in which E is represented as an effect of C.

Figure 15.23 provides our first illustration. In the "true" model, dimensions C and D are mutual causes of each other. The estimated model does not include the path from D to C, represented in the figure by the broken line. The estimated model provides a poor fit. Although the adjusted goodness of fit index is .911, both the chi square (186.53) and the focused measures of fit suggest a lack of good fit. The normalized residuals (Table 15.24) are large when the indicators of dimension D are paired with the indicators of dimension A or dimension B. The modification indices (Table 15.25) point to correlated error between C and D, or to missing paths either from A to D, or from B to D. It is reasonable to assume that a better fit could be achieved, but it is not immediately evident which parameter should be freed in any re-

Table 15.23. Significant modification indices for estimating model of Figure 15.22

Parameter	Modification index
$B \rightarrow E$	63.82
$C \leftrightarrow E$	62.10
$C \leftarrow E$	15.51

specification. The most likely candidate would seem to be the correlated error between C and D based on the modification index. Making this respecification and reestimating yields an adequate fit even though it does not duplicate exactly the "true" model. The respecified model is "correct" in some sense, but it is incomplete.

In the example above, the estimating model designates the direction of effect as being from C to D. This is opposite to the direction of the larger of the two reciprocal effects in the "true" model. We next specify a "true" model in which the effect of C on D is greater than that of D on C, so that the direction of effect specified in the estimating model will be in the same direction as the larger of the two reciprocal effects. In this instance, the estimating model, although misspecified, adequately fits the data. The diagnostics do not indicate a need to change any feature of the estimating model. Correlated error between dimensions C and D is weakly suggested, but the modification index suggesting it is too small to be judged significant by the usual criteria. Hence, as illustrated in this second example, a recursive assumption specification error is likely to be discerned by our goodness of fit measures if the direction specified is opposite to the larger of the two reciprocal effects, but not otherwise.

Figures 15.24 and 15.25 illustrate two additional models used to explore recursive assumption specification errors. In these models, the path represented by a broken line was omitted in the estimated model. Neither of these estimated models fits the covariance structures adequately. In Fig. 15.24, the instrumental nature of A and B provides sufficient information to identify the parameters of the reciprocal effects between C and D. However, the

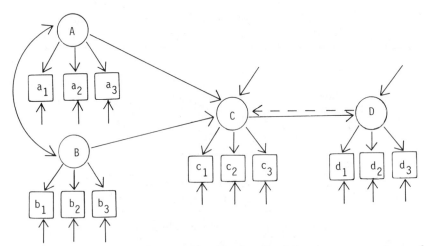

Figure 15.23. "True" model (including broken line) that is nonrecursive and estimating model (omitting broken line) that represents the process as recursive.

Table 15.24. Normalized residuals matrix for estimating model of Figure 15.23

	c_1	c_2	c_3	d_1	d_2	d_3	a_1	a_2	a_3	b_1	b_2	b_3
c_1	.000											
c_2	.020	−.000										
c_3	−.005	−.004	.000									
d_1	.037	.038	.014	.000								
d_2	.038	.038	.014	.000	.000							
d_3	.038	.039	.015	.000	.000	.000						
a_1	.056	.058	.043	−1.630	−1.637	−1.650	.000					
a_2	.060	.061	.046	−1.751	−1.760	−1.772	−.000	.000				
a_3	.063	.063	.048	−1.815	−1.826	−1.838	−.000	.000	−.000			
b_1	.071	.074	.055	−2.092	−2.100	−2.114	.000	.000	.000	.000		
b_2	.080	.081	.061	−2.338	−2.350	−2.363	−.000	−.000	−.000	−.000	−.000	
b_3	.072	.074	.056	−2.132	−2.142	−2.156	.000	.000	.000	.000	.000	.000

Table 15.25. Significant
modification
indices for
estimating model
of Figure 15.23

Parameter	Modification index
$C \leftarrow\!\rightarrow D$	88.59
$B \rightarrow D$	47.35
$A \rightarrow D$	28.39
$C \leftarrow D$	7.93

diagnostic indices suggest the inclusion of paths from B to C and from A to D, and this respecification provides an adequate fit even though it is incorrect.

In Fig. 15.25, the diagnostic indices are again confusing. The normalized residuals for indicators of D and C are relatively large when these are paired with indicators of the exogenous variables. The modification indices point to correlated error between dimensions C and D, or to direct paths from dimensions A to D, or from B to D. Freeing the correlated error between C and D provides an adequate fit, and it is, again, "correct" in some sense, but incomplete. In both of these models, there is no clue in the diagnostic indices that a reciprocal effect or nonrecursive process should have been included in the estimating model. As in the direction of effect misspecifications, with

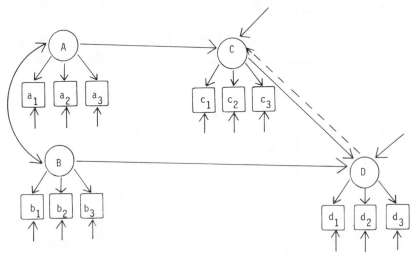

Figure 15.24. "True" model (including broken line) that is nonrecursive and estimating model (omitting broken line) that represents the process as recursive.

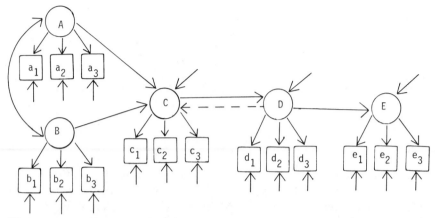

Figure 15.25. "True" model (including broken line) that is nonrecursive and estimating model (omitting broken line) that represents the process as recursive.

recursive assumption specification errors, the investigator has limited information from the diagnostic indices, and the information they give may lead to a respecification that provides a better fit without replicating the "true" model.

The simulations explored here point toward considerable difficulty in diagnosing misspecifications in the structural part of a multiple indicator model. If there is an omitted path in the structural model, or correlated error between dimensions, the diagnostic indices will point to such a specification error. However, the same diagnostic clues that point toward omitted paths or correlated error also appear when the problem is a direction of effect specification error or a recursive assumption specification error. This weakens our ability to diagnose correlated error or omitted paths at the structural level, since these types of specification errors imply, but are not implied by, diagnostic clues of certain kinds. It should be evident that great care should be exercised in respecifying the structural part of the model on the basis of the diagnostic indices.

A SUPPLEMENTARY STRATEGY: EXAMINING SUBMODELS

Glymour [4] has suggested a strategy for discerning misspecifications by examining the consistency of results from various submodels. This is similar to the use of submodels by Costner and Schoenberg [3], but they focused entirely on submodels that included different sets of indicators, while Glymour includes consistency between submodels at the structural level. Wimsatt [15] has discussed a related strategy that requires consistent estimates of

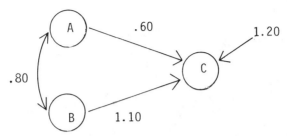

Figure 15.26. Submodel of Figure 15.22.

structural parameters from different sets of measures. In a different but related vein, Land and Felson [10] have used the term *sensitivity analysis* to refer to the estimation of models specified in slightly different ways to see the effect on parameters of major interest. With the exception of Land and Felson's "sensitivity analysis," all of these strategies point to discrepant estimates, when the model would imply identical estimates, as evidence for misspecification. Starting with a different purpose, Land and Felson do the reverse, that is, they try various specifications to examine their effect on parameter estimates instead of examining different parameter estimates to discern their implications for a different specification.

We now consider the potential utility of submodels in discerning mis-specifications in the structural model that were not pinpointed adequately by the overall or focused measures of fit. We refer again to Figs. 15.21 and 15.22. Recall that when we respecified the model shown in Fig. 15.22 to include correlated error between dimensions C and E, as suggested by the focused measures of fit, we achieved an adequate fit even though this was not the correct respecification. The submodel strategy we now explore is an alternative to dependence on the focused measures of fit. We begin by formulating various submodels based on the estimating model of interest, in this instance, Fig. 15.22. Three such submodels are shown in Figs. 15.26, 15.27 and 15.28. Each of these submodels is implied by Fig. 15.22. In the first submodel, for example, dimensions D and E have been omitted, but, if Fig. 15.22 is the "true" model, those omissions should have no effect on the

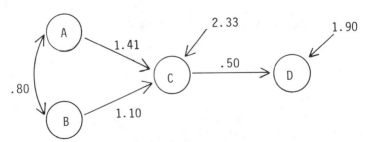

Figure 15.27. Submodel of Figure 15.22.

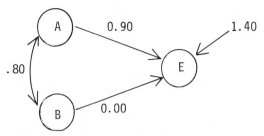

Figure 15.28. Submodel of Figure 15.22.

parameters for the effects of A and B on C. Similarly, the second submodel, omitting E, should have no effect on the parameters for the effects of A and B on C, provided the model of Fig. 15.22 is the "true" model. Hence, assuming Fig. 15.22 to display the "true" model, we should expect that the parameter estimate for the path from A to C should be identical when estimated on the basis of either of these two submodels. Similarly, the parameter estimate for the path from B to C should be the same in the two submodels. As shown in Figs. 15.26 and 15.27, the estimates for the path from A to C differ between the two submodels. We infer that there must be some source of covariance between A and C in addition to that shown in these submodels, and hence in addition to that shown in the estimating model of Fig. 15.22. We next turn to the submodel shown in Fig. 15.28. This submodel is also implied by the larger, estimating model of Fig. 15.22. In the larger model, there are two paths (direct and indirect through C) from A to E, and these should be combined in the submodel shown. There is also one path (indirect through C) from B to E, and the estimate in the submodel should be the product of these two paths. But, as shown in Fig. 15.28, there is in the data no support for the supposition of an effect—direct or indirect—of B on E. Again, the submodel suggests a misspecification. One must now exercise some ingenuity in finding an alternative specification that would (1) introduce into the model an additional source of covariance between A and C, and (2) eliminate from the model any effect of B on E. One way to accomplish both purposes is to reverse the postulated direction of effect between C and E. There is, of course, no "diagnostic index" that tells us to make that change, but it becomes evident by trial and error that this change will indeed correct the errors highlighted by the submodels.

In large and complex models, the number of potential submodels will be large and their estimation and comparison may be tedious. Alternatively, in relatively simple models, too few submodels may be implied to make possible the adequate assessment of the specification in this way. Furthermore, it will not always be evident what respecification is required to correct for inconsistent parameter estimates. However, submodels provide an investigator with an additional check on the adequacy of the specification and, hence, may either increase one's confidence in the model as specified or suggest that more work needs to be done to achieve a specification compatible with the data.

MULTIPLE SPECIFICATION ERRORS: RETURN TO THE ORIGINAL SIMULATION EXAMPLE

Having examined the utility of the diagnostic indices in a series of relatively simple models, we turn now to an exploration of their usefulness when the problems are not so oversimplified. Will we be able to use the insights suggested above to ferret out specification errors in a model that includes several such errors of more than one type? Our results above suggest that the focused measures of fit are better able to identify problems of specification at the measurement level than at the structural level. But the overall ability of the focused measures to lead directly to the misspecification is disappointingly low. Will we find that they are even less useful with a more complex pattern of misspecification. We return to the first illustration (Fig. 15.6) and apply our "educated judgment."

Following our strategy, we first specify a confirmatory factor analysis model for Fig. 15.6, and this was shown in Fig. 15.8. Estimating this model, we find a poor fit, the chi square is 196.83, suggesting that part of the problem in the original specification is due to a poorly specified measurement model. The adjusted goodness of fit index (.744) and the Q-plot confirm the inability of this measurement model to represent adequately the "true" measurement model. The normalized residuals and the significant modification indices for this initial specification are presented in Tables 15.26 and 15.27. Among the normalized residuals, a group of large values stand out; these are the residuals between indicator b_3 and the indicators of dimension A. The larger modification indices are for the parameters representing a path from dimension A to b_3 (47.13) or a correlated error between indicator b_1 and a_2 (40.03). This pattern of residuals and the large modification index in the previous simulations have suggested an omitted path misspecification. The indicator b_3 appears to be a "shared" indicator of A and B; we reestimate the model freeing this parameter and achieve a much improved fit. The chi-square value is still large enough to reject the model, and the focused indices still show major deviations. A large normalized residual occurs between indicators a_2 and b_1. This residual is a definite outlier. As shown earlier, this is a typical pattern found in the residuals matrix in the presence of spurious association misspecification between indicators of different dimensions. The modification index also shows that a large reduction in chi square would occur if correlated error were permitted between these two indicators. The parameter representing correlated error between b_1 and a_2 was freed, with a resulting significant decrease in the chi-square value.

Table 15.26. Normalized residuals matrix for confirmatory factor analysis of estimating model of Figure 15.6

	c_1	c_2	c_3	d_1	d_2	d_3	a_1	a_2	a_3	b_1	b_2	b_3
c_1	.001											
c_2	.380	.001										
c_3	−.136	−.081	.001									
d_1	.059	.099	.729	−.000								
d_2	−.400	.371	.234	−.079	−.000							
d_3	−.376	−.344	.307	−.008	.021	−.000						
a_1	−.211	−.188	.272	.362	−.062	−.018	−.001					
a_2	−.293	−.270	.169	.246	−.167	−.132	−.073	−.000				
a_3	−.208	−.182	.299	.397	−.045	.000	.099	−.054	−.000			
b_1	−.539	−.508	.177	−.496	−.878	−.904	−1.717	1.727	−1.784	.002		
b_2	−.149	−.102	.690	−.230	−.664	−.665	−1.669	−1.720	−1.729	.747	.001	
b_3	−.593	−.560	.171	1.660	1.197	1.342	2.790	2.592	2.936	−.615	−.182	.001

Table 15.27. Significant modification indices for confirmatory factor analysis of estimating model of Figure 15.6

Parameter	Modification index
$A \rightarrow b_3$	47.13
$a_2 \leftrightharpoons b_1$	40.03
$A \rightarrow b_2$	25.57
$a_3 \leftrightharpoons b_3$	16.37
$a_3 \leftrightharpoons b_1$	15.33
$b_1 \leftrightharpoons b_2$	14.92
$a_2 \leftrightharpoons b_2$	14.00
$D \rightarrow b_3$	11.58
$a_1 \leftrightharpoons b_3$	11.00
$a_1 \leftrightharpoons b_1$	10.49
$c_1 \leftrightharpoons c_2$	9.68
$b_1 \leftrightharpoons b_3$	6.95

In three steps we have reformulated the measurement model for Fig. 15.8 to achieve an excellent fit between the model and data. After having included the path from dimension A to b_3 and the correlated error between b_1 and a_2, the chi square is a nonsignificant 24.2. The adjusted goodness of fit index is .95 and the root mean square residual is .12, quite small in comparison to the covariances in the data. The model can easily be accepted as an adequate representation of the "true" measurement model. The focused measures indicate that little improvement could be achieved by freeing other parameters. The normalized residuals show no large deviations and the largest modification index is a relatively small 9.896 for correlated error between c_1 and c_2. We conclude that we have fit the data very well with the measurement model as respecified. We now move to an examination of the structural part of the model. We include the new measurement specifications and the information concerning the fit of the data to help determine the adequacy of the structural model.

Estimating the structural model as depicted in Fig. 15.6 but incorporating the new measurement specifications does not produce a set of parameter estimates that fit the observed covariance structure. We therefore conclude that there are structural misspecifications in the model. The normalized residuals and modification indexes are found in Tables 15.28 and 15.29. The chi square is 107.0 and the adjusted goodness of fit index is .884 — lower than the adjusted index of .950 found for the final measurement model. The

Table 15.28. Normalized residuals matrix for estimating model of Figure 15.6 incorporating measurement model changes—A to b_3 and correlated error a_2 and b_2

	c_1	c_2	c_3	d_1	d_2	d_3	a_1	a_2	a_3	b_1	b_2	b_3
c_1	.001											
c_2	.448	.001										
c_3	−.116	−.056	.001									
d_1	−.087	−.048	.544	.000								
d_2	−.555	−.527	.037	−.058	.000							
d_3	−.555	−.526	.082	−.005	.014	.000						
a_1	−.474	−.457	−.033	4.784	4.461	4.889	−.000					
a_2	−.459	−.439	−.031	4.614	4.302	4.716	−.000	−.000				
a_3	−.498	−.479	−.034	5.033	4.693	5.138	−.000	.000	.000			
b_1	−.358	−.320	.336	.616	.252	.304	.000	−.000	.000	−.000		
b_2	−.405	−.360	.379	.715	.292	.350	.000	.000	.000	.001	.001	
b_3	−.493	−.454	.240	2.843	2.397	2.628	.000	.000	.000	.001	.001	.001

Table 15.29. Significant modification indices for estimating model of Figure 15.6 incorporating measurement model changes — A to b_3 and correlated error a_2 and b_2

Parameter	Modification index
$A \rightarrow D$	53.09
$c_1 \overset{\curvearrowleft}{} c_2$	12.65
$C \overset{\curvearrowleft}{} D$	10.44
$C \leftarrow D$	5.33

pattern of normalized residuals is fairly straightforward. The indicators of dimension A (a_1, a_2, a_3, and b_3) all have larger than expected positive values with the indicators of dimension D. Such a pattern suggests an omitted path between dimensions A and D. The modification index confirms this diagnosis. The largest modification index (53.09) suggests including a structural path from A to D as well. The other large indices specify correlated error between C and D, a path from D to C, or freeing correlated error between indicator c_1 and c_2. The last named suggestion can be dismissed since we

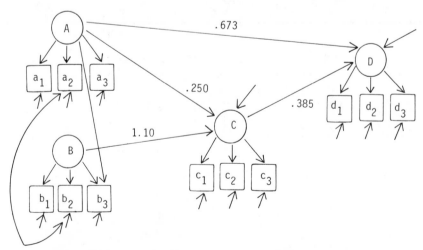

Figure 15.29. Respecification of the model of Figure 15.6.

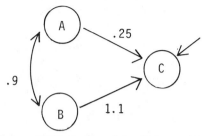

Figure 15.30. Submodel of Figure 15.29.

have already respecified the measurement model. Most of the information points to freeing the path from dimension A to dimension D.

The respecified model including the path from A to D provides an excellent fit of the model and data. The chi square is 24.29 and the adjusted goodness of fit index is .95. The Q-plot is reasonably linear with no major deviations and has a slope much greater than 1.0. The modification indices and the normalized residuals point out no major deviations. As in the measurement model, the correlated error between c_1 and c_2 is the only exception, and its impact seems to be minimal. We conclude that the model, as respecified in Fig. 15.29, is as adequate a representation of the "true" model as we can obtain.

However, given the tenuous nature of the diagnostics at the structural level, it seems best to test a few submodels to gain greater confidence in our respecification of the structural model. Figures 15.30, 15.31, and 15.32 show a set of submodels that will help provide a more thorough examination of the final specification in Fig. 15.29. Our expectation in Fig. 15.30 is that the paths from the exogenous dimensions A and B to dimension C should not change from their values in Fig. 15.29. This proves to be the case. In Fig. 15.31, the paths from A and B to D should be a simple function of the parameters shown in Fig. 15.29. The path from A to D should be $.673 + (.250)(.385) = .769$. The path from B to D should be $(1.1)(.385) = .4235$. Both expectations are confirmed. Finally, in Fig. 15.32, we expect the value of the path from dimension A to C to increase when we have removed

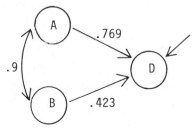

Figure 15.31. Submodel of Figure 15.29.

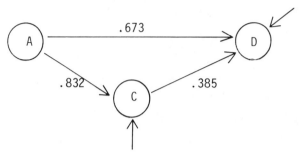

Figure 15.32. Submodel of Figure 15.29.

dimension B, but the direct path from A to D should be stable. These expectations are also supported. Although this does not guarantee that all misspecifications have been corrected, the exploration of submodels suggests considerable confidence in our final specification.

In this instance, we are in the unusual circumstance of being able to compare our final specification to the "true" model that generated the data. Our final specification is accurate at the structural level. At the measurement level, we failed to include correlated error between c_1 and c_2 and between d_2 and d_3. The former was suggested by the modification index in our final measurement specification. Had we included it, we would have provided a still better fit of the model. The clues, however, would not suggest including the correlated error between d_2 and d_3. But the diagnostic clues for these specification errors were not clear. The decision not to include these features of the "true" specification in our attempt to reach a final model based on the diagnostic indices was based on the adequacy of the fit. Given the artificial nature of the simulated data, it might have been reasonable to attempt to attain perfect agreement between model and data, but such a standard is unrealistic for nonsimulated data. Leaving out these two features of the "true" model would be "normal" with real data since they did not greatly interfere with the achievement of a good fit.

In the context of a more complicated set of specification errors than those considered previously, we have demonstrated the utility of the diagnostic indices. These indices are not seriously affected by the number of misspecifications present in a model, it seems. However, a warning is necessary; the "more complicated" model used in this illustration did not include any of the misspecifications found difficult to discover — misspecifications such as errors in the direction of effect or violations of the recursive assumption. Some caution is appropriate, and compatibility with substantive theory and intuitive reasoning should always supplement the guidance provided by the diagnostic indices.

REFERENCES

[1] Burt, Ronald S. "Interpretational Confounding of Unobserved Variables in Structural Equation Models." *Sociological Methods and Research* 5 (1976): 3–52.

[2] Byron, R. P. "Testing for Misspecification in Econometric Systems Using Full Information." *International Econometric Review* 13 (1972): 745–756.

[3] Costner, Herbert L., and Schoenberg, Ronald. "Diagnosing Indicator Ills in Multiple Indicator Models." In *Structural Equation Models in the Social Sciences,* edited by Arthur S. Goldberger and Otis Dudley Duncan. New York, Seminar Press, 1973.

[4] Glymour, Clark. *Theory and Evidence.* Princeton: Princeton University Press, 1980.

[5] Jöreskog, Karl G. "A General Method for Analysis of Covariance Structures." *Biometrika* 57 (1970): 239–251.

[6] Jöreskog, Karl G. "Introduction to LISREL V." *DATA* 1 (1981): 1–7.

[7] Jöreskog, Karl G., and Sörbom, Dag. *LISREL V: Analysis of Linear Structural Relationships by the Method of Maximum Likelihood; User's Guide.* University of Uppsala, 1981.

[8] Jöreskog, Karl G., and van Thillo, M. "LISREL: A General Computer Program for Estimating a Linear Structural Equation System Involving Multiple Indicators of Unmeasured Variables." *Research Bulletin* 71–1. Princeton: Educational Testing Service, 1972.

[9] Kalleberg, A. L. "A Causal Approach to the Measurement of Job Satisfaction." *Social Science Research* 3 (1974): 299–321.

[10] Land, Kenneth C., and Felson, Marcus. "Sensitivity Analysis of Arbitrarily Identified Simultaneous-Equation Models." *Sociological Methods and Research* 6 (1978): 283–307, and this volume.

[11] Long, J. Scott. "Estimation and Hypothesis Testing in Linear Models Containing Measurement Error." *Sociological Methods and Research* 5 (1976): 157–206.

[12] Saris, W. E., Pijper, W. M., and Zegwaard, P. "Detection of Specification Errors in Linear Structural Equation Models." In *Sociological Methodology 1979,* edited by Karl F. Schuessler. San Francisco: Jossey-Bass, 1979.

[13] Sörbom, Dag. "Detection of Correlated Errors in Longitudinal Data." *British Journal of Mathematical and Statistical Psychology* 28 (1975): 138–151.

[14] Stolzenberg, Ross M. "Estimating An Equation with Multiplicative and Additive Terms, with An Application to Analysis of Wage Differentials between Men and Women in 1960." *Sociological Methods and Research* 2 (1974): 313–331.

[15] Wimsatt, William C. "Robustness, Reliability and Overdetermination." In *Scientific Inquiry in the Social Sciences,* edited by Marilyn B. Brewer and Barry E. Collins. San Francisco: Jossey-Bass, 1981.

Multiple Indicators and Complex Causal Models

John L. Sullivan

16

Several authors have recently pointed out the usefulness of multiple indicators. Curtis and Jackson [3] argue that the use of each indicator separately has certain advantages over combining them into an index. It increases the number of predictions made by a particular model, enables the careful researcher to determine the existence of an unknown spurious cause, increases confidence in the validity of the indicators, and guides conceptual reformulation. Other authors, such as Siegel and Hodge [4], Costner [2], and Blalock [1] have illustrated the use of multiple indicators to determine the existence and nature of measurement error. This chapter addresses itself to the former use of multiple indicators, extends the implications of Curtis and Jackson to a more general model, and introduces a method that deals with the problems arising as a result of this extension.

A common problem encountered in testing complex causal models is the selection of indicators. A common solution is the selection of one indicator (from among perhaps three or four) that seems "on its face" to best represent the construct we wish to measure. This involves a loss of information and accuracy, as the two or three indicators that are thrown out are likely to have some validity, and their addition may produce a more correct representation of the construct. A second solution is that of combining the three or four indicators into an index in an attempt to represent the construct more accurately than any one (even the "best") of the indicators could possibly do. This seems preferable, yet as Curtis and Jackson note, this has certain costs involved as well, thus, their argument in favor of using each indicator separately. As will be noted later, factor analysis is not deemed an acceptable solution to the problem.

EXTENSION TO THE GENERAL MODEL

The use of each indicator separately has certain practical limitations whenever we are dealing with a large number of variables, hence an even larger number of indicators. The number of tests of the predictions, without even attempting to assess the assumptions of the model, quickly becomes unmanageable. In Fig. 16.1, for example, six predictions are made, but there are 106 tests of these predictions, a number that would increase rapidly with every additional indicator. If one wishes to check the correlational assumptions of the model, the number of tests becomes even more unmanageable.

Must we, then, lose the advantages of multiple indicators precisely in those situations where we need them the most? As our models become more complex and meaningful, must we fall back upon selecting the "best" indicator? Hopefully, the answer is no. As a sensible compromise, I would suggest the use of multiple–partial correlation coefficients to test complex

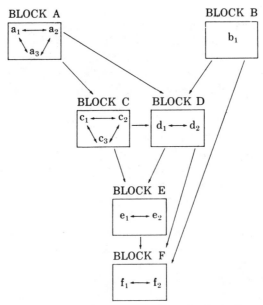

Figure 16.1. Causal model with multiple indicators.

causal models. This involves using the indicators of the dependent[1] variable separately, but allowing the indicators of the independent and the control variables to operate as a block.[2] Thus, we have more tests of each prediction (the same number as the number of indicators of the dependent variable in each test) but also a more accurate representation of the theoretical constructs. For example, in Figure 16.1, the prediction that $r_{AE \cdot CD} = 0$ now has two rather than 36 tests

$$r_{e_1(a_1a_2a_3) \cdot (c_1c_2c_3d_1d_2)} = 0$$

and

$$r_{e_2(a_1a_2a_3) \cdot (c_1c_2c_3d_1d_2)} = 0$$

That is, we allow all of the indicators of the control variables to wipe out as much variation as they can in the dependent variable, and then see how much of the remaining variation is explained by all of the indicators of the independent variable. One could argue that "of course, any partial will drop to or near zero if we allow so many variables to first wipe out variation in the dependent variable." However, I propose that one also use this method to check the assumptions upon which the model is based, thus looking for non-zero multiple–partials. For example, the model in Figure 16.1 assumes causal connections between D and E, D and F, and E and F. If this is true, then D and F ought to be related, even controlling for E. Therefore, the following two "tests" or "checks" of that assumption can be made:

$$r_{f_1(d_1d_2) \cdot (e_1e_2)} \neq 0$$

and

$$r_{f_2(d_1d_2) \cdot (e_1e_2)} \neq 0$$

---

[1] "Dependent" is used advisedly in this case. It is a dependent variable only in terms of one particular test of the model, not in terms of the entire model.

[2] Of course, the use of this procedure means that the empirical equations for each test of a single prediction will differ. The weights given to various indicators shift because the slope coefficients shift as the various indicators of the "dependent" variable are used. Since we are interested in the "total direct effects" of one block of variables on another block, this probably is not a serious problem, although our estimates will vary.

Standard *F*-tests can be used to determine whether the multiple-partial is significantly greater than zero.

AN EMPIRICAL ILLUSTRATION

A brief example of this technique follows, more as an illustration of its use than as a test of this particular model. Table 16.1 lists the constructs and the indicators corresponding to Fig. 16.1. The unit of analysis is the state, with an N of 35, as there were 35 states with senatorial elections in 1966, and one of the constructs, electoral choice, is dependent upon questionnaires administered in that election. The results of the tests of the predictions and assumptions of the model are presented in Tables 16.2 and 16.3. The small N implies large sampling errors, hence these results are merely suggestive.

Table 16.2 lists the six predictions and the fourteen tests of these predictions. All fourteen tests are successful at the .01 level of confidence. This makes it difficult to reject the null hypothesis that the correlation is equal to zero. If the .05 level of confidence is used, two of the thirteen tests are

Table 16.1. Constructs and indicators of Figure 16.1

Block A. Resources
      a_1 percentage home ownership
      a_2 median income
      a_3 percentage white-collar occupation
Block B. Heterogeneity
      b_1 weighted sum of the standard deviations (by congressional district) of
          a_1, a_2, c_1, c_3, d_1, percentage urban and percentage foreign stock
Block C. Sophistication of electorate
      c_1 percentage with a college education
      c_2 median education
      c_3 percentage functionally illiterate
Block D. Discrimination
      d_1 percentage black
      d_2 difference between the ratio of home ownership/renter occupied
          dwellings for whites versus nonwhites
Block E. Party competition
      e_1 percentage for presidential candidate who carried the state in 1960
      e_2 mean difference, winner's percentage versus loser's percentage of the
          vote, by congressional district
Block F. Electoral choice
      f_1 senatorial challenger versus incumbent attitudinal differences on
          issues, 1966[a]
      f_2 mean attitudinal differences on issues, challenger versus incumbent, by
          congressional district

---

[a] From a 16-point scale developed from questionnaires given to all candidates in the 1966 Congressional and Senatorial elections by NBC. See Robert O'Connor and John Sullivan, *Systemic Determinism or Popular Choice?*, unpublished mimeo, University of North Carolina.

Table 16.2. Predictions

	.05 level	.01 level
1. $r_{AB} = 0$		
(1) $r_{a_1b_1} = .13$	NS	NS
(2) $r_{a_2b_1} = -.14$	NS	NS
(3) $r_{a_3b_1} = .27$	NS	NS
2. $r_{BC} = 0$		
(1) $r_{c_1b_1} = .12$	NS	NS
(2) $r_{c_2b_1} = -.25$	NS	NS
(3) $r_{c_3b_1} = -.13$	NS	NS
3. $r_{BE \cdot D} = 0$		
(1) $r_{e_1(b_1) \cdot (d_1d_2)} = .10$	NS	NS
(2) $r_{e_2(b_1) \cdot (d_1d_2)} = .16$	NS	NS
4. $r_{CF \cdot DE} = 0$		
(1) $r_{f_1(c_1c_2c_3) \cdot (d_1d_2e_1e_2)} = .14$	NS	NS
(2) $r_{f_2(c_1c_2c_3) \cdot (d_1d_2e_1e_2)} = .32$	NS	NS
5. $r_{AE \cdot CD} = 0$		
(1) $r_{e_1(a_1a_2a_3) \cdot (c_1c_2c_3d_1d_2)} = .18$	NS	NS
(2) $r_{e_2(a_1a_2a_3) \cdot (c_1c_2c_3d_1d_2)} = .43$	Sign	NS
6. $r_{AF \cdot CD} = 0$		
(1) $r_{f_1(a_1a_2a_3) \cdot (c_1c_2c_3d_1d_2)} = .28$	NS	NS
(2) $r_{f_2(a_1a_2a_3) \cdot (c_1c_2c_3d_1d_2)} = .46$	Sign	NS

unsuccessful, whereas approximately one is expected to be unsuccessful by chance.

Fifteen of the 33 tests of the nine assumptions are not met at the .01 level. Again, however, with such a small N, the .01 confidence level is highly rigorous. If the .05 level is substituted, 22 of these tests are successful, while eleven are unsuccessful. Of these eleven, five occur as the only unsuccessful test of a certain assumption, whereas four occur in the six tests of assumption three, and the other two are the only two tests of assumption two. It appears that these two assumptions may in fact be false.

The model assumes that a high degree of sophistication causes a low degree of discrimination. Four of the six "tests" of this assumption indicate that the data do not fit. However, two of the three tests using percentage black as an indicator of discrimination indicate that sophistication and discrimination are very strongly related. The failure of the three tests using the difference between the ratio of home ownership/renter occupied dwellings for whites versus nonwhites as an indicator of discrimination may reflect its inadequacy as an indicator. This method allows us to explore and explain such inconsistencies. In any case, the size of the correlations for the two successful tests indicates that the arrow from sophistication to discrimination in Fig. 16.1 should not be removed until further evidence is presented.

Table 16.3. Assumptions

	.05 level	.01 level
1. $r_{AC} \neq 0$		
(1) $r_{a_1c_1} = .61$	Sign	Sign
(2) $r_{a_1c_2} = .74$	Sign	Sign
(3) $r_{a_1c_3} = -.74$	Sign	Sign
(4) $r_{a_2c_1} = .10$	NS	NS
(5) $r_{a_2c_2} = .55$	Sign	Sign
(6) $r_{a_2c_3} = -.63$	Sign	Sign
(7) $r_{a_3c_1} = .76$	Sign	Sign
(8) $r_{a_3c_2} = .71$	Sign	Sign
(9) $r_{a_3c_3} = -.62$	Sign	Sign
2. $r_{BD} \neq 0$		
(1) $r_{b_1d_1} = .33$	NS	NS
(2) $r_{b_1d_2} = .17$	NS	NS
3. $r_{CD} \neq 0$		
(1) $r_{c_1d_1} = -.29$	NS	NS
(2) $r_{c_1d_2} = 0$	NS	NS
(3) $r_{c_2d_1} = -.78$	Sign	Sign
(4) $r_{c_2d_2} = .08$	NS	NS
(5) $r_{c_3d_1} = .93$	Sign	Sign
(6) $r_{c_3d_2} = -.13$	NS	NS
4. $r_{DE} \neq 0$		
(1) $r_{d_1e_1} = -.46$	Sign	Sign
(2) $r_{d_1e_2} = .65$	Sign	Sign
(3) $r_{d_2e_1} = -.47$	Sign	Sign
(4) $r_{d_2e_2} = -.03$	NS	NS
5. $r_{EF} \neq 0$		
(1) $r_{e_1f_1} = .46$	Sign	Sign
(2) $r_{e_1f_2} = .22$	NS	NS
(3) $r_{e_2f_1} = -.65$	Sign	Sign
(4) $r_{e_2f_2} = -.79$	Sign	Sign
6. $r_{AD \cdot C} \neq 0$		
(1) $r_{d_1(a_1a_2a_3) \cdot (c_1c_2c_3)} = .37$	Sign	NS
(2) $r_{d_2(a_1a_2a_3) \cdot (c_1c_2c_3)} = .70$	Sign	Sign
7. $r_{BF \cdot DE} \neq 0$		
(1) $r_{f_1(b_1) \cdot (d_1d_2e_1e_2)} = .43$	Sign	NS
(2) $r_{f_2(b_1) \cdot (d_1d_2e_1e_2)} = .34$	NS	NS
8. $r_{CE \cdot D} \neq 0$		
(1) $r_{e_1(c_1c_2c_3) \cdot (d_1d_2)} = .27$	NS	NS
(2) $r_{e_2(c_1c_2c_3) \cdot (d_1d_2)} = .39$	Sign	NS
9. $r_{DF \cdot E} \neq 0$		
(1) $r_{f_1(d_1d_2) \cdot (e_1e_2)} = .55$	Sign	Sign
(2) $r_{f_2(d_1d_2) \cdot (e_1e_2)} = .42$	Sign	NS

The model also assumes that a high degree of heterogeneity causes a low degree of discrimination. As Table 16.3 indicates, this assumes that $r_{BD} \neq 0$, and both tests of that assumption are unsuccessful. However, the relationship between the index of heterogeneity and percentage black appears to be significantly nonlinear.[3] Despite that fact, $r_{b_1 d_1} = .33$. Using an unrestricted model and relating the index to percentage black, an $E^2 = .56$ ($E = .75$) is obtained. That is, over half of the variance in percentage black is "explained" by the index, if the assumption of linearity is removed. Further research may specify the exact form of this relationship, but for present purposes it is sufficient to demonstrate that it exists.

CONCLUSION

The use of multiple–partial correlation coefficients allows the researcher to make ample use of multiple indicators while at the same time retaining a manageable number of predictions. The result is better measurement of underlying constructs and parsimony in evaluating a specific model. It has other advantages. Unlike factor analysis, the indicators may be causally related to one another apart from the spuriousness produced by the assumption that both (or all) indicators are at least partially caused by the underlying factors. No assumptions about the causal interrelations of any one block of indicators need to be made. They may take any form, including reciprocal causation, as indicated in Fig. 16.1. We do not have to know which variables in Block A affect which variables in Block C, but we do assume that no variables in Block C affect Block A. That is, we assume a block-recursive system and attempt to assess relationships between, but not within, the blocks. This is a common situation, especially in macrolevel studies. This procedure also forces the decision of which indicators "indicate" which construct to be made on theoretical rather than empirical grounds. This forces clearer conceptualization, avoids the false impression

[3] The use of causal modeling assumes linearity and no interaction. Using the .01 level of confidence, three of 69 tests for linearity are significantly nonlinear. One is between percentage black and percentage home ownership and is only one of six tests among indicators of the constructs discrimination and resources. Another is between percentage college educated and percentage home ownership, one of nine tests among indicators of sophistication and resources. The last is one of two tests among the indicators of heterogeneity and discrimination. Only in the latter case is theoretical nonlinearity suspected.

The assumption of no interaction was also tested, using the .01 level of confidence. There are 187 tests, of which only 10 show significant interaction. Five of the ten interactions involve only one of between eight to eighteen tests among indicators of the theoretical constructs involved. Hence, theoretical interaction is highly doubtful. The remaining five show the possibility of interaction between resources and discrimination, *vis à vis* choice, and sophistication and discrimination *vis à vis* party competition. The former involves 3 of 12, the latter 2 of 12, tests.

that empirical techniques can do our theory building for us, and results in a more valid representation of the constructs.

As Curtis and Jackson note, different indicators of the same construct need not be highly associated with one another and may in fact be negatively associated. They may explain different portions of the variance in the construct, having little common variance. This makes empirical determination (on the basis of the degree of association among indicators) of which indicators represent the same construct at best tenuous, at worst misleading, antitheoretical, and a deterrent to our goals and purposes as social scientists.

REFERENCES

[1] Blalock, H. M., Jr. "Multiple Indicators and the Causal Approach to Measurement Error." *American Journal of Sociology* 75 (1969): 264–272.

[2] Costner, Herbert L. "Theory, Deduction and Rules of Correspondence." *American Journal of Sociology* 75 (1969): 245–263.

[3] Curtis, Richard F., and Jackson, Elton F. "Multiple Indicators in Survey Research." *American Journal of Sociology* 68 (1962): 195–204.

[4] Siegel, Paul M., and Hodge, Robert W. "A Causal Approach to the Study of Measurement Error." *Methodology in Social Research*, edited by H. M. Blalock and Ann B. Blalock. New York: McGraw-Hill, 1968.

Problems of Aggregation

Michael T. Hannan

17

The thrust of the majority of chapters in this volume is to demonstrate the advantages of regression-based linear causal models for theory testing and construction in the social sciences. As has already been discussed, such approaches to making causal inferences from nonexperimental data place a number of constraints on the analyst. In particular, he must make explicit most of the assumptions underlying both his model and analysis operations. Among other things, he must close the theoretical model and make assumptions about the influences of outside variables, distinguish between measured and unmeasured variables, and specify the relations between theoretical constructs and indicators. All such constraints seem likely to improve the quality of nonexperimental research and enhance the possibilities of cumulation in the social sciences.

It is crucial in any discussion of these techniques to keep in mind the restrictiveness of the assumptions justifying their use. Given the state of theory and data in the nonexperimental social sciences, it is highly unlikely that all of the assumptions underlying any one of the techniques will be met in any substantively interesting application. In other words, complications are almost certain to arise when substantive specialists employ the proposed techniques. The advocacy of the use of linear causal model-testing procedures demands an examination of the impact of those complications that are thought to be most likely to arise in specific areas of application. A number of the econometricians and biometricians who have pioneered in the development of linear causal analysis have devoted considerable attention to complications. Among the complications that have received the most attention are errors in variables (including measurement error), errors of specification, multicollinearity, identification problems, problems of autocorrelation, the introduction of unmeasured variables, and changes in units of analysis.

As sociologists increasingly make use of the techniques under discussion, they will undoubtedly encounter many of the same complications found in economic and biological applications. In such cases, we should be quick to seek out and adopt formulations that have proved fruitful in other disciplines, if such formulations exist. It seems likely, however, that as a consequence of differing theoretical predispositions and problem foci, sociologists will also face some complications not previously addressed in other fields. In this case, we may be able to modify procedures created to handle quite different problems, or we may have to begin *de nouveau.* I would suspect that most of the complications sociologists will face will turn out to be analogous to those arising in quite different applications in other fields, with a component specific or unique to sociological research designs. If this proves to be the case, a useful set of methodological tasks would involve: (1) the identification of complications most likely to lead to faulty inference in the sociological applications of linear causal techniques; (2) a search for problem formulations that have been created to identify (and hopefully, to resolve) the same or analogous complications in other disciplines; and (3) critically evaluating the feasibility and fruitfulness of employing such formulations in sociological designs.

This is not to suggest some new line of methodological inquiry. Sociologists have been engaged in such a process almost continually since they became aware of the model-testing strategies. A notable early example is Blalock's [Chaps. 4 and 5] discussion of complications likely to arise in sociological applications of what has come to be known as the Simon – Blalock technique. In this chapter, I continue and expand on a discussion presented there concerning the effects of changes in units or levels of analysis. In this case, I borrow quite heavily from a highly technical body of literature in economics. Much of the focus will be on reformulating mathematical and statistical arguments in causal terms more familiar to sociologists and in trying to demonstrate that an explicit causal perspective elucidates some issues that tend to be hidden in the technical complexity of the original arguments.

Sociologists are undoubtedly more familiar with that aspect of the problem of changing units labeled *disaggregation.* The attention of the discipline was forcefully directed to the problem of disaggregation by W. S. Robinson's [30] discussion of the "ecological fallacy." He demonstrated the wide potential divergence between measures of association computed on individual level data and those computed on aggregate data. The ecological fallacy, then, consists of inferring individual relationships from calculated aggregate relationships. Sociologists are just becoming aware of the issues involved in the converse problem of *aggregation*. Much of the relevant commentary on past and present positions in the social sciences on aggregation and disaggregation is summarized in an excellent symposium collection of quantitative ecological analysis (Doggan and Rokkan, [9]), which will almost certainly stimulate additional thinking on the issues.

The available discussions of the issues tend in large measure to stress fallacious technique. Thus, we see reference to the ecological fallacy, the individualistic fallacy (arising in the aggregation of individual units), and the more inclusive term: the fallacy of the wrong level (Galtung [13]). I agree that at some point in the development of a methodology it is useful to stress possible fallacies. But, it seems that once a complication has been brought to the attention of a discipline, a continued stress on fallacies *per se* may tend to stifle research rather than improve research practice or extend the range of research possibilities.[1] The establishment of a more sophisticated methodology would seem to require movement from such categorical thinking to a consideration of likely magnitude of errors and consequent faulty inference associated with specified procedures under a variety of situations. This seems particularly true with respect to some of the problems involved in changing units of analysis. Robinson's [30] paper was directed at discouraging what he saw as faulty research practice. It has become clear, however, that the problems of changing units of analysis in research designs do not arise solely as a consequence of poorly conceived research. There are a number of situations in which the researcher has no choice but to compromise his design by employing data gathered on the "wrong" level or to abandon an interesting and perhaps important line of investigation.

In other words, problems of aggregation and disaggregation arise largely as a consequence of *missing data*. There are, no doubt, contemporary examples of the uncritical shifting of units that Robinson criticized. This is not the focus of this analysis, however. An early and still important example of missing data problems is presented by Durkheim's analysis of suicide. His analysis presents a clear example of the problem of disaggregation of empirical relations. Doggan and Rokkan [9, pp. 5–6] suggest a wide variety of situations in which missing data problems are likely to arise:

> There may be data for *individuals* across a range of areal units but no way of identifying the characteristics of the proximal community contexts of their behavior (for an ecological analysis). This is frequently the case in secondary analysis of nationwide sample surveys; for reasons of secrecy, or economy, or

[1] A forceful statement of this position is presented by Allardt [2].

sloppy administration, there may no longer be any possibility of allocating individual respondents to any known set of primary sampling units.

The contrary situation is even more frequent: no individual data are at hand, but *aggregate* distributions have been established for territorial units at different levels. This is the case for a wide variety of official statistics: the primary individual data have either been kept secret from the outset, as in elections or referenda, or cannot be made available for administrative or economic reasons, as will often be the case for census data, school grades, tax records, criminal statistics.

Problems of missing data are likely to be particularly acute in historical or "backward-looking" longitudinal research.[2] Here the researcher does not have the option of collecting the appropriate data; he must work with what is available or abandon the project.

To this point, I have been considering changes of units or levels within a single design. Another aspect of the problem of changing units has received relatively little attention. This involves the problem of comparing research results (e.g., path coefficients, partial regression coefficients) of studies formulated on different levels of analysis employing the "same" models. In such attempts at comparison, it becomes crucial to find a way to separate out the effects that are due to aggregation or disaggregation bias (to be defined below) from "true" differences at different levels. The Doggan and Rokkan collection evidences increasing interest in this issue stimulated by an interest in cross-national comparisons of results at different levels and the beginnings of data banks that will facilitate such comparisons. Allardt [2, p. 45], for example, evidences this concern: "The real issue in methodological discussions of today does not seem to be the ecological fallacy but the techniques used in analyzing data from many levels of social organization." This is another example of a concern with the effects of changing units of analysis that is not concerned with poor research technique but with extending the range of possibilities of existing methods.

The aggregation–disaggregation issues involved in comparing research results from different levels are not so obvious when the results are stated in verbal form. The same seems to be true when the results are stated in more or less descriptive form using percentage differences or coefficients of association. This perhaps explains the relative lack of attention paid to this issue. However, as attention shifts to comparing regression coefficients and similar measures and to mathematical formulations, I would expect the complications I will outline in this chapter to become potentially more troublesome.

[2] Linz [24] presents an excellent discussion of some of the cross-level analysis problems faced in such research. We will see below that time-series or longitudinal research is subject to peculiar aggregation–disaggregation problems.

THE CONSISTENCY FORMULATION

There are a number of perspectives from which sociologists might address the issue of changing levels of analysis. At the most abstract level, there is an admixture of theoretical and methodological issues. Although the distinction between theoretical and methodological concerns is often arbitrary and artificial, a chapter of this nature cannot hope to address both sides of the issues. For this reason, I will not address the many theoretical issues involved in this problem. Let it suffice to say that the import of the technical results will depend largely on the reader's position on the question of "theoretical scope" (to use Wagner's [35] term). Elsewhere (Hannan [19], I have developed a distinction suggested by Wagner between a *homology approach*, which supports the free translation of theoretical propositions across levels of analysis, and an *inconsistency position*, which rejects such a notion. I would suspect that most sociologists would reject both extreme positions and place themselves somewhere in the middle of the continuum suggested by the terms. At any rate, readers whose positions are closer to the inconsistency end of the continuum are likely to be tempted to reject aggregation and disaggregation out of hand. But it is not so simple as this. It appears that even some of those social scientists who on the theoretical level would reject the homology assumption make simplifying assumptions either in gathering data (e.g., using simple random samples in an attempt to obtain "structural" information) or in making abstract theoretical statements (e.g., constructing mathematical models that are assumed to hold for more than one level) that imply homology. In any case, the seriousness with which the reader views the conclusion that changes in levels of analysis, especially in longitudinal research, tend to produce "artificial" changes in statistical parameter estimates will depend largely on the degree to which he thinks this is natural and appropriate (due to the presence of an ontological gap between levels, "emergent properties," etc.).

The earliest concerns with problems of aggregation and disaggregation, at least in economics, arose in the context of exact or deterministic models. A number of early discussions raised serious issues of lack of consistency between analogous macro- and microfunctions when both were composed of exact relationships. Our concern is more directed at stochastic models. The statistical estimation aspects of the problem of changing units followed logically from the deterministic results. Sociologists focused from the beginning on coefficients of association, while economists have tended to restrict their attention (with a few notable exceptions discussed below) to regression estimates. In both disciplines, methodological concern seems to have been aroused when it was found that aggregation and disaggregation produced results that could not be accounted for by the homology model. This is implied in the sociological literature and made explicit in the economic discussions. The economists who early examined the issue for-

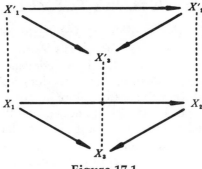

Figure 17.1

mulated an operational analog of the homology model labeled the *consistency criterion*. This was crucial since it provided a criterion against which to evaluate the effects of actually changing levels with empirical data on the statistical estimates on the level of interest. That is, this formulation permitted the identification of *aggregation – disaggregation bias*. The consistency criterion allows one to identify the changes in estimates that are due solely to aggregation – disaggregation. In other words, if the model specifies that the properties of models at different levels are identical, then any deviation at different levels can be attributed to the biasing effect of changing units. This is precisely how the problem was formulated for sociologists by Robinson [30] and for economists by Klein [20], May [26], and Theil [32].

Let us focus on the economists' formulation. We are concerned with three kinds of relationships: *microrelations*, composed of microvariables, *macrorelations*, composed of macrovariables and formed by analogy with the micromodel, and *aggregation relations*, which define macrovariables in terms of "corresponding" microvariables.[3] For a simple three-variable recursive model, we can represent the situation as in Fig. 17.1. Here X_1, X_2, and X_3 are microvariables, X'_1, X'_2, and X'_3 are the corresponding macrovariables. We draw dashed lines from microvariables to corresponding macrovariables to indicate a systematic functional relationship to which we do not attach any causal significance. All of the other relations expressed in the model can be thought of as causal.

Before addressing the analytic issues raised by this formulation, I will present two examples, one from economics and one from sociology. Both are meant to be suggestive and illustrative rather than to be serious substantive models. Allen [3] suggests the following situation. Suppose we had a well-established micro demand function that stated that the individual household's demand for tea was a function of household income, the price

---

[3] This is the appropriate formulation for considering problems of aggregation. A concern with disaggregation would have us begin with a macromodel, a micromodel formed by analogy with the macromodel, and a set of disaggregation relations.

of tea, and the prices of several related commodities. By analogy, we might construct a macromodel in which the aggregate demand for tea was a function of national income, the price of tea, and a general price index. The aggregation relations are obvious, the macro demand variable is the sum of the individual household demands, national income is the sum of all the incomes of households, etc.

A sociologist might be more interested in a model like the following. Consider a micromodel that takes the individual academic achievement (ae_i) as a function of student's parents' education (pe_i), father's occupation (fo_i), and student's measured academic aptitude (a_i):

$$ae_i = f_i(pe_i, fo_i, a_i)$$

$$(i = 1, \ldots, N)$$

(17.1)

The subscripting of the symbol denoting the function indicates that we are not requiring that all individuals "behave" alike with respect to changes in the explanatory variables. The question of whether or not all individuals in the sample do behave alike with respect to the variables included in the model becomes a significant issue in the discussion of aggregation in k-variable models (below). Second, consider a macromodel formed by analogy to the micromodel:

$$AE = F(PE, FO, A)$$

(17.2)

This model can be assumed to hold for classes, homerooms, entire schools, etc. Finally, we must define the functional relations between microvariables and macrovariables:

$$AE = g(ae_1, \ldots, ae_n)$$

$$PE = h(pe_1, \ldots, pe_n)$$

$$FO = l(fo_1, \ldots, fo_n)$$

$$A = m(a_1, \ldots, a_n)$$

(17.3)

These relations might express the macrovariables as the arithmetic means of the (scaled) microvariables for each homeroom, or on the other hand, we might define the macrovariables as distribution or dispersion measures of the corresponding microvariables, etc. For the present time, we do not specify anything about the form of the three kinds of relations except to note that "noncorresponding" microvariables do not enter into the function defining macrovariables. In Fig. 17.1, this is indicated by the absence of a dashed line drawn connecting any macrovariable with any noncorresponding microvariable.

Now that we have considered examples of the three kinds of relations, we can consider the issue of consistency. In any analysis of aggregation and disaggregation, we must concern ourselves with the interrelations of all three kinds of relationships. The analytic problem arises as a result of the functional dependence of the macrovariables on the microvariables. This functional dependence (expressed in the aggregation relations) produces more relationships than can be chosen independently as inspection of Fig. 17.1 would suggest.

Such a situation raises the possibility that the relations may be defined in such a way as to be inconsistent with each other. Although the parallels are not completely clear, this seems much like the situation of overidentification of models, which allows the possibility that the model be inconsistent with certain sets of data. Here the possibility that the three kinds of relations may be defined in ways such that, when the three are taken together, inconsistencies arise, motivates the study of the conditions under which they will be consistent. The economists have defined consistency very concretely. Green [17, p. 35], for example, states the following: "Consistency means that a knowledge of the "macro-relations" . . . and of the value of the aggregate independent variables would lead to the same value of the aggregate dependent variable as a knowledge of the micro-relations and the values of the individual independent variables." Perhaps reference to a specific simple model will help clarify the meaning of this definition. What does it require of the model drawn in Fig. 17.1? We are concerned with the equality of two different methods for generating predicted values of the macro dependent variable, X_3' in this case. The first is the straightforward regression (since I have defined the model as a linear recursive one) of observed values of X_3' on X_1' and X_2'. We will denote the resulting prediction $\hat{X}_3'$. The second method involves obtaining predicted values of X_{3i} for all microunits using the microregressions and then aggregating the X_{3i} according to the aggregation rule (relation) defining X_3'. We will denote this value $\hat{X}_3''$. By the definition presented above, aggregation will be consistent for this model if and only if $\hat{X}_3' = \hat{X}_3''$. If this equality does not hold, the difference $\hat{X}_3' - \hat{X}_3''$ is said to be due to the presence of *aggregation bias* in the macroparameters. We will be able to show below what such bias terms are likely to be. It is the presence of varying magnitudes of aggregation bias in parameter estimates that I am proposing as one potential complication in the testing and evaluating of causal models.

Returning to the general statement of the problem of consistency, we can define any two of the three relations and analyze the restrictions that the third must meet for aggregation to be consistent. Since we are dealing with aggregation, we expect the micromodel to be well defined. Two broad strategies thus present themselves.[4] First, we can define the micromodel and an analogous macromodel and attempt to find the class of aggregation relations that will produce consistent aggregation. This approach has been suggested by Klein [20] and Theil [32]. A second approach would be to define the micromodel and employ some widely used or theoretically appropriate aggregation relations and then search for consistent macromodels. This approach has been advocated by May [26]. This presentation will pursue the first course simply because the available statistical literature follows Theil's strategy.

To this point, I have framed the problem in terms of aggregation. Sociologists are more familiar with the problems of disaggregation. However, examination of the sociological analyses of the disaggregation problem shows that the problem of aggregation must be dealt with (at least implicitly) before conclusions about disaggregation can be drawn. In this chapter, I will restrict myself to the logically prior problem of aggregation. Elsewhere (Hannan [19]), I have summarized the existing literature on disaggregation and related the problem to what is being considered here.

AGGREGATION BIAS AS MEASUREMENT ERROR?

Before moving on to the more technical analysis of aggregation bias, we should examine the implications of the close similarity of the problem of aggregation and conventional treatments of measurement error. One of the central aspects of this formulation of the aggregation problem is the requirement that corresponding microvariables and macrovariables be functionally related. That is, we require that there be unambiguous rules that assign unique values to the macrovariables on the basis of the values taken by the corresponding microvariables. First of all, how much does this restrict the generality of this analysis? Consistency as defined here is meaningful only when the macrovariables are all mathematical transformations of corresponding microvariables. Thus, we can restate the above question to ask how often sociologists use macrovariables of this nature.

Lazarsfeld and Menzel [23] in a well-known discussion of the properties of measures at different levels of analysis label these kinds of macrovariables *analytical*. They are distinguished primarily from *global* macrovariables, which are defined without reference to the properties of the microunits that comprise the aggregate or collectivity. Analytical variables are typically means, proportions, standard deviations, etc. Examples of global

---

[4] This oversimplifies the variety of approaches adopted at one time or another. A useful classification of such approaches together with an up-to-date bibliography from the economics literature is found in Nataf [29].

variables usually characterize a collectivity on the basis of presence or absence of a characteristic like money as a medium of exchange or achievement motive in folk tales. However, Lazarsfeld and Menzel also characterize density measures as global variables.

It is my impression that the macrovariables employed in sociological research are more frequently analytical than global. In fact, one of the most pressing measurement problems facing the discipline is precisely a general inability to construct macro- or "structural" measures that mirror the meaning of macrotheories, (i.e., measures that are more than simple transformations of individual properties or behaviors). If this assessment is accurate, the empirical problems of changing units in sociological research when macrovariables are analytical does not involve rare or special cases but may represent the modal case.

Unfortunately, this analysis will not help clarify the fundamental measurement problem mentioned above. Still, the aggregation problem as developed to this point seems to have much in common with conventional treatments of measurement error. As I have already pointed out, problems of aggregation–disaggregation arise in sociology primarily as a consequence of absence of data at the level of interest. If we consider the observed macrovariables as imperfect measures of "structural" variables (e.g., measured per capita income as an imperfect measure of "true" per capita income), we can conceptualize the intersection of the issues of measurement error and aggregation bias in a rather straightforward way.

Consider the more complicated model drawn in Fig. 17.2. Here we introduce a set of theoretical macrovariables, $\overline{X}_1$, $\overline{X}_2$, and $\overline{X}_3$. We now consider the aggregated microvalues (what we are calling the macrovalues) to be measured values of the theoretical macrovariables. Following current prac-

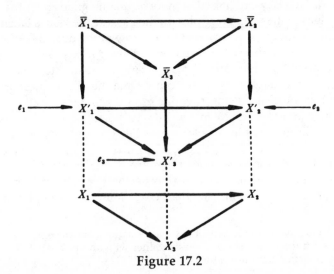

Figure 17.2

tice in causal analyses of measurement error, we assume that the measured values are produced by the joint action of the "true" values and a number of other influences (e.g., response set, coding error, censorship) lumped together as measurement error. Thus, in this model we have causal arrows drawn from macro theoretical variables to corresponding measured variables, as well as arrows from the symbols e_1, e_2, e_3 representing the measurement error terms. Note that for purposes of simplicity we are assuming that the micromodel is perfectly measured. We wouldn't, of course, expect this in any application. But the presence of measurement error in the micromodel is not relevant to this brief excursus.

We can use the model drawn in Fig. 17.2 to point out the main difference between the two concerns. In considering measurement issues, we are by definition dealing with the correspondence between observed and unmeasured values (in this example, $\overline{X}_1$ and $\overline{X}'_1$, etc.). In aggregation analyses, we must concern ourselves with measured quantities at two levels of aggregation. Further, we do not usually specify a causal relation between corresponding microvariables and macrovariables. Despite this fundamental difference, there remain numerous similarities between the two concerns. And Theil [32] has suggested that it might prove useful to consider measurement error and aggregation bias as two special cases of *specification error*.[5] The reader who wishes to pursue this formal similarity is urged to consult this provocative but rather technically difficult work.

AGGREGATION BIAS: THE BIVARIATE CASE

Sociologists have tended to confine their attention *vis-à-vis* problems of changing units to relationships between two variables. Thus, we will start with this case and later discover that the bivariate case is rather a special case in the sense that some of the most troublesome problems of aggregation do not arise. The attention of sociologists has been generally even further restricted to problems concerning changes in obtained coefficients of correlation. I will begin with the problems arising in grouping or aggregation affecting correlation coefficients and then consider the same problem with respect to regression coefficients.

It was noticed as early as 1939 (Thorndike [34]) that measures of association increased with the grouping of units into aggregates. Gehkle and Biehel [14] reported similar results. It was not until 1950 (Robinson [30]; Yule and Kendall [36]) that the problem received analytical treatments. Robinson presented an example that is useful as a point of reference. In the 1934

---

[5] The term *specification error* refers to any difference between the assumptions of a statistical model and the real properties it is assumed to represent. Below, in discussing Grunfeld and Griliches' [18] paper, we use the term in a more restrictive sense to refer to the omission of variables from regression functions that have some independent explanatory (in the statistical sense of prediction) power.

census, he found that the individual correlation between race and literacy was .20. However, the correlation computed on state marginals (percentage white, percentage literate, etc.) was .77, and the same statistic computed on the eight census district marginals was .94. As in this case, it often appears that when microobservations are grouped on areal basis, that measures of association increase as a monotonic function of the level of aggregation.

There have been a number of explanations offered for this phenomenon. It is useful to briefly consider each in turn, since they begin to shed some light on the general problem of aggregation.

Yule and Kendall [36, p. 312] proposed that the problem was one of *modifiable units:*

> This example serves to bring out an important distinction between two different types of data to which correlation analysis may be applied. The difficulty does not arise when we are considering the relationship, say, between heights of fathers and sons. The ultimate unit in this case . . . is a unique non-modifiable unit.
>
> On the other hand, our geographical areas . . . are modifiable units, and necessarily so. . . . A similar effect arises whenever we attempt to measure concomitant variation extending over continuous regions of space or time.

This distinction has *intuitive* appeal and seems to convey something of the difficulty involved in attempting to shift levels with empirical data. However, I am not convinced that this is a useful practical distinction. As Galtung [13, p. 45] correctly notes, individuals, which in most sociological analyses of the problem are seen as the unique nonmodifiable units, can be considered as boundaries for a variety of subunits, such as psychological syndromes, status behaviors, behaviors in a time sequence, etc. Even the most elementary familiarity with social psychology (not to mention behavioral and physiological psychology) suffices to make this point. It would seem that even single behavioral acts are modifiable according to the measurement procedure employed to record it. Yet, in defense of Yule and Kendall, I think their point is clear: Units that are grouped on some areal or temporal criterion are liable to be highly arbitrary aggregates. It would seem to make sense to begin to discuss degrees of modifiability of data sets. Granting this, nothing specific in this perspective helps us understand the technical problem.

Yule and Kendall did not content themselves with pointing to the modifiability problem but also suggested a more specific explanation. They pointed to the effect of the presence of random components in variables on correlation coefficients. This attenuation effect has become well known to sociologists in the context of the effects of random measurement error. They suggest that by the effect of "a sampling effect"—which, unfortunately, is not clearly discussed—the correlation coefficient increases with grouping. In a later chapter on sampling, they discuss "dependent sampling," sampling from a population in which values of the variables sampled are highly correlated. Consider the development they present. We have two variables, x and y, each of which is composed of a "true" component and a random component:

$$x = x' + e$$

$$y = y' + f$$

We assume a causal connection between x' and y', but assume that e and f are uncorrelated with x' and y' and with each other. In this case, we can easily show that:

$$\text{var } x = \text{var } x' + \text{var } e$$

$$\text{var } y = \text{var } y' + \text{var } f$$

$$\text{cov } (x, y) = \text{cov } (x', y')$$

We can further demonstrate that the correlation between x and y (the observed values) that we denote by r is related to the correlation of the "true" components, r', as follows:

$$r = \frac{r'}{\sqrt{1 + \dfrac{\text{var } e}{\text{var } x'}} \sqrt{1 + \dfrac{\text{var } f}{\text{var } y'}}} \tag{17.4}$$

Since variances are by definition nonnegative, we can see that r' cannot be less than r. To the extent to which r is less than r', we say that the correlation between the true values has been *attenuated* by the presence of random components.

The impact on these correlations of aggregation is easiest to show using the case in which the aggregates are simple sums.[6] We must ask what happens to the variances of the true components and the random components as units are grouped and sums taken (since it is ratios of such quantities that determine the attenuation effect). Consider a pair of random variables x_1 and x_2. An elementary result gives us:

---

[6] The extension to other linear combinations of microvalues is trivial. But, as we will see below, we cannot easily handle nonlinear aggregation relations.

$$\text{var } (x_1 + x_2) = \text{var } x_1 + \text{var } x_2 + 2 \text{ cov } (x_1, x_2)$$

The extension to the k-variable case is straightforward. This result tells us that the variance of a sum of random variables depends on the sign of the covariance term (which can take on either positive or negative values). Since we have assumed that e and f are to be distributed randomly in the population, we would expect the covariance terms for sums of e's and f's to be approximately zero, and, as a result, for each random component the variance of the sum of random variables will be approximately equal to the sum of the variance of the random variables entering the sum. This would also be true of the systematic variables x' and y' if their values were distributed randomly in space or time (i.e., if they were not systematically related to the aggregation or grouping criterion).[7] Yule and Kendall invoke the assumption (or, perhaps, empirical law) that adjacent units in any spatial or temporal distribution tend to be alike on most variables relative to the heterogeneity of any larger population (i.e., homogeneous units tend to cluster). As a result, areal or temporal aggregation will tend to group positively intercorrelated values[8] of x' and of y'. This is usually called the *clustering effect.*

In areal or temporal aggregation, then, the usual case will be that the variance of the sums (recalling the restriction to summation aggregation relations) will be greater than the sums of the variances. It is in this sense that "systematic effects represented by x' and y' will be cumulative, whereas the random effects represented by e and f will tend to cancel out — the larger the number of units we have, the less, relatively speaking, will their total be affected by erratic fluctuations" (Yule and Kendall [36, p. 314]). As Yule and Kendall point out, the denominator in (17.4) will be reduced in such a situation as we increase the size of the groupings, and thus if r' is a constant, r will continually increase toward its ceiling of 1.0.

The reader who consults the work cited above will find that a number of issues that underlie Yule and Kendall's presentation but that remained implicit were introduced explicitly in the above presentation. These modifications are suggested by the analyses that have been motivated by the analysis of covariance analog suggested by areal or temporal aggregation. If area or time period is taken as the nominal covariable, the importance of the clustering effect becomes obvious. Robinson [30] was the first to make this demonstration. I will briefly summarize the argument, since excellent treat-

---

[7] This statement relies on Blalock's formulation, which will be treated below. Earlier analyses like the one considered here tended to restrict attention to areal proximity grouping and did not make clear that it is the presence of a systematic connection between the values of the variables and the grouping procedure that produces this inflation effect.

[8] Here and elsewhere in this chapter we will use the term *intercorrelation of values of a variable* to mean one of two things. If we have data at a single point in time, this will refer to a high positive intraclass correlation. If we have observations at numerous points in time, this will refer to a serial correlation of the variables.

ments are available (particularly Duncan, Cuzzort, and Duncan, [11] and Alker [1]). The following discussion and the necessary formulas presented in Table 17.1 are adapted from Alker's comprehensive analysis.[9]

The theorems employed in the analysis of covariance all begin with a decomposition like the following:

$$(X_i - X_.) = (X_i - X_{.r}) + (X_{.r} - X_.) \qquad (17.5)$$

which simply states that the difference between a unit's value and the grand mean can be expressed as the sum of the difference between the unit's value and its subgroup (or regional) mean and the difference between the subgroup mean and the grand mean. By multiplying expression (17.5) by a comparable expression for the ith unit's Y value and taking expectations, we arrive at the "fundamental theorem" of the analysis of covariance:

$$C_{XY} = WC_{XY} + BC_{XY} \qquad (17.6)$$

This tells us that the covariance of X and Y for some N units can be represented as the sum of a within-region covariance and a between-region (or "ecological") covariance. As the definitional formulas of Table 17.1 make clear, the within-region covariance is a population weighted sum over all regions of the covariance of individual X and Y values within each region, and the ecological covariance is a population weighted average product of regional deviations in X and Y. We can prove that the two components on the righthand side of (17.6) are independent.

Our aim is to transform (17.6) into an expression in correlation coefficients. We can do this by successively dividing (17.6) by three standard deviation terms

$$(\sqrt{C_{XX}C_{YY}}, \quad \sqrt{WC_{XX}WC_{YY}}, \quad \sqrt{BC_{XX}BC_{YY}})$$

to obtain correlational expressions as follows:

$$R_{XY} = \frac{WC_{XY} + BC_{XY}}{\sqrt{C_{XX}C_{YY}}} = \frac{WC_{XY}}{\sqrt{WC_{XX}WC_{YY}}} \cdot \sqrt{\frac{WC_{XX}WC_{YY}}{C_{XX}C_{YY}}}$$
$$+ \frac{BC_{XY}}{\sqrt{BC_{XX}BC_{YY}}} \cdot \sqrt{\frac{BC_{XX}BC_{YY}}{C_{XX}C_{YY}}}$$

[9] I am restricting this analysis to areal aggregation. Alker concisely illustrates the formal similarity between temporal and areal aggregation by adding a subscript denoting time period of observation to each entry in Table 17.1 and then aggregating both across time periods and across units.

Table 17.1. Statistical elements of covariance theorems

A1. X_{ir}, Y_{ir}		X (or Y) value for unit i in region r
A2. $X_{\cdot r}, Y_{\cdot r}$	$\dfrac{1}{N_r} \displaystyle\sum_{i=1}^{N_r} X_{ir}; \dfrac{1}{N_r} \displaystyle\sum_{i=1}^{N_r} Y_{ir}$ Where N_r is the number of units in region r.	Average X (or Y) value for units i in region r
A3. $X_{\cdot\cdot}, Y_{\cdot\cdot}$	$\dfrac{1}{N} \displaystyle\sum_{i=1}^{N} \sum_{r=1}^{R} X_{ir},$ etc.	Average X (or Y) value for all units in all regions
B1. C_{XX}	$\dfrac{1}{N} \displaystyle\sum_{i=1}^{N} (X_{ir} - X_{\cdot\cdot})^2$ (similar term defined for Y)	Universal variance of X
B2. C_{XY}	$\dfrac{1}{N} \displaystyle\sum_{i=1}^{N} (X_{ir} - X_{\cdot\cdot})(Y_{ir} - Y_{\cdot\cdot})$	Universal covariance of X and Y
B3. WC_{XX}	$\dfrac{1}{N} \displaystyle\sum_{i=1}^{N} (X_{ir} - X_{\cdot r})^2$ (similar term defined for Y)	Within-group variance of X
B4. WC_{XY}	$\dfrac{1}{N} \displaystyle\sum_{i=1}^{N} (X_{ir} - X_{\cdot r})(Y_{ir} - Y_{\cdot r})$	Within-group covariance of X and Y
B5. BC_{XX}	$\dfrac{1}{R} \displaystyle\sum_{r=1}^{R} (X_{\cdot r} - X_{\cdot\cdot})^2$ (similar term defined for Y)	Between-group ("ecological") variance of X
B6. BC_{XY}	$\dfrac{1}{R} \displaystyle\sum_{r=1}^{R} (X_{\cdot r} - X_{\cdot\cdot})(Y_{\cdot r} - Y_{\cdot\cdot})$	Between-group covariance of X and Y
C1. E_{XR}^2	$\dfrac{BC_{XX}}{C_{XX}}$	Correlation ratio of X and R
C2. E_{YR}^2	$\dfrac{BC_{YY}}{C_{YY}}$	Correlation ratio of Y and R
D1. R_{XY}	$\dfrac{C_{XY}}{\sqrt{C_{XX}C_{YY}}}$	Universal correlation between X and Y
D2. WR_{XY}	$\dfrac{WC_{XY}}{\sqrt{WC_{XX}WC_{YY}}}$	Within-group correlation of X and Y
D3. BR_{XY}	$\dfrac{BC_{XY}}{\sqrt{BC_{XX}BC_{YY}}}$	Between-group ("ecological") correlation of X and Y

Noting that $WC_{XX} = C_{XX} - BC_{XX}$, and $WC_{YY} = C_{YY} - BC_{YY}$, and using the definitions of the correlation ratios, we arrive at the following result:

$$R_{XY} = WR_{XY}\sqrt{1 - E_{YR}^2}\,\sqrt{1 - E_{XR}^2} + BR_{XY}E_{YR}E_{XR} \qquad (17.7)$$

Robinson [30] employed a slightly different version of (17.7):

$$BR_{XY} = R_{XY}/E_{XR}E_{YR} - WR_{XY}\sqrt{1 - E_{XR}^2}\,\sqrt{1 - E_{YR}^2}/E_{XR}E_{YR} \qquad (17.8)$$

From this, Robinson deduced what I will call the *consistency condition for correlation coefficients:*

$$WR_{XY} = R_{XY}(1 - E_{XY}E_{YR})/\sqrt{1 - E_{XR}^2}\,\sqrt{1 - E_{YR}^2} \text{ or } WR_{XY} = kR_{XY} \qquad (17.9)$$

The individual or micro and the macro or "ecological" correlations are equal only if this identity holds. The minimum value of the quantity represented by k in the abbreviated expression is unity. This is the case in which the within-area correlations are all exactly equal to the individual (total) correlation. But Robinson argues that an examination of expression (17.9) shows that the inflation effect depends on the number of subareas employed and on the level (inclusiveness) of aggregation.

As smaller areas are consolidated, two things happen. First, the within-areas correlation increases as a result of increasing heterogeneity of subareas. This effect tends to diminish the ecological correlation, since the proportion of variance accounted for by an area is equal to $1 - WR_{XY}^2$. Second, the values of the correlation ratios, E_{XR} and E_{YR}, decrease as a consequence of the decrease in the heterogeneity of the values of X and Y in the subareas. This tends to increase the ecological correlation. But Robinson [30, pp. 356–347] argues:

> These two tendencies are of unequal importance. Investigation of [our (17.8)] with respect to the changes in the values of E_{XR}, E_{YR}, and WR_{XY} indicates that the influence of the changes of the E's is considerably more important than the influence of changes in the value of WR_{XY}. The net effect of changes in the E's and WR_{XY} taken together is to increase the numerical value of the ecological correlation as consolidation takes place.

What is crucial is that grouping units somehow changes variation in the variables of interest. The narrow focus on areal aggregation (the "ecological fallacy") deflected attention from the more general problem of aggregation. Exactly how grouping or aggregation affects variation in such variables became much clearer when Blalock [5] raised the issue in the context of linear causal analysis.

Before shifting to an explicitly causal treatment, we can, following Grunfeld and Griliches [18], more precisely specify the dependence of the inflation of correlation coefficients on clustering. Grunfeld and Griliches [18, p. 4] prefer to speak of a *synchronization effect,* defined as follows: "The higher the correlation between the independent variables of different individuals or behavior units, *ceribus paribus,* the higher the R^2 of the aggregate equation relative to the R^2's of the micro-equation." This term may be preferable to clustering effect, since the latter term suggests a concern only with areal aggregation.

Grunfeld and Griliches did not introduce the grouping criterion explicitly into their analysis but rather employed a simple regression model: $y_i(t) = bx_i(t) + u_i(t)$. Both variables are assumed to be measured as deviations about their means, and the t in parentheses denotes the time period of observation. The aim is to demonstrate that the inflation effect depends on the intercorrelations of the observed values of the explanatory variables and of the disturbance term. Note that we have in this model a time series of observations for all microunits. We can thus compute product-moment correlations of arrays of microvalues for pairs of microunits.[10] Grunfeld and Griliches assume that such correlations for values of the explanatory variable and disturbance term, ρ_x and ρ_u, respectively, are constant for all pairs of microunits. If we make two more simplifying assumptions: that all microunits can be represented by a micromodel with constant coefficient (i.e., $b_i = b_j$ for all i, j),[11] and that the variances S_{xi}^2 and S_{ui}^2 are constants, we can arrive at the following result:[12]

$$\frac{R^2\text{macro}}{R^2\text{micro}} = \frac{b^2 S_{xi}^2 + S_{ui}^2}{b^2 S_{xi}^2 + S_{ui}^2 \left[\dfrac{1 + (N-1)\rho_u}{1 + (N-1)\rho_x}\right]} \tag{17.10}$$

where N is the number of microunits.

It is clear from (17.10) that the inflation of correlation coefficients with aggregation depends on the relative magnitude of the two synchronization coefficients, ρ_x and ρ_u. All that is required for such inflation is that ρ_x be larger in magnitude than ρ_u. An important contribution of this analysis is the demonstration that the inflation effect is dependent on the completeness of the micromodel. In a perfectly specified micromodel,[13] we would expect the disturbance term to be composed of a large number of "random shocks." As a consequence, we would expect ρ_u to be approximately equal to zero. In

[10] Grunfeld and Griliches [18, Appendix E] develop the parallel result for the case where we have a set of microvalues at a single point in time. In this case, the appropriate measure of synchronization is the intraclass correlation coefficient.

[11] In "Aggregation Bias," p. 426, we will develop the significance of this restriction. Simply, if this condition does not hold, we would observe a peculiar type of "time series aggregation bias."

[12] Grunfeld and Griliches systematically relax these assumptions in a series of appendices that demonstrate that this general result is not contradicted.

[13] Here we are using the term *specification error* in the narrow sense: failure to include one or more causally important variables in the micromodel.

such a case, even a very slight positive synchronization of values of the explanatory variable would produce inflation of the macrocoefficient of determination. The point is that the size of the coefficient ρ_u depends on the size of the specification error in the general case. This will become clearer in the analysis motivated by causal modeling considerations.

By now it has become clear that we must concern ourselves with the effects of aggregation on the variation in the variables of interest. Blalock [5] has made this issue explicit and has made it the focus of his concern with the problem of changing units. He argues that [5, p. 98] "in shifting from one unit of analysis to another we are very likely to affect the manner in which outside and potentially disturbing influences are operating on the dependent and independent variables under consideration." The way in which this is likely to occur is that variation in X or Y or both may be manipulated in such a way that their effects become confounded with the effects of other variables. The key insight is that the effect of shifting levels on the behavior of correlation and regression coefficients depends on the manner in which units have been put together. If we explicitly consider the aggregation or grouping criterion as a variable in the system, we can see that the effect of a given kind of grouping on the variation of other variables in the system depends on the systematic connections between the grouping criterion and each other variable in the system. We have already noted, for example, that grouping by area of residence will maximize variation in most social variables (i.e., will increase between-group variation relative to within-group variation). As Blalock points out, however, we seldom have precise information on the connections between the aggregation criterion and the variables in the model.

In order to gain some insight into the likely effects of several grouping procedures on bivariate relations, Blalock examined four types of groupings: random grouping, grouping that maximizes variation in X, grouping that maximizes variation in Y, and grouping by physical propinquity. The first three are artificial groupings, the fourth corresponds to the case we have been concerned with so far. Figures 17.3, 17.4, and 17.5 illustrate the situation modeled in each of the first three grouping procedures. Figure 17.3 represents random grouping—the situation in which the grouping criterion is not systematically connected with any variable in the model. Blalock's manipulation of county level data showed that both correlation and regression coefficients remained fairly stable across a series of more and more

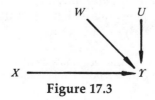

Figure 17.3

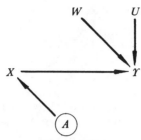

Figure 17.4

inclusive groupings. Those fluctuations that did arise are attributed by Blalock to sampling error.[14]

The second procedure, grouping so as to maximize variation in X (Fig.
17.4) is essentially random with respect to those independent causal factors
that are not systematically related to X (W and U in our example). As a result,
this type of grouping will increase the between-group variation in X relative
to that in these other variables. This will lessen the "nuisance" effect of such
variables, meaning that X will "explain" a larger proportion of the variation
in Y. This effect should increase with the level of grouping (as more consolidation takes place, the systematic effects of A will increase the between-
group variation relative to the within-group variation in X). This argument
could be rephrased in terms of the clustering effect, since the effect of A is to
group positively correlated[15] values of X. The data support this argument; r_{xy}
increases consistently (monotonically) with the level of grouping. However,
we would not expect that the reduction of nuisance effects would change
the nature of the relationship between X and Y. In this case, the slope b_{yx}
remained remarkably constant across levels of aggregation. But the reverse
slope (which is not of direct interest since we are taking Y as dependent in
the model) increased with grouping. Blalock notes that this is a mathemati-
cal artifact, since $r_{xy}^2 = b_{yx}b_{xy}$ and r_{xy}^2 increases as b_{yx} remains constant.

The third procedure, grouping that maximizes variation in Y (Fig. 17.5),
produces symmetrical results. We find that r_{xy}^2 increases with grouping (this
is expected since the correlation coefficient is symmetric with respect to X
and Y). In this case, it is the slope b_{xy} that remains nearly constant, so that b_{yx}
(the slope of interest) increases with level of grouping. This is a serious
complication and can be explained causally by noting that maximizing
variation in Y confounds variation in Y due to X, with that due to other
factors such as W and U. If, for example, all three causal factors are related
positively to Y, then grouping units with high Y values will very likely group
units which have high X, W, and U values. As Blalock notes [5, p. 108],

---

[14] But see the discussion of Cramer's analysis below.
[15] Since this analysis is cross-sectional, this comment refers to increasing intraclass correlations.

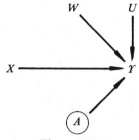

Figure 17.5

"Almost any variable that is even slightly related to Y initially becomes a good predictor of Y under such a grouping procedure because of the fact that its effects are being confounded with those of the other variables related to Y."

The fourth procedure is closer to what concerns us in this chapter (i.e., it is empirically a more likely case). Blalock's proximity grouping appeared to be somewhat intermediate between procedures one and two. However, in any instance, it is an empirical question whether or not proximity (including temporal proximity) grouping will have more effect on X or Y. I have implied above that most of the variables of interest to social scientists are likely to be found clustered to some degree. If this is the case, a more realistic model is probably more like that presented in Fig. 17.6, which is a composite of the models of Figs. 17.4 and 17.5 with additional arrows drawn from A to the other causal factors. The issue then becomes which of these relations are the strongest. In Blalock's proximity grouping, it is apparently the case that the relation of A to X is stronger than the relations from A to W, U and Y.

The model drawn in Figure 17.6 raises an extremely interesting issue: namely, the possibility of the grouping criterion producing a spurious relationship between X and Y. Consider the models drawn in Fig. 17.7. Here we would expect to find no relationship between X and Y at the microlevel; but at the macrolevel, such a relationship could appear as a result of the grouping process. This somewhat obvious possibility has not been discussed in the aggregation-disaggregation literature. Alker [1] has raised this issue and suggests that the literature on ecological correlations would have developed

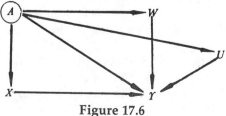

Figure 17.6

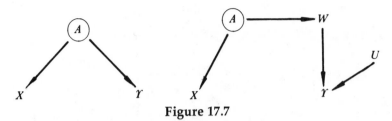

Figure 17.7

quite differently if those who addressed the problem had considered the issue of spurious relations. Alker illustrates this possibility using Robinson's race–literacy example. Here the individual level correlation was .20 and the correlation computed on the eight census district marginals was .95. Without exception, the previous analyses of the ecological fallacy problem took the observed individual correlation as a "pure" datum. The individual level correlation was implicitly considered a substantively interesting and reliable benchmark against which to compare aggregation–disaggregation effects.

Alker suggests a perspective from which this observed correlation of .20 can be considered spurious. Due to the presence of the clustering or synchronization effect, we expect that a portion of the variance in both X (being nonwhite) and Y (being illiterate) will be ecologically determined. Suppose we had reason to suspect that approximately a quarter of the variation in each variable was ecologically determined (i.e., $E_{XR}^2 = E_{YR}^2 = .25$). The correlational version of the covariance theorem for this example is as follows:

$$.20 = WR_{XY} \sqrt{1 - E_{XR}^2} \sqrt{1 - E_{YR}^2} + .95 \, E_{XR} E_{YR}$$

Substituting the values assumed for the correlation ratios, we obtain

$$.20 = .75 \, WR_{XY} + .24$$

from which we deduce $WR_{XY} = -.05$. Thus, in a sense, the grouping criterion (which Alker suggests was operating as level of industrialization and/or urbanization) "partials out" the correlation between X and Y. If we had three interval variables, this situation would be exactly analogous to the basic three-variable prediction equation of the Simon–Blalock strategy, $R_{XY \cdot R} \cong 0$.

This presentation is meant to be heuristic. We cannot, of course, differentiate the three-variable model drawn in Fig. 17.7A from a number of alternative three-variable models. Further, situations in which the grouping criterion is operating as a confounding variable in this sense can be identified only in the context of a theoretical formulation that relates the micro-variables to the grouping variable (e.g., level of industrialization). Alker argues convincingly that this analysis problem points to the need for cross-level theories.

Before moving to that discussion of aggregation of k-variable models, I would like to briefly mention an important result bearing on the previous discussion that to this point has not been mentioned in the sociological literature. Cramer [8][16] has addressed himself to the effects of various kinds of grouping procedures on bivariate correlation and regression coefficients. With respect to the former, his analysis provides an alternative procedure for demonstrating that the inflation of correlation coefficients depends on the degree of consolidation. If we hold the number of microunits constant, we can group units into successively larger and more inclusive aggregates. I am (following Robinson) referring to this process as consolidation. Blalock, for instance, grouped units in pairs, fives, tens, and, finally, fifteens. When a variation in a variable is being maximized in grouping, one places those units that are most alike on X values in the same group. As consolidation increases, both between-group and within-group variation increases. Be-tween-group variation increases because the means diverge further and further apart. Within-group variation increases because fives will be less homogeneous than pairs, etc. This is essentially the analysis problem Robin-son discussed in the framework of the analysis of covariance. The demon-stration of the effect of clustering (which is here produced by the grouping criterion) suggests that the increase in between-group variation is more important than the increase in within-group variation. Cramer offers an alternative method of demonstrating that the inflation effect is a monotonic increasing function of the level of consolidation for any fixed number of microunits. Similarly, the inflation effect also increases monotonically with the number of microunits employed in the grouping.

It is the discussion of the effects of grouping on the regression estimates that bears on an issue not previously addressed. Restricting himself to Blalock's types 1 and 2 grouping procedures, Cramer proves that for a simple bivariate model,[17] the regression computed on the aggregates gives an unbiased estimate of the population slope parameter.[18] However, the aggregate slope is less efficient than the slope computed on the microdata.

---

[16] I would like to thank Arthur S. Goldberger for recommending this paper, which for some reason seems to have been overlooked in the economics literature as well.

[17] The development of the argument is somewhat laborious and will not be attempted here. The reader is referred to the original paper. A less mathematically demanding version is presented in Hannan [19, Chapter 2].

[18] He does not mention the possible complications introduced by inadvertent manipulation of the dependent variable in aggregation.

For the simple bivariate case, the loss of efficiency can be expressed as a ratio of the between-group sum of squares of X to the total sums in X.

For random grouping, Cramer shows that the reduction of efficiency is an increasing function of the level of consolidation. Blalock's data conform to this observation. The case of grouping that maximizes variation in X is more problematic, as discussed above. Cramer's formulation demonstrates that as the between-group sums of squares in X relative to the within-group sums of squares increases, the efficiency of the estimator increases. Blalock's data show no variation in the slope b_{yx} with this type of grouping across several levels of consolidation. What we would need for comparison would be a grouping procedure that did not perfectly maximize variation in X as Blalock did. He suggests that his proximity grouping approximated this situation (i.e., combined the features of random grouping and grouping that maximizes X). We see that b_{yx} increasingly deviated from the true values with increasing consolidation. This, then, is an additional argument for the need of precisely specifying the relationship between the grouping criterion and the variables in the model. The ideal case seems to be that in which A (the grouping criterion) is not related to Y or to any other causes of Y but is related to X in such a way that grouping results in great increases in the between-group variation in X.

AGGREGATION BIAS: THE K-VARIABLE CASE

The approach presented in the last section can be extended to apply to more complex models. The potentially most troublesome models are those in which the grouping criterion is systematically related to a number of variables in the model. In such cases, we are likely to encounter some combination of maximizing variation in dependent variables, controlling for prior variables in causal chains, and spurious macrorelations. General statements are difficult here as verbal statements very quickly become inadequate as the potential difficulties become more numerous and complex. However, a special aggregation complication arises when we have time-series observations on K-variable models that have been extensively developed by a number of econometricians. I would like to develop the argument briefly here. After considering this special case of aggregation bias, I will attempt to relate it to the causal concerns already expressed.

Before considering the time-series problem, I should note that there is a rather extensive literature in economics dealing with the consistency conditions for *deterministic* K-variable micromodels (where as above we allow each microunit to be represented in a separate model). A series of theorems[19] proves that consistency as we have defined it is attainable only

[19] Green [17] presents a more precise statement along with necessary theorems and associated lemmas. In general, they involve a mathematical restatement (in terms of partial derivatives and differentials) of the consistency formulation presented above.

when all relations (micro-, macro-, and aggregation relations) are *linear*.[20] The implications of this result are quite extraordinary given that we are discussing exact relations. For the sociologist, one practical conclusion is that consistency is ruled out by the use of distributional measures as macrovariables. This is troublesome when homogeneity – heterogeneity is a relevant dimension at both microlevels and macrolevels.[21] This conclusion, of course, holds *a fortiori* for stochastic aggregation problems.

Yet special problems arise in the case for stochastic K-variable models even when all relations are linear. To appreciate one of the more serious and unusual aggregation bias problems, we must shift our attention to relations that are estimated from a time-series of observations (i.e., situations in which we have measures for all microunits on the variables of concern at several points in time).[22] The use of such time-series, so basic to economic analysis, is still relatively rare in sociology. There seems to be an increasing emphasis on undertaking such studies, however. Examples of such studies are the cohort studies of high school or college classes where all individuals in the study are followed up and remeasured at regular time intervals. More common, perhaps, are the "synthetic cohort" studies in which retrospective measures of previous states are employed in the time-series (cf. Blau and Duncan, 1967).

The discussion must now become relatively concrete. We follow Theil's [32] formulation.[23] We are requiring that all relations be linear. Consider the following *micromodel*:

$$y_i(t) = \alpha_i + \sum_{k=1}^{K} \beta_{ki} x_{ki}(t) + u_i(t)$$

$$(i = 1, \ldots, N)$$

(17.11)

where each microunit is permitted to behave idiosyncratically with respect to changes in the K micro explanatory variables, the t in parentheses denotes the time period of observation, and the disturbance is assumed to have zero mean. We will restrict our attention to the simplest case of aggregation in which all macrovariables are the sums of corresponding microvariables (the

[20] Nonlinear functions that are easily transformed into linear functions (e.g., logarithmic functions) present no particular difficulties, however.

[21] Boudon [7] develops the implications for disaggregation of formulations in which this dimension is introduced. He argues that nonlinear formulations are generally appropriate in attempts at disaggregating data that have been grouped on some areal proximity basis.

[22] We are concerned with a cross-sectional model estimated at numerous points in time rather than with dynamic formulations in which time plays a central role.

[23] Excellent summary statements of Theil's analysis prepared for economists are found in Allen [3, Chapter 20] and Fox [12, Chapter 14].

extension to arithmetic means and proportions is straightforward):

$$y(t) = \sum_{i=1}^{N} y_i(t)$$

$$x_1(t) = \sum_{i=1}^{N} x_{1i}(t) \tag{17.12}$$

$$\cdot$$
$$\cdot$$
$$\cdot$$

$$x_K(t) = \sum_{i=1}^{N} x_{Ki}(t)$$

We assume that a *macromodel* is defined analogously to (17.11):

$$y(t) = \alpha + \sum_{k=1}^{K} \beta_k x_k(t) + u(t) \tag{17.13}$$

The analytic problem arises when we use least squares to estimate the macromodel directly[24] and compare the obtained coefficients with those obtained from estimating the micromodel. The notion of simple consistency would suggest that we would want the macrocoefficients to be simple sums of the corresponding microcoefficients. We can use the consistency formulation to see this. Consistency is achieved when the two method of generating predicted macro dependent variable values discussed above are identical. We can express changes in the macromodel as follows:

$$\Delta y(t) = \alpha + \sum_{k=1}^{K} \beta_k \Delta x_k + u(t) \tag{17.14}$$

For each equation in the micromodel, change is defined as follows:

$$\Delta y_i(t) = \alpha_i + \sum_{k=1}^{K} \beta_{ki} \Delta x_{ki} + u_i(t) \tag{17.15}$$

Consistency requires that:

[24] Here we see the necessity of time-series observations. Since by definition only a single macrovalue corresponds to the array of microvalues of each corresponding variable at each time period, there would be no variation in the macromodel in a single cross-section.

$$\Delta y(t) = \sum_{i=1}^{N} \Delta y_i(t) = \sum_{i=1}^{N} \alpha_i + \sum_{i=1}^{N} \sum_{k=1}^{K} \beta_{ki} \Delta x_{ki} + \sum_{i=1}^{N} u_i(t) \qquad (17.16)$$

In other words, when the aggregates are simple sums, aggregation is consistent if and only if:

$$\alpha = \sum_{i=1}^{N} \alpha_i \qquad \beta_k = \sum_{i=1}^{N} \beta_{ki} \qquad u(t) = \sum_{i=1}^{N} u_i(t) \qquad (17.17)$$

There is nothing obscure in these definitions of the macroparameters. This is simply what intuition would suggest would be the appropriate set of definitions. We can see that the substitution of the values defined in (17.17) into the original macromodel (17.11) will yield (17.16) as required.

Theil's analysis is intended to demonstrate that we are not usually so fortunate as to obtain the required macroparameters. To pursue this and the resulting contradiction between micromodels and macromodels we need to define a new set of functions. We define a set of *auxiliary regressions* that express each microvalue (of explanatory variables) as a linear function of the whole set of macro explanatory variables. More precisely, they are the least squares estimates of the "time paths" of changes in the microvariables as functions of the macrovariables:

$$x_{ki}(t) = A_{ki} + B_{ki,1} x_1(t) + \dots + B_{ki,K} x_K(t) + V_{ki}(t) \qquad (17.18)$$

where $V_{ki}(t)$ are residuals that have zero mean and are uncorrelated with the T values taken by the macrovariables $x_{k'}$ ($k' = 1, \dots, K$). We need not assign any substantive meaning to these auxiliary regressions since they are simply a formal device by which we demonstrate the existence of aggregation bias.

Theil's Theorem 1 demonstrates that the parameters of (17.11) the macromodel are determined as follows (when the aggregates are simple sums):

$$\alpha = \sum_{i=1}^{N} \alpha_i + \sum_{k=1}^{K} \sum_{i=1}^{N} A_{ki} \beta_{ki}$$

$$\beta_k = \sum_{k'=1}^{K} \sum_{i=1}^{N} B_{k'i,k} \beta_{k'i} (k = 1, \dots, K) \qquad (17.19)$$

$$u(t) = \sum_{i=1}^{N} u_i(t) + \sum_{i=1}^{N} \sum_{k=1}^{K} \beta_{ki} V_{ki}(t)$$

Thus the macroparameters have some rather unusual properties:

1. The macrointercept is the sum of the corresponding microintercepts plus a series of terms involving noncorresponding parameters.
2. The macro regression coefficients that can be rewritten

$$\frac{1}{N}\sum_{i=1}^{N}\beta_{ki} + \sum_{i=1}^{N}(B_{ki,k} - \frac{1}{N}\beta_{ki} + \sum_{i=1}^{N}\sum_{k'\neq k}B_{k'i,k}\beta_{ki})$$

are composed of the sum of the arithmetic mean of the corresponding microparameters, plus a weighted mean of the corresponding microparameters plus a sum of weighted arithmetic means of noncorresponding coefficients.
3. The macrodisturbance is also composed of a sum of the sum of corresponding disturbances plus a weighted sum of noncorresponding parameters.

It is the presence of the terms involving noncorresponding parameters (Allen calls these "cross-effects") that is disturbing. Theil labels all terms other than

$$\sum_{i=1}^{N}\alpha_1, \frac{1}{N}\sum_{i=1}^{N}\beta_{ki} \text{ (or } \bar{\beta}_{ki})$$

and

$$\sum_{i=1}^{N}u_i(t)$$

as *aggregation bias terms*.

We can further examine these aggregation bias terms by employing the second half of Theil's Theorem 1, which states that the following restrictions on coefficients of the auxiliary regressions must hold:

$$\sum_{i=1}^{N}A_{ki} = \sum_{i=1}^{N}V_{ki}(t) = 0; \qquad \sum_{i=1}^{N}B_{k'i,k} = \begin{array}{l}1 \text{ if } k' = k \\ 0 \text{ otherwise}\end{array} \qquad (17.20)$$

We want this requirement to be met since we have defined

$$x_k = \sum_{i=1}^{N}x_{ki}$$

and the summation of expression (17.18) will not give this result unless the restrictions are met. Using these restrictions, we can rewrite the expressions of (17.19) so as to represent the aggregation bias terms as covariances between microparameters and auxiliary regression coefficients:[25]

$$\alpha = \sum_{i=1}^{N} \alpha_i + N \sum_{k=1}^{K} \text{cov} (A_{ki}, \beta_{ki})$$

$$\beta_k = \bar{\beta}_{ki} + N \sum_{k'=1}^{K} \text{cov} (B_{k'i,k}, \beta_{ki}) \qquad (17.21)$$

$$u(t) = \sum_{i=1}^{N} u_i(t) + N \text{cov} \sum_{k=1}^{K} (V_{ki}(t), \beta_{ki})$$

Thus, we see that the three macroterms (the two coefficients plus the disturbance) are equal to linear combinations of corresponding microterms apart from certain covariance corrections. *These covariance corrections represent the aggregation bias in the macromodel.* In other words, it is the presence of these covariance corrections that will give rise to inconsistencies or contradictions between micromodels and macromodels. The reader may convince himself of this by pursuing our usual two-procedure method of generating predicted values for the macro dependent variable employing the coefficient values presented above for the macroparameters.

There are a number of special cases in which such aggregation bias vanishes. I will mention only one.[26] If we can write the micromodel with constant coefficients, all aggregation bias will disappear, since the covariance of a constant with any random variable is by definition zero. In other words, if all microunits behave exactly alike with respect to the variables in the model over the time period defined by the observations, one need not concern himself with aggregation bias. Theil proposed that it is highly unlikely that this condition would be realized in economic studies in which consumers or households are microunits. Economists expect wealthy families to behave differently economically from poor families. Sociologists would expect the same result and would stipulate further that many distinctions, rural–urban, black–white, conservative–liberal, etc., would be likely to produce similar kinds of differences. Malinvaud [25] suggests that if there is only one dimension (e.g., wealth) which is suspected to be responsi-

[25] I will not derive these results here as the intervening steps are fully spelled out by Theil [32, pp. 15–17].

[26] The other cases are those in which the microvariables are all exact functions of the macrovariables (i.e., the auxiliary regression is deterministic rather than stochastic) or when we have additional knowledge that allows us to posit that the covariances are zero for the time span of concern.

ble for large differences in microparameters, one should stratify his sample on this dimension and run separate analyses. Within each subsample, we might then reasonably expect the microparameters to be nearly constant. Sociologists are used to addressing the same problem in a very different context. The problem of variable microparameters can be seen as a classical statistical interaction problem: Other variables are changing the nature of the relationship across subsamples.

It is interesting to note that sociologists concerned with disaggregation have arrived at the identical conclusion about the aggregation problem. Goodman [16], for example, arrives at a formulation for the two-variable cases in which disaggregation is consistent if the grouping criterion does not interact with the relationship of interest. That is, if the microparameters are reasonably constant across ecological areas, then aggregation is consistent and Goodman's disaggregation strategy will give reasonably good estimates of the microvalues. Boudon (1964), however, has argued vigorously that it is highly unlikely that such interactions will be weak or missing in most interesting cases.

What about the case in which the microparameters vary? Is aggregation bias an inevitable complication? Theil argues that it is not and proposes a strategy of "perfect aggregation." He advocates a peculiar sort of *fixed-weight aggregation,* which produces constant microparameters. Consider the following weighted aggregates (in place of simple sums):

$$y'(t) = \sum_{i=1}^{N} s_i y_i(t) \tag{17.22}$$

$$x'_k(t) = \sum_{i=1}^{N} w_{ki} x_{ki}(t) \tag{17.23}$$

where s_i and w_{ki} are fixed-weights (constant across time periods). We can write the micromodel in the form:

$$\begin{aligned} s_i y_i(t) &= s_i \alpha_i + \sum_{k=1}^{K} \frac{s_i \beta_{ki}}{w_{ki}} x_{ki}(t) + s_i u_i(t) \\ &= \alpha'_i + \sum_{k=1}^{K} \beta'_{ki}[w_{ki} x_{ki}(t)] + u'_i(t) \end{aligned} \tag{17.24}$$

and see that it is equivalent to our earlier unweighted formulation. To see how fixed-weight aggregation can produce constant coefficients in the micromodel, we set the weights s_i equal to unity and assign the following values to w_{ki}:

$$w_{ki} = \frac{\beta_{ki}}{c_{ki}} \text{ for all pairs } (k, i)$$

where the c's are arbitrary constants. In other words, we weight microvariables by the corresponding microcoefficients. In this case, (17.24) is reduced to:

$$y_i(t) = \alpha_i + \sum_{k=1}^{K} c_k[w_{ki}x_{ki}(t)] + u_i(t) \qquad (17.25)$$

so that the microcoefficients β'_{ki} of (17.24) are all identically equal to c_k. "This implies that if aggregation is performed such that all microvalues of $x_{ki}(t)$ are weighted proportionately to their microparameters β_{ki}, both the intercept α and the rates of change β_k of the macro equation depend on microparameters corresponding only (Theil, [32, p. 18]).

This strategy, while mathematically sound, presents some obvious difficulties from the theory tester's perspective. Lancaster [22] has presented perhaps the most compelling criticism of this approach in commenting on the issues in the context of the Keynesian Savings Function [22, p. 18]:

> The definition of Y (the macro explanatory variable) is not a "natural" definition; it does not conform to any standard statistical series; it is a solution only to the problem in hand and has no special use for any other aggregate income problem; it depends on the exact distribution of income at each period; but it makes the micro- and macro-relationships always consistent.

In other words, the macrocoefficients are dependent on the exact distributions of the microvariables in the population under study for the time period specified. For this reason, the macrocoefficients are sensitive to changes in the distributions of the microvariables across time periods. Similarly, this characteristic of "perfect aggregation" makes it extremely difficult, if not impossible, to compare macroresults obtained from different populations or in the same population at different time periods.

The theorist operates in terms of general and abstract relationships that are neither time nor place specific. Of course, all operationalizations are to some extent time and place specific. Yet, this strategy seems more time and population bound than most procedures. In any particular application, "perfect aggregation" is likely to seem highly *ad hoc* and to have little intuitive or substantive appeal. None of these criticisms apply, however, to pure prediction problems. They are specific to attempts at theory testing or construction.

In an extremely interesting analysis, Boot and deWit [6] estimated the aggregation bias in the coefficients of an investment model. They found

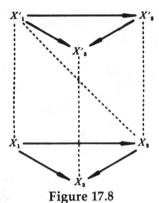

Figure 17.8

such bias to be relatively small. Fox [12], perhaps on the basis of this paper, suggests that in most economic applications aggregation bias should not exceed 10% of the parameter values.[27] This is an encouraging result, since it suggests that until we operate in terms of highly precise models, aggregation bias will represent only a minor inconvenience.

It is interesting to recast this argument in causal terms, following Blalock's example. Examination of (17.21) shows that the presence of aggregation bias in the macroparameter estimates depends on non-zero coefficients in the auxiliary regressions. More specifically, it depends on non-zero coefficients associated with *noncorresponding* macrovariables. To this point, we have followed the example of the economists and have treated the auxiliary regressions as purely formal. But it appears that insight into the aggregation problem arising in time-series can be gained by intensive examination of these relations. The question is, under what conditions are we likely to find microvariables associated across time with noncorresponding macrovariables?

Theil suggests that this is likely to result from multicollinearity of the macrovariables. But, since the corresponding macrovariable is included in the auxiliary regression along with the noncorresponding macrovariables, multicollinearity should not produce strong "partial" relationships of the kind required here. Reference to a simple specific model is helpful here. We can revise the first model presented in Fig. 17.1 to include an arrow connecting one of the microvariables (X_1) with a noncorresponding macrovariable (X_2'), as in Fig. 17.8. The multicollinearity explanation suggests that there is no direct causal link between the two variables, but that it is spurious due to high collinearity of X_1' and X_2'. I have already suggested that most of this effect would be "credited" to X_1' in the auxiliary regression. Three

---

[27] Boot and deWit found positive aggregation bias terms. It is, of course, possible for the covariance terms to be negative so that aggregation bias would deflate the parameter estimates.

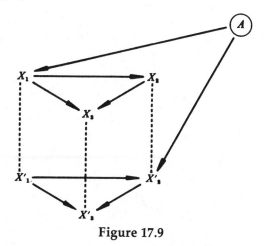

Figure 17.9

possibilities remain: (1) X_1 is systematically producing changes in X'_2 across time; (2) X'_2 is systematically producing changes in X_1 across time; (3) there is some omitted variable jointly affecting X_1 and X'_2 across time (the relationship is spurious owing to the effects of some "lurking" variable).

Consider a model in which individual income and individual political radicalism are both included as explanatory variables. We might suspect that variations in individual radicalism will vary across time with variations in community (or national) income if there is enough variability in the latter during the time period of observation. Such a situation will give rise to aggregation bias if we estimate the analogous model across time with community radicalism and community income as explanatory variables. The third case listed above comes closest to the situation we have already analyzed. We see here a possibility of a temporal spuriousness at the macrolevel analogous to the areal spuriousness pointed out by Alker. In the model we have used, this spuriousness does not relate variables that are otherwise unrelated but simply inflates or deflates the microrelationship. The introduction of a time dimension raises the possibility of a kind of spuriousness that is not apparent when we restrict our attention to static models estimated at one point in time. The situation I am describing is pictured in Figure 17.9 for our three-variable model.

What of the other possibilities? It seems more plausible that a direct causal impact would flow from macrovariables to microvariables rather than *vice versa*, although there is no basis on which to rule out the possibility of direct impact of changes in microvalues on macrovalues (particularly given the simple definition of aggregate variables employed here).[28] Ad-

[28] When such simple aggregates are used, some of the problems raised can be thought of as part-whole correlation problems. Recent discussions of this problem (Bartko and Pettigrew [4]) do not seem to shed too much light on the problem that concerns us here.

dressing this issue in any specific case demands the formulation of cross-level theories. I would strongly endorse the statement made by Fox [12, p. 496] relevant to aggregation problems in economics:

> If we are to avoid a naive empiricism in dealing with economic aggregates we must first do some rather careful bridge-building between the *microvariables* and *microrelationships* associated with individual consumers or firms and the *macrovariables* and *macrorelationships* associated with large aggregates of consumers or firms at a national or regional level.

Quite a few sociologists are willing to make strong arguments for the necessity of creating theories that include variables from different levels of social structure. There is no point in elaborating on this as the case is strongly made in many of the essays collected in the Doggan and Rokkan volume referred to previously. The analysis of some of the technical problems involved in changing units of analysis with empirical data (when the simplest of aggregates are used as macrovariables) brings us to the same conclusion. I would suggest that an examination of the work done to this point on the aggregation problem strongly suggests that progress in resolving the complex issues arising in attempts at changing levels of analysis is heavily dependent on theoretical advance in specifying crosslevel relationships.

ADDITIONAL COMPLICATIONS

I have indicated at numerous points that this analysis considers only the simplest cases of aggregation problems. In particular, we have limited our concerns to very simple (or simplistic) macrovariables, to perfectly specified recursive micromodels. I have already referred to the issues raised by the choice of simple aggregates. The literature referred to in the presentation of the aggregation problem contains analyses of the effects of specification error in aggregation and the aggregation problem relevant to nonrecursive systems of simultaneous equations. Additional problems discussed in this literature include the effects of introducing lagged endogenous variables in the micromodel and the effect of simultaneously aggregating over individuals (as we have done) and over variables (creating fewer macrovariables than microvariables). Space considerations preclude consideration of such complications. The reader is referred to the references listed above. Theil [32], Green [17], Nataf [29], and Hannan [19] are the best sources of references on specific complications.

CONCLUSIONS

Two major conclusions emerge from this overview of aggregation problems. The first concerns the novelty of aggregation problems. We have pointed to specific statistical problems arising from the aggregation of functions or relationships. But the thrust of this presentation was to argue that the aggregation complications previously analyzed can be seen as special cases of more familiar causal modeling complications. The key insight is that when data sets are aggregated nonrandomly (i.e., when the aggregation criterion is systematically related to one or more variables in the model), the relative variation in the variables tends to be affected. The most serious and also the most likely outcome of this process is the confounding of the variation of sets of variables. A special case of confounding effects is the production of spurious relationships at the aggregate level. Further, we noted that different but analogous mechanisms operate in cross-sectional and time-series analyses to produce similar effects. Unfortunately, many of these aggregation effects are only vaguely understood and considerable additional analytical and empirical analyses seem needed.

The second major conclusion is that the resolution of the kind of aggregation problems pointed to here seems to demand advances in cross-level theorizing. This conclusion emerges from the analysis of the aggregation bias mechanisms discussed above. The successful handling of any particular case of aggregation problems demands two things. First, one must specify the factors operating in the usually rather vague aggregation criterion (e.g., physical propinquity) to systematically affect variation in the variables of concern. Second, one must be able to specify the relationship of these factors to the variables in the model. Both tasks would seem to require theories that include in their scope both the microvariables being aggregated and the (presumably) macrolevel factors operating in the grouping or aggregation process. A theory seems required because of the extremely large number of possible factors that might be assumed to be operating in the type of grouping criteria usually encountered and the difficulties faced in attempting to specify the differential relationship of such factors to the variables in the model.

Finally, I would like to stress that this presentation of aggregation problems is in no way exhaustive. In economics, for example, aggregation problems have been raised and analyzed in a wide variety of theoretical models. There seems to have been a recent upsurge of renewed interest in aggregation problems in theoretical economics. Much of this recent work suggests interesting possibilities for the handling of the cross-level theory problems just discussed.

ACKNOWLEDGMENTS

I would like to acknowledge the helpful comments of H. M. Blalock, Jr., Allan Mazur, John Meyer, and Francesca Cancian on an earlier draft and the editorial and typing assistance of the Laboratory for Social Research, Stanford University.

REFERENCES

[1] Alker, Hayward R., Jr. "A typology of ecological fallacies." In *Quantitative Ecological Analysis in the Social Sciences*, edited by M. Doggan and S. Rokkan. Cambridge, Mass.: MIT Press, 1969.

[2] Allardt, Erik. "Aggregate analysis: The problem of its informative value." In *Quantitative Ecological Analysis in the Social Sciences*, edited by M. Doggan and S. Rokkan. Cambridge, Mass: MIT Press, 1969.

[3] Allen, R. G. D. *Mathematical Economics*. London: Macmillan, 1956.

[4] Bartko, John J., and Pettigrew, Karen D. "A note on the correlation of parts with wholes." *The American Statistician* 22 (1968): 41.

[5] Blalock, H. M., Jr. *Causal Inferences in Non-Experimental Research*. Chapel Hill: University of North Carolina Press, 1964.

[6] Boot, J. C. G., and DeWit, G. M. "Investment demand: An empirical contribution to aggregation problem." *International Economic Review* 1 (1960): 3–30.

[7] Boudon, Raymond. "Propriétés individuelles et propriétés collectives: Une probleme d'analyse ecologique." *Revue francaise de sociologie* 4 (1963): 275–299.

[8] Cramer, J. S. "Efficient grouping: Regression and correlation in Engel curve analysis." *Journal of the American Statistical Association* 59 (1964): 233–250.

[9] Doggan, Mattei, and Rokkan, Stein, eds. *Quantitative Ecological Analysis in the Social Sciences*. Cambridge, Mass.: MIT Press, 1969.

[10] Duncan, Otis Dudley, and Davis, Beverly. "An alternative to ecological correlation." *American Sociological Review* 18 (1953): 665–666.

[11] Duncan, Otis Dudley, Cuzzort, Ray P., and Duncan, Beverly. *Statistical Geography*. Glencoe, Ill.: The Free Press, 1961.

[12] Fox, Karl. *Intermediate Economic Statistics*. New York: John Wiley & Sons, 1968.

[13] Galtung, Johan. *Theory and Methods of Social Research*. New York: Columbia University Press, 1967.

[14] Gehkle, C., and Biehel, K. "Certain Effects of Grouping upon the Size of the Correlation Coefficient in Census Tract Material." *Journal of the American Statistical Association Supplement* 29 (1934): 169–170.

[15] Goodman, Leo. "Ecological regression and the behavior of individuals." *American Sociological Review* 18 (1953): 663–664.

[16] Goodman, Leo. "Some alternatives to ecological correlation." *American Journal of Sociology* 64 (1959): 610–625.

[17] Green, H. A. John. *Aggregation in Economic Analysis*. Princeton, N.J.: Princeton University Press, 1964.

[18] Grunfeld, Yehuda, and Griliches, Zvi. "Is aggregation necessarily bad?" *Review of Economics and Statistics* 42 (1960): 1–13.

[19] Hannan, Michael T. *Problems of Aggregation and Disaggregation in Sociological Research*. Chapel Hill, N.C.: University of North Carolina, Institute for Research in Social Science, 1970.

[20] Klein, Lawrence R. "Remarks on the theory of aggregation." *Econometrica* 14 (1946): 303–312.

[21] Klein, Lawrence R. *A Textbook of Econometrics*. Evanston, Ill.: Row, Peterson, 1953.

[22] Lancaster, Kelvin. "Economic aggregation and additivity." In *The Structure of Economic Science*, edited by Sherman Kropp. Englewood Cliffs, N.J.: Prentice-Hall, 1966.

[23] Lazarsfeld, Paul, and Menzel, Herbert. "On the relations between individual and collective properties." In *Complex Organizations*, edited by Amitai Etzioni. New York: Holt, Rinehart and Winston, 1965.

[24] Linz, Juan J. "Ecological analysis and survey research." In *Quantitative Ecological Analysis in the Social Sciences,* edited by M. Doggan and S. Rokkan. Cambridge, Mass.: MIT Press, 1969.

[25] Malinvaud, E. *Statistical Methods of Econometrics.* Chicago: Rand McNally, 1966.

[26] May, Kenneth O. "The aggregation problem for a one-industry model." *Econometrica* 14 (1968): 285–298.

[27] Naroll, Raoul. "Some thoughts on comparative method in cultural anthropology." In *Methodology in Social Research,* edited by H. M. Blalock and Ann Blalock. New York: McGraw-Hill, 1968.

[28] Nataf, André. "Sur la possibilité de construction de certains macromodels." *Econometrica* 16 (1948): 232–244.

[29] Nataf, André "Aggregation." In *International Encyclopedia of the Social Sciences,* edited by D. Sills New York: Macmillan and the Free Press, 1968.

[30] Robinson, William S. "Ecological correlations and the behavior of individuals." *American Sociological Review* 15 (1950): 351–357.

[31] Searle, S. R. "Correlation between means of parts and wholes." *The American Statistician* 23 (1969): 23–24.

[32] Theil, Henri. *Linear Aggregation in Economic Relations.* Amsterdam: North Holland Publishing Company, 1954.

[33] Theil, Henri. "The aggregation implications of identifiable structure macro-relations." *Econometrica* 27 (1959): 14–29.

[34] Thorndike, Edward L. "On the fallacy of imputing the correlations found for groups to the individuals or smaller groups composing them." *American Journal of Psychology* 52 (1939): 122–124.

[35] Wagner, Helmut R. "Displacement of scope: A problem of the relationship between small-scale and large-scale sociological theories." *American Journal of Sociology* 69 (1964): 517–584.

[36] Yule, G. Udny, and Kendall, Maurice G. *An Introduction to the Theory of Statistics.* London: Charles Griffin, 1950.

Author Index

Numbers in italics indicate page where complete reference is given.

Subject Index

860030584